THE RISE AND FALL OF THE DC-10

ALSO BY JOHN GODSON

UNSAFE AT ANY HEIGHT
RUNWAY
PAPA INDIA – THE TRIDENT TRAGEDY
AN EYE FOR AN EYE
THE AIRLINE PILOT
DESIGN BY ACCIDENT

THE RISE AND FALL OF THE DC-10

John Godson

NEW ENGLISH LIBRARY
TIMES MIRROR

To all those who died in the world's worst air disaster – and to their relatives.

First published by New English Library,
Barnard's Inn, Holborn, London EC1N 2JR in 1975

Printed and bound in Great Britain by
the Pitman Press, Bath

45002355 9

INTRODUCTION

THIS is the story of a well reputed aircraft manufacturing company, its designers and engineers; the story of an administrator of the Federal Aviation Administration, the person so-called responsible for civil aviation safety within the United States of America; the story of a smallish international airline that was sold an aircraft, unwanted by the original orderer; and it culminates with the story of a group of ordinary airline passengers who, because of a series of faults and inefficiences all along the line, were to be together in the French *Forêt d' Ermenonville* for their appointment with fate.

The publishers would like to thank the National Transportation Safety Board of America for permission to reproduce their photographs; also *Le Figaro* for the centre-spread illustration; and Air and General Photos for the photograph of the Turkish Airline DC-10, taken at Heathrow Airport.

CHAPTER 1

Donald W. Douglas was a young man back in the days when the brothers Wilbur and Orville Wright were experimenting and indeed, finally succeeded in performing the advent of powered flight. The publicity generated by the Wright brothers' successes at Kitty Hawk, grabbed the imagination of many a young man then as space flights and trips to the moon do with the youngsters of today.

Douglas was an engineer by trade, and he was impressed by the possibilities that powered flight could bring. For instance, people, one or two of them at a time, could be carried rapidly over long distances, such as 50 miles or so, at a much faster speed than was then possible by any other means of transportation.

In the early months of the Douglas Aircraft Company, which was founded on the 23 July 1920, Douglas played around with certain plans until he finally designed and then sold to the United States Navy an idea which he had already been working on for a land-based aeroplane that could fly out over the waters that surrounded the country, and drop a torpedo on any war-like craft seen approaching the American shores. The Navy was impressed and placed an order for three of the model which was called the DT–1.

The machine had quite fantastic characteristics for those days. It had a range of 270 miles (in still air); weighed 7,293 lbs. when fully laden for action; was 37 feet 8 inches long and 50 feet from wing-tip to wing-tip; and it carried a crew of two, together with one torpedo, at its maximum speed of 99 miles per hour.

Once delivered, both the Navy and Douglas realized that there was one major problem. As it was a land-based aircraft with one engine and with two wheels on short, stubby, fixed legs underneath, and as it was designed to fly out over the water and drop its deadly weapon on some marauding enemy that was presumably about to attack the Confederation, what would happen if the single, 400 horse-power Liberty powerplant gave trouble?

The answer to that question meant the sudden end to the DT–1 and the introduction of the DT–2 in 1921. The difference between them was that the latter had an interchangeable undercarriage system – from wheels to floats – meaning that the aircraft could then land on either solid ground or water, dependent upon the availability at the time. The DT–2 was just what the U.S. Navy needed. They ordered 63 of them.

With the tiny production lines busy, Donald Douglas tried his hand at the next project. Realizing that people had actually seen and experienced the power of air supremacy during the latter part of the First World War, his thoughts turned to another military possibility, this time one for the U.S. Army that could be used as either an observation aircraft, or as a fighter, or as a bomber. In all, the Army purchased 153 of these O–38s from the Douglas Aircraft Company – a machine that could fly around on its single Pratt & Whitney R–1690–5 radial engine for up to 2 hours and 10 minutes at the enhanced maximum speed of 150 miles per hour, sporting its menacing two 0.30 inch Browning machine guns and with its four 100 lbs. bombs.

Then, because of its success, in 1922 Douglas started designing a further development which also interested the U.S. Army, and was designated the O–43A. With an order for 25, the factory went to work on this single engined observation aircraft with its one fixed and one flexible 0.30 inch Browning machine guns.

These were followed by the TBD "Devistator", with a vastly increased range over the earlier models, and the B–18, the first Douglas machine to sport two engines and was capable of carrying up to 6,500 lbs. of bombs, as well as being equipped with three Browning machine guns.

By this time, civil aviation was starting to catch on in both Europe and the United States, and in January 1932, a carrier called Transcontinental and Western Air (which was quite a mouthful, and was therefore shortened to T.W.A.), approached the Douglas Aircraft Company and asked whether they would care to compete for an order for a three-engined civil aircraft T.W.A. were thinking of buying. The Douglas designers and engineers were interested and looked at the possibilities of making an all-metal, large capacity, passenger carrying machine. As a result of this interest, a prototype was built to test out the all-metal theories, and also to show that a higher-powered two-engine design was equally as good as a lower-powered three-engine model, but with the one big advantage of saving fuel, and therefore increasing the distances over which it could fly. This prototype was designated the DC–1 (DC standing for Douglas Commercial), and its maiden flight was in the middle of July 1933.

The DC–1 was so far advanced over previous types of planes that virtually every time it took to the air, some flight record was broken. T.W.A. were very impressed with the two-engined machine, even though they had been convinced that their idea had needed three engines, that they immediately signed a contract with the Douglas Aircraft Company. But the DC–1 was only made experimentally to see whether this rather radical designers and engineers idea worked.

While it was flying around the country breaking records during tests and demonstrations, Wright reported that they had been able to make further advances in developing increased power from their engines. The Douglas engineers took advantage of this to enlarge the DC–1 and also to extend its flight range, with the result that the order for 20 aircraft for T.W.A. were called DC–2s.

This new model was capable of carrying 14 passengers (two more than in the DC–1), and it also had a greater non-stop range (3,700 miles, compared with the DC–1's 3,000 miles). The first flight made by a DC–2 was in May 1934.

At around the time of its maiden flight, the Dutch airline K.L.M. ordered a few of these machines, and later that year, they entered one of them in the world famous MacRobertson race from London, England, to Melbourne, Australia. The DC–2 was lined up against 20 high-speed airplanes, some of which had been especially built for the prestigious competition. A certain amount of mocking was afforded the K.L.M. entry, and this turned into relative hilarity when the airline announced that the DC–2 would also be carrying 14 passengers during the flight. However, it managed to be placed second in the speed contest, and was first in its own category, that of "transport aircraft".

The world-wide press coverage of the MacRobertson race, and possibly especially of this ridiculous all-metal competitor carrying 14 passengers, stimulated further orders for the DC–2, and by 1938 the company had produced 133 of these airplanes. During the early part of its production however, Douglas was approached by American Airlines who suggested that if the fuselage was widened, it could accommodate some sleeping berths and thus permit the airline to offer a continuous service from the east to the west coast – a flight which at that time, had to be accomplished in easy stages with stop-overs here and there because night-flying was extremely perilous without sophisticated instruments, and also in order to let both passengers and crew get some sleep.

As a result of this inquiry, the Douglas Aircraft Company came up with the idea for the DST – the Douglas Sleeper Transport – first flown in December 1935, and with added instrumentation, was classed as the most advanced commercial aircraft in the sky.

But with the DST production lines started, the engineers at Santa Monica, California, reasoned that the wide fuselage would also permit an increase in the number of seats that could be contained inside the passenger cabin, from 14 to 21. At around the same time, Pratt & Whitney approached Douglas with their new 1,200 horse-power engine design, and this was quickly seen to be a further improvement over the DC–2's 875 horse-power Wright powerplants.

The new machine became the legendary Douglas DC–3. Because of its combination of speed, payload and range, the aircraft soon dominated the field of commercial air transportation throughout the world, and there was one stage in its life where more than 90% of all civil air passengers were being transported solely in DC–3s.

However, with war clouds looming darkly over Europe, the American military was also interested in the DC–3. Between 1936 and 1939 the Douglas Aircraft Company built 10,925 of these machines; 10,123 for military use, and designated the C–47. At the close of World War Two approximately half of the military versions were converted for civilian operations, and as the air force's requirements changed, additional C–47s became DC–3s.

Once the DC–3 design had been settled and the factorys' production lines began producing the model, the designers and engineers turned their interest to an advancement yet again. This time, they were thinking of a much bigger four-engined aircraft which ultimately became the DC–4. A prototype, designated the DC–4E ("E" for Experimental), was first flown in June 1938, and during its demonstration flights to the different big cities in the United States early in 1939, it was seen as the "Jumbo" transport of that period. It represented quite an appreciable advance in both concept and construction over its predecessors, incorporating such innovations as underwing fuelling, retractable tricycle landing gears, and a redesigned cabin heating system that permitted a more stable temperature to be maintained.

Then came the DC–5 – of which only 12 were built – a high-winged monoplane with many of the elements and systems taken from the DC–3. But because a high-winged configuration had the effect of reducing both the capacity and the range, in the long run it proved to be less economical than the DC–3. After its first flight in February 1939, one was ordered by an American businessman as an "executive aircraft", 4 were purchased by K.L.M., and the remaining 7 were delivered to the United States Marine Corps.

On the basis of the DC–3 alone, Douglas merited its reputation as the leader in civil air transportation. Donald Douglas Sr. never lost sight of the prime importance of retaining this leading position in *civil* air transportation; other companies, after the Second World War and the endless cold war that followed it, became inextricably involved in the artificial economy of the blooming complex of defence industries. Douglas sought to retain its leadership *and* its independence by building on the continually increasing demand for aircraft designed for peace, not war. Thus it became fully tragic to see Douglas trapped in the bursting soap-bubble of an inflated, "permanent wartime economy", staking in great measure not only their reputation but their actual survival on the airplane which is the subject of this book. The DC–10, unlike other jet air transports, was the end product of the evolutionary development of the most successful series of civil air transport aircraft in world history. It is true that Douglas built thousands of military aircraft up to the time of the merger with McDonnell but the great dreams of Donald Douglas are best embodied in the DC–3, the DC–4, the DC–6, the DC–7, the DC–8 and 9 . . . and the DC–10.

At the start of World War Two, the DC–4 was about to go into production for the major American airlines. After a postponement because of the urgent need to get military aircraft into the air, it was first flown in March 1942. Then, when production really started, it was built as the military C–54 or as the R–5D Skymaster. A total of 1,163 DC–4s were delivered to the U.S. Army and Navy, and they established an enviable record especially in the field of trans-Oceanic travel by a normally land-based aircraft, and after the war, 79 more were produced solely as civilian machines, while in the meantime, 250 surplus military models also found their way into airline services as 44-passenger DC–4s.

Douglas engineers next designed the DC–6, drawing heavily on the proven aerodynamics and structure of the DC–4, with the most important changes being once again, the substitution of more powerful engines and the addition of cabin pressurization and air conditioning to the "stretched" fuselage. It was in February 1946, that the DC–6 made its maiden flight. In service, it carried 50 passengers and flew at more than 300 miles per hour. After the first 174 were delivered to the airlines, the plane was again enlarged by extending the cabin length, and this "new model" was called the DC–6A – the first modern airplane to be produced solely for the purposes of carrying cargo.

With the enlarged fuselage, the airlines rapidly realized that this would permit them to carry even more passengers, and so Douglas, always ready to oblige, made some slight alterations in the design and called it the 54-seater DC–6B. Its maiden flight was in February 1951, and while the production lines were making the 583 DC–6s ordered, the Wright Company advised the designers that they had developed an even more powerful engine than the 2,500 horse-power Pratt & Whitneys being used on the DC–6 series – their 3,250 horse-power Wright R–3350 – and this resulted in the return to the drawing boards, and by a further "stretching" the fuselage and increasing the fuel tank sizes, they ended up with the DC–7.

For the first time in American civil aviation history, the DC–7 made it possible for non-stop flights from the east to the west coast of the continent. The 64-seater, which flew for the first time in May 1953, offered what was classed as luxurious and fast transportation, speeding its way from New York to Los Angeles at the incredible rate of 365 miles per hour, and while they were being manufactured, the designers made some minor aerodynamic improvements and slightly increased the fuel tank capacities which brought about the introduction of the DC–7B.

Once these aircraft were seen to be capable of making the long and laborious journey from one side of the United States to the other, a number of the international airlines approached the Douglas Aircraft Company, wondering whether an even longer ranged version would be possible. The Douglas engineers found that the most efficient way of carrying the additional fuel required for pan-Atlantic and pan-Pacific flights, would be to add a section in the area where the wings joined on to the fuselage, so they added about 5 feet to the inboard wing and this not only permitted longer flight hops, but it also increased the passenger seating from 62 to 99 people. This model, called the DC–7C, made its first crossing in December 1955, and cut hours off the Trans-Atlantic flights.

However, by now, commercial war clouds were looming. In 1952, the British de Havilland Aircraft Company delivered its first jet-turbine powered airliner to the British Overseas Airways

Corporation, and for the next nine months or so, B.O.A.C. was using these machines on certain of its world routes. The Comet 1 proved to be much faster, much smoother in flight, much quieter in the cabin, even though it could carry less than half the number of passengers in a DC–7C. But it had problems. After three disastrous crashes killing 110 people, the British government, who had presumably tested it, found it acceptable, and issued it with a certificate of airworthiness, had to shame-facedly turn around and revoke its permission, and this effectively saved people's lives by ridding the industry of the Comet 1. But, even when the Comet was dead and gone, in the few short months it had been in service, it had demonstrated that passengers would accept the new type of engines, and preferred the speed and comfort that they had supplied.

Soon after the Comet 1 was introduced, the Douglas Aircraft Company, realizing that if they did not make a full study of this newly powered machine, could see the end of their leadership in the civil aircraft manufacturing market. In June 1952, they set up a "special projects" office to look into the possibility of adding these jet-turbine engines to their DC-series. In early 1953, their design was sufficiently advanced to offer a pure-jet transport to their airline customers. In reality, this would have been only the third passenger carrying plane that they actually had to sit down and work out from scratch; as we have seen, the DC–1 begat the DC–2, the DC–3 and the DC–5. And the DC–4 was the foundation for the DC–6 and DC–7 series. But then, early in 1955, orders started coming in for their all new, all jet-powered DC–8.

On 30 May 1958, DC–8 production number one flew for the very first time. They were surrounded by opposition (the de Havilland Company was in the process of flight testing the re-jigged design for their soon-to-be-introduced Comet 4, and the Boeing Aircraft Company was already flight testing their first major post-war contender in the civil aircraft field, their model 707), but Douglas held back taking the decision on whether or not to build a jet-powered transport until the very last minute, and then decided that in order to try and catch up with their competitors, no prototype machine would be built. Their reasoning behind this must have mainly been based on their past record. Everyone knew and had more than likely already flown in a Douglas commercial aircraft, whereas hardly anyone had heard of Boeing, and naturally, everyone had read of the de Havilland Company and *their* past "efforts" with the Comet 1.

So in 1957, with both the Boeing 707 and the Comet 4 being tested in the air, Douglas made the fateful decision to bypass the prototype stages and commence straight-in with the production aircraft. This was practically their downfall, taking the company close to bankruptcy, because at the stages of final assembly and initial flight testing, the DC–8 proved to have many problems, all of which had to be corrected during production, costing heavily in both time and money. With the assembly lines tooled up and working, as the problems were uncovered, major re-tooling had to be effected, and the aircraft that were advanced in the production stages ahead of the area where the fault had been uncovered, had to be taken apart and modified. Of the three major contenders for jet civil air transportation, the DC–8 was the last in the air, the last into service and had lost a large number of potential customers to the Boeing Company's more advanced project.

However, it did have one small advantage in its tardiness. Douglas were able to increase its speed, payload and range over those offered by its two major competitors, which brought back a few of their past airline customers. In all, 294 DC–8s, with a mixed-class seating capacity of up to 159 passengers, were built.

With the introduction of the jets, civil aviation came into the second big period of advancement. (The first was immediately at the end of the war when the different countries had the tasks of returning their fighting men to their own home territories.) In 1959, with jet transportation seen to be quicker, cleaner, cheaper and less tiring, all the world was at the airlines' ticket counters. Following Douglas's time-proven formula of "stretching" fuselage designs of their successful transports in order to increase both capacity and range, the company came up with three distinctly new versions of the DC–8 designated the Super Sixty Series. Its main reason for returning to old habits was because it was being seen to be soundly beaten in the "order book" department by the Boeing Company, and on making a careful study of the Model 707, Douglas engineers realized that because of its short and stubby undercarriage, in order to stretch the 707, the whole underneath and wing area would have to be completely redesigned. They, on the other hand, could – as usual – easily "stretch" the DC–8 because of its higher undercarriage which, like so many other things, could be traced back to the 1937-designed DC–4.

So along came the DC–8 Super 61. This model accommodated up to 251 passengers in a fuselage that was 36 feet longer than the original DC–8 preceding it. And with the enormous advance in size, many of its operational systems and electrical and electronic circuits had to be completely redesigned.

Next came the DC–8 Super 62, possibly the most successful of all Douglas-built jet aircraft, developed mainly for extremely long-range operations. The fuselage was stretched by only 6 feet from the original DC–8s, but the big differences were the wing-tips and the increased wing-span; newer, more powerful engines; and a newly designed underwing engine pylon (onto which the engines were attached) which greatly reduced "drag". The DC–8 Super 62 became the longest-ranged jet in the air with a seating capacity for 189 passengers.

The final one in this series was called the DC–8 Super 63, combining the wing-engine-pylon arrangement of the Super 62, with the long-range fuselage of the Super 61. This resulted in a large-capacity transport that was capable of flying intercontinental routes non-stop.

The next major scheme in the DC saga began with market studies in 1962, which showed that the airlines needed a jetliner to replace the smaller propeller driven aircraft still serving many communities not large enough to support the operations of the DC–8, Boeing 707 class of plane. The twin-jetted DC–9 was developed to fill this void.

The first flight took place in February 1965, and the T-tailed DC–9 then started to take over on routes where DC–3s and Viscounts were the mainstay of air travel. However, with the Douglas Aircraft Company just starting to break even on their DC–8 design, little beknown to them, they were letting themselves in for something that was nearly to be the beginning of the end.

So in talking about the Douglas Aircraft Company, we are discussing one of the foremost aircraft manufacturing concerns; possibly one that was too complacent, depending to a great deal on their past good name revolving around the foresight and development capabilities of Donald W. Douglas, back in the year 1932, when he decided to make radical steps in aircraft construction by moving away from the cardboard and string machines of those days and entering the world of the all-metal airplane.

Possibly the best way of showing the great strides made in the evolution of aircraft design and advancement is to list the Douglas products manufactured prior to this time, which, by comparison, shows the build-up of the era. This list appears as the Appendix.

CHAPTER 2

DURING the latter half of 1965, engineers at the Douglas plant at Long Beach were working on a new, secret project. It was called the "DC–10"; the tenth in the line of Douglas commercial air transport vehicles – a double-decker, 400-passenger, long-range development of a mixture of their DC–8 and the huge Lockheed C–5A military machine. The C–5A as may be remembered, is the biggest jet turbine-powered aircraft in the world. It is in fact the biggest ever built and flown, capable of carrying over 1,000 troops or an enormous amount of military equipment. The project nearly bankrupted the Lockheed Aircraft Corporation in the late-1960s, when the government suddenly discovered that it was being "over-charged", and demanded quite a chunk of its money back. Douglas's idea was to "split-the-difference" between its quite successful DC–8 design and that of the huge Lockheed military C–5A.

Ironically, at the time of writing, the McDonnell Douglas Corporation is caught in the same net that previously affected Lockheed. An inquiry is being held into reportedly "excessive profits" made by the Corporation in the financial years 1967 and 1968 ($131 million and $125 million respectively). The investigative committee has realized that the Douglas part of the Corporation was not a good profit maker, and it therefore has been assumed that these returns came from the McDonnell military contracts side. However they have complained that McDonnell Douglas has not supplied them with an exact breakdown of details from which divisions these profits were obtained, and this seems to be the cause of the trouble.

It has been pointed out that the five-man Renegotiation Board, which originally conducted an inquiry into these huge returns and found them practically "acceptable", was appointed by the then President Nixon – the election campaign for which McDonnell Douglas bought advertisements, and its employees handed over just on $100,000 to the "re-election" campaign funds in 1972.

The new senatorial inquiry will be looking into the aspects of why this Renegotiation Board accepted that just under a quarter of a billion dollars profits in two years was not excessive, deeming that all that was necessary to "readjust the balance" was for McDonnell Douglas to hand back $5 million.

In the light of the Watergate and related conspiracies, the determinations of this new inquiry will be watched with interest.

At that time, none of the major manufacturers had formally made up their minds about their "next generation" aircraft. Boeing had stated that they were "stretching" their 727 and calling it the 727 Series 200, in order to directly compete with the then proposed Anglo-French Airbus. In the meantime, they were also considering the prospects of going ahead with their large 747, long-haul, low-cost, 400-passenger transport. However, no official decision on these plans had been made even though the Boeing Company had approached some of the larger airlines with the idea.

On 3 January 1966, the Douglas Aircraft Company formally announced the details of their DC–10. The "final decision" on the shape had been determined in November 1965, when the company planned to directly compete with the earlier released details of the Boeing 747. No technical information was made public at that time; no firm commitment to build the design was offered by Douglas; all they had done was to draw some outlines on pieces of paper and then sent their sales force around to the major carriers in order to sound them out.

The DC–10 was then being considered as two separate ideas. There was the smaller DC–10 D–950 with two passenger decks; 194 feet 4 inches in overall length, 60 feet 10 inches from the ground to the top of the tail, and 167 feet 6 inches from wing-tip to wing-tip. This was again broken down into three separate possibilities; the DC–10 D–950 Series 10 would have carried 376 passengers plus 3,400 cubic feet of space underneath the lower deck for cargo; the DC–10 D–950 Series 30 would have a passenger carrying capacity of up to 457 places, with an added 4,290 cubic feet for cargo; and the DC–10 D–950 Series 50 was for 524 passenger seats plus 5,025 cubic feet of cargo space.

The fuselage design itself was unique for civil machines. It was a "double-bubble", in the shape of an "8", with the upper deck being in the smaller circle of the "8", which itself would have been similar in diameter to the fuselage of the DC–8. The lower passenger deck was to have been situated in the upper part of the much bigger lower section of the "8", with the cargo compartments and wing-through-fuselage areas in the lower part. Enginewise, Douglas were planning on slinging two high-by-pass ratio turbo-fans under each wing, and were looking for something similar to the ones Boeing were going to use on their 747s, each of which would give off in the vicinity of 40,000 lbs. of static thrust.

The second idea was for the DC–10 D–952, again being considered in two sizes carrying up to 563 passengers and, because it would be higher off the ground, it had greater possibilities of being further "stretched" at a later date. This version was bigger than their proposed D–950; it was 212 feet in overall length, 75 feet from ground level to the top of the tail, and 206 feet 9 inches from wing-tip to wing-tip. It was even bigger than the planned 747.

Both Boeing and Douglas continued their efforts to interest the airlines in their ideas. Boeing had gained some support for their "smaller of the two" planes. However, Douglas, without any sales luck to that time, approached American Airlines early in April 1966, with their "double-bubble" DC–10, and it just so happened that American were looking for a new aircraft for the 1970s – one that would be capable of taking a large number of passengers along their Chicago to Los Angeles route, and one that could also carry a full payload out of the short runway at La Guardia, New York, over to Chicago.

American Airlines thought it was looking for a large passenger capacity jumbo-type fuselage, but with only two engines instead of Douglas's intended four. This of course, compounded the already confused situation, and at the annual shareholders' meeting on 1 May 1966, Donald W. Douglas Jr., the company's president, indicating that they may wait until 1975 before making a decision to build the proposed 500-passenger DC–10, added that since Pan American had placed an order for the Boeing 747, they were not prepared to rush into production: "the company would wait until it felt the market was ready", he said. Douglas also pointed out that the availability of the 251-passenger DC–8 Series 60s would limit the demand for these larger aircraft: "for some time to come".

Opposition plans were taking shape in the form of competitive, cheaper, more economical planes than the DC–8 Series 60. In Europe, British, French and West German government representatives met to discuss the proposed European Airbus and in order to offer "return competition", in mid-May 1966, Jackson R.

McGowen, the Douglas group vice-president, revealed an idea for a short-haul, twin-engined 250–350 seater aircraft which would be in service by late 1970 or early 1971.

As soon as Boeing had announced the provisional go-ahead for their Model 747, Douglas were looking more at the Airbus idea instead of attempting to compete directly with their Jumbo Jet. Obviously, if the decision was made for Douglas to go ahead with the two-engined DC–10, it would make a considerable difference to the Europeans' plans for an Airbus.

While one section of Douglas's "advanced project design group" was studying the two-engined jet, another had put forward the suggestion that the 251-passenger DC–8 Series 60 aircraft (which themselves were "stretched" versions of the smaller DC–8), be "stretched" again to what could be known as the DC–8 Series 83, and which would offer some direct competition to the Boeing 747, with the one advantage that Douglas would have their machine in the air first. The DC–8 Series 83 was going to be a wider-cabined, 19 feet longer version of the DC–8 Series 63, capable of carrying up to 350 passengers, and for which Douglas salesmen had rapidly visited the major long-haul airlines that were not already committed to the Boeing 747, in order to try and drum up some interest. So as to make it sound like a new project and in the light of the earlier announced decision at the annual shareholders meeting not to go ahead with the "double-bubble", this new "stretched" and widened DC–8 was re-designated the DC–10, however, after being unable to interest any of the air carriers, this idea too was dropped.

The competition amongst the different aircraft builders became fierce. In the latter half of 1966, there was a variety of machines available for the short-haul low-capacity needs of the airlines, however everyone could see that trouble was brewing ahead, as a number of the world's major airports were starting to get "clogged up" at peak periods, with planes having to fly around in circles in stacks for anything up to an hour and a half, awaiting their turns to land. There seemed to be only three answers to this evidently fast developing problem: (1) build more airports; (2) start cutting down on the numbers of aircraft; or (3) come up with much bigger designs in order that more people could be carried in less machines.

It was in this aura of uncertainty that the whole civil air transport business operated. The biggest aircraft then available was the Douglas DC–8 Series 63 with a capacity of 251 passengers. Boeing had made a firm commitment to go ahead with their 747 design, capable of taking from 350 to 500 people. And now Lockheed were proposing to make a civil version of their military C–5A, with up to 750 passenger seats.

Other rumours in the background included the Europeans being interested in a 250 to 300-seater Airbus. Douglas's problem was to find the right niche in the market in order to be able to design an aircraft of which they could build sufficient numbers to make it a profitable proposition.

But by mid-November 1966, the Douglas Aircraft Company was on the verge of financial crisis. They started exploring the possibility of a merger with several major manufacturers. Two of these companies, North American and General Dynamics, had admitted being involved in the preliminary talks, but it was soon realized that a third, the McDonnell Company, seemed to be the most likely bidder. McDonnell had been very active in the military field though they had never built a civilian plane; whereas Douglas was in the main a civil constructor, after having been a "bit-of-each" up to the end of the Second World War with their A–26 Invaders.

Paradoxically, the news of Douglas searching around for a partner came at a time when the company already had a huge backlog of civil aircraft orders – but the fact was that they had been in financial difficulties since about 1963 and had been unable to make their DC–9 machines pay.

During the talks, the big American banks were understood to have been opposed to a merger with either North American or McDonnell because of the lack of experience of these companies in commercial aircraft development and production. Douglas needed $75 million from the banks, together with credit facilities from its supplies of a further $25 million, *and* some advanced payments from its airline customers in order to be able to continue. At that time they had already received a large number of orders for their then-being-manufactured aircraft; with 342 outstanding for the loss-making DC–9, and 156 for the DC–8 series.

Some of the banks did come to their immediate aid. The plane makers desperately needed assistance. Also, the McDonnell Company, as part of its merger proposals, was arranging to obtain a further $300 million credit through a group of ten banks headed by the huge Chase Manhatten.

In the middle of the talks, near disaster struck when, on 24 February 1967, a Northeast Airlines DC–6B made an emergency landing at Kennedy Airport, New York, after a ten-foot section of the right-hand side of the cabin had been blown out at altitude.

Two of the nine passengers on board were slightly injured and some luggage and clothing was sucked out of the hole. The DC–6B was on a scheduled flight from Philadelphia to Boston, and was out over the Atlantic Ocean at the time the explosive decompression occurred. The Federal Aviation Agency (as it was then called) ordered all DC–6 and DC–7 aircraft to be flown unpressurized until the problem was resolved.

This, of course, was no good for the Douglas image, and it tended to compound the financial difficulties. However, it was later determined that metal fatigue had caused the huge panel to separate, and after certain inspections and maintenance had been carried out on all these aircraft, they were allowed to return to normal flight conditions.

On 1 May 1967, the McDonnell Douglas Corporation was born. The new combine started in its headquarters in St. Louis, Missouri, with the McDonnell Company and the Douglas Aircraft Company operating separately as its principal components. Prior to the formation, Douglas had announced a loss of $27,560,000 for the year ended 30 November 1966, compared with a net earnings of $14,598,000 for the previous twelve months. Once the monetary problems had been resolved, it was back to the drawing boards for the designers. Within one month of the merger, Douglas and Lockheed were at loggerheads, each working on a 250 to 350 seat, 1,800 to 2,500 nautical mile range aircraft.

Both of them were offering "families" – Douglas with the generally larger three- and four-engined versions, bearing the type number DC–10 (for the fourth time), and Lockheed with the two- and three-engined ideas being called the CL 1011. Both aircraft, in their tri-jet versions, featured an engine under each wing, with a third buried in the machine's tail. However, the new DC–10, even though much smaller than the original, was still being thought of as a partial double-decker. There would be normal, high-density seating on the main deck, with the flight crew in an upstairs cockpit in a similar, but smaller "humped" arrangement as was being planned by Boeing for their 747s. And there would be another passenger area, accommodating 38 seats, below the forward part of the main deck, between the front of the wings and the nose.

On 1 November 1967, the McDonnell Douglas Corporation authorized further design and development work to be done on their DC–10 idea. Following more talks with American Airlines, they had chosen the three-engined version as the one most likely to succeed, though some potential customers had indicated that with the forward lower deck being used as a passenger cabin, this did not leave too much room for the carriage of baggage and extra freight.

However, by December 1967, both Douglas and Lockheed had changed their minds about their designs *yet once again*. After conferences with their prospective customers, both of them chose longer fuselages which increased the Lockheed seating capacity to around the 300-mark, and reduced the DC–10's to about the same figure as now Douglas were considering all passenger seats to be on the one main level, with the lower forward deck reverting to the accommodation of cargo and baggage.

In January 1968, McDonnell Douglas salesmen canvassed the European airlines with their newer versions of the DC–10B. This time, the passenger seats had been increased to a maximum of 345, with five doors on either side of the fuselage. Three main cargo compartments were to be located beneath the passenger cabin floor, two of which were to be equipped with mechanized handling systems for the rapid loading and unloading of pre-packed containers. The aft cargo compartment was designed for the carrying of bulk cargo and animals. Access to it was to be via a door, built into the left-hand side of the fuselage, below the cabin floor level. And between the forward and middle compartments there was to be one large downstairs galley.

On 19 February 1968, American Airlines committed themselves to the biggest order in aviation history when they signed a contract for 25 McDonnell Douglas DC–10s together with an option to buy a further 25, with the combined value of more than $800 million (£323,000,000 – FFr3,553,000,000 – SFr2,422,500,-000 – Y218,025,000,000). This was the first and at that time the only airline to favour the new generation of large-capacity tri-jets.

American Airline's choice came as quite a surprise in view of the fact that it had been in closer contact with the Lockheed 1011 project than the McDonnell Douglas DC–10, and its order stipulated a delivery date for the first aircraft in the latter half of 1971. However the contract was not completely binding as Douglas had officially announced that they would not go ahead with the idea until such a time as they could get other air carriers interested.

"The DC–10 is the aircraft of the future," George A. Spater, president of American Airlines said when the contracts were being announced. "We do not foresee it will be superceded or obsoleted by any aircraft on the horizon, supersonic, or otherwise."

On 25 April 1968, at the United Air Lines annual shareholders' meeting, the president, G. E. Keck, announced the company's intentions of signing up for 30 DC–10s, with a further 30 aircraft on option, as long as the machine was powered by the General Electric CF6–36 engines.

With the orders starting to come in, the new generation jets were taking a more positive shape. The only thing that was needed was a firm commitment from the Douglas Aircraft Company to go ahead with the project.

CHAPTER 3

On 25 April 1968, after nearly three years of submitting and then retracting designs, McDonnell Douglas Corporation officially announced that the DC–10 was to go ahead. The arrangements for managing its development and production were also fixed, and within a couple of weeks of the announcement, Douglas published the specifications and performance details for their idea. A full-scale mock-up including a cabin section and a flight deck had been completed at Long Beach, and it was then seen that not since the days of the Boeing 707 and Douglas DC–8 rivalry had there been a pair of such closely similar aircraft as the DC–10 and the Lockheed 1011. Although the 1011 appeared to have a slight advantage in power, Douglas was claiming that the DC–10 had more seating capacity. However in all other respects the rival claims were virtually identical.

The DC–10 was being designed for a 3-man flight crew. A second officer or flight engineer was to be behind the first officer on a swivelling tracked seat which could face forwards from behind the centre console, or sideways at his own systems panel. There was also an arrangement for two observers' seats in the large cockpit area, and the overall layout was based on a "scaling up" of the 2-man, DC–9 flight deck.

The passenger cabin of the DC–10 was to be 18 feet 9 inches wide with a ceiling 8 feet 3 inches above the floor. The economy-class section was designed for eight 19 inch-wide seats running across the cabin in pairs, with two 20 inch-wide aisles. In first-class, it was to be six abreast paired seating. Douglas was also planning for larger cabin windows than in any previous machine: 11 inches deep by 16 inches wide, and spaced just 9 inches apart.

Then there was the problem of feeding 345 hungry passengers, realizing that flights would be started around normal meal times, and yet their DC–10 was, relatively speaking, a short-haul aircraft. In the first week in December 1968, at Long Beach, trials in the mock-up were conducted to see whether 250 people could be served with drinks and a multi-coursed meal in under $1\frac{1}{4}$ hours. This limitation was selected as being short enough to allow a similar "real-life" service on the New York to Chicago run.

In the tests there were 46 first-class passengers served by 2, and 204 coach passengers served by 4 stewardesses, with 2 others employed in the food storage and preparation galley on the lower deck. Observers monitored the operations both upstairs and down, through 10 strategically placed closed-circuit television cameras, with video-taping for later study by time-and-motion specialists.

The test began in a reception area adjoining the mock-up of the aircraft, where passengers "checked-in" and were assigned seat numbers. Stewardesses verified the tickets, directed them to their places and then took orders for pre-luncheon drinks. Except in first-class, passengers paid for these drinks with "Monopoly money" which required the attendants in some cases, to provide change. In first-class, there was a three-course meal with a choice of three entrées, and wine was also served. The coach-class eats were somewhat smaller.

All meals and drinks, except coffee from the main-deck coffee-makers, came from the below-deck food-preparation galley, a new and key element in the arrangements. Food was provided in 8 modular containers, each holding 4 serving carts. Within each cart there were stacked trays for drinks, with complete meal trays – or for first-class, separate entrées. Where required, refrigeration was contained within the modules. Hot food was warmed and cooked in 6 high-temperature electric ovens lined along one wall of the galley above a preparation bench.

The attendants who prepared the meals downstairs, sent the carts loaded with trays up to a "service centre" on the main passenger cabin floor level in two electrically operated lifts, one of which was tall enough to carry a stewardess.

Following the tests, further studies and modifications were made over the ensuing months, with Douglas setting on the finished aircraft early in 1969, and then sending the design plans around to the major sub-contractors while in the meantime the salesmen were in force visiting the world's airlines to try and get more of them interested in their machine. On 5 June 1969, Douglas announced that they had negotiated an order for 14 of the slightly larger DC–10–30s with the KUSS group of European carriers, made up of K.L.M., U.T.A., Scandinavian Airlines System and Swissair.

However, even further changes in the fuselage were being required by them. They wanted an intercontinental machine, which meant one of an all-up greater capacity and weight, and this in turn required Douglas to add a third main undercarriage leg to the system below the centre of the aircraft. Other major revisions concerned the interior layout. The group did not want galleys on the lower deck, but needed more cargo area, so the feeding arrangements for them were moved up to the passenger cabin level.

On 25 June 1969, initial assembly started on DC–10 fuselage number 1 at the Santa Monica, California factory, when cockpit window frames were joined in the very first step of constructing the nose section. It had taken Douglas just on 14 months from the time they announced the official go-ahead, up to the phase where construction was started.

The build-up of the workforce for the DC–10 was rapid. In Santa Monica and Long Beach, 10,000 people were employed on the project by May 1969. By December 1969, the number had risen to about 14,000 and they were then predicting that a further 5,000 bodies would be needed before long.

CHAPTER 4

OVER on the other side of the American continent, in Washington D.C., it was a bright but chilly Monday morning, the 24 March 1969, for the arrival of the "new boy" who was taking up his position as the head of the Federal Aviation Administration. He was John Hixon Shaffer, a fifty-year-old political appointee of the then United States President, Richard Milhouse Nixon.

As the press handout from the F.A.A. at the time said:

John Hixon Shaffer is serving as the Administrator of the Federal Aviation Administration during one of the most dynamic periods in aviation history. With his background and experience as an officer of one of America's largest and diversified corporations, jet pilot and aeronautical engineer, he has a comprehensive understanding of the many problems facing aviation and the leadership capability to direct the efforts to solve them.

Shaffer was born in Everett, Pennsylvania, on 25 February 1919. His secondary schooling was gained at the Franklyn and Marshall Academy in Lancaster, Pennsylvania, from which he obtained an appointment to the United States Military Academy at West Point, where he played Varsity football and graduated in the top half of his class. He secured his pilot rating while still a cadet at the Academy.

During World War Two, Shaffer flew in 46 combat missions as a B–26 pilot with the U.S. 9th Air Force in Europe. After the war, he took an aeronautical refresher course at the engineering school of the Wright-Patterson Air Force Base at Dayton, Ohio, graduating in 1945. He then went on to achieve a Master of Science degree at the Columbia University in 1946.

In June 1946, Shaffer was named production project officer of the Air Force's B–50 programme, and he served in this capacity until September 1948, when he was appointed weapons systems manager of the B–47 project. He remained in that post until 4 January 1954, resigning his Air Force commission as a lieutenant colonel to become general production and assistant plant manager of the Mercury assembly plant of the Ford Motor Company at Metuchen, New Jersey.

He joined Thompson, Ramo and Woolridge Inc. as assistant to the group vice-president, automotive, on 1 January 1957. In June 1958, Shaffer was named vice-president of marketing and sales for the whole of the equipment group of companies. In 1962, he was elected a corporate vice-president (consumer relations) and was later assigned to Washington D.C. with responsibility for the company's advanced technical capability; to broadening requirements and changing national priorities; specific sales assignments on major programmes; plus administrative responsibility for the Washington operations which involved some 500 engineers and scientists. He resigned from that position after being recommended to that of the head of the F.A.A.

Following his nomination . . . he played a key role in the drive for the most comprehensive aviation legislation ever enacted into Federal Law – the Airport and Airways Development Act of 1970. Under his leadership, the F.A.A. has taken prompt action to implement the new programs authorized by the Act.

In America, civil aviation comes under the direct control of two government departments and one independent agency. There is the Civil Aeronautics Board, which is responsible for the licensing of companies wishing to operate air transport as well as granting rights to specific carriers for specific flight sectors, and authorizing the amounts of money they may charge passengers, goods and mail for transportation over such routes.

There is the F.A.A., responsible for the whole of the technical and operational side of aviation, including the rule-making and policing of such rules in order to maintain certain minimum standards of safety, which the F.A.A. itself has determined as being necessary.

And finally, there is the independent National Transportation Safety Board, charged under the Department of Transportation Act of 15 October 1966, with the investigating and determining of the probable causes of all aviation accidents, and where specific faults have been seen to have been involved, to make recommendations to the Federal Aviation Administration in order to avoid recurrences in the future.

Over the years, the F.A.A. has come to think of itself as the sole responsible body for civil aviation – not only at home in the Americas, but also theoretically around the world, as American manufactured aircraft are dominant in most other countries.

And it has, over the past few years, felt an inward revulsion at the fact that it seems to have been continually pestered by this "junior body" the National Transportation Safety Board, showering it with pieces of paper called "recommendations" which have no substance in law.

This interdepartmental in-fighting has probably been responsible for many unnecessary air passenger deaths over the years, and may be traced back to the Federal Aviation Administration's dislike of the fact that practically every recommendation made by the National Transportation Safety Board points directly to an area of failure within the F.A.A.'s own regulatory or technical activities.

During Shaffer's reign as Administrator, according to the records of the Safety Board, the rejection of N.T.S.B. recommendations increased with each passing year. This detestation may be seen in a comparison of the figures released by these departments, supposedly being annual totals for the number made by the National Transportation Safety Board. For example, in 1970, the N.T.S.B. said it issued 121 safety recommendations; 23 of these were totally complied with by the F.A.A.; a further 22 received "essential compliance"; 17 others were "partially complied with"; while 12 were rejected outright; and the remaining 47 are still "pending".

When an approach was made to the F.A.A. asking for "their version" of the same figures for the same period, they said that the N.T.S.B. issued 115 safety recommendations. Of these, 105 were "totally complied with" and the remaining 10 were rejected.

The same thing may be seen in 1971. The number of recommendations (70, according to both the N.T.S.B. and the F.A.A.) resulted in 25 being totally complied with according to the N.T.S.B. (62 according to the F.A.A.); 18 being complied with in essence according to the N.T.S.B.; a further 9 being partially complied with; 4 being rejected outright (whereas the F.A.A.'s figure is 8); and 14 still pending.

Of all the years during which Shaffer was administrator, 1972 was the worst for the F.A.A.'s rejection of N.T.S.B. recommendations. The Safety Board said it issued 234 (the F.A.A. says the figure was 202); the Board says that only 35 of them were given total compliance (whereas the F.A.A. maintains it adopted 184); 21 were complied with in essence; a further 19 received partial compliance; 27 others were rejected outright (the F.A.A.'s figure for rejection is only 18); *and 132 of these recommendations are "still pending"*.

On 2 January 1973, a press-release from the F.A.A. read in part:

He (Shaffer) also had been in the forefront of many other innovative efforts to make the National Aviation System safer and more efficient. These include: developing new policies and programs for environmental quality; implementing a comprehensive anti-hijacking system and improving the overall air transport security program; advancing the automation program in air traffic control including establishing new management techniques in systems engineering; restructuring F.A.A. regional organizations; establishing new management and supervisory training programs; stepping up recruitment of air traffic controllers; providing new centralized flow control of the nation's air traffic to reduce congestion and delays; and establishing a general aviation accident prevention program to replace added emphasis on standardization of training to reduce accidents.

Unfortunately, this hand-out does not deal with an aspect of Shaffer's administration that was to prove disastrous in the long run: the rejection, during his tenure, of most of the National Transportation Safety Board's recommendations.

With regard to Shaffer's interests in implementing N.T.S.B. safety recommendations, some of the correspondence tells its own story.

On 2 May 1970, a Douglas DC–9 owned by Overseas National Airways Inc., and operated as the Dutch Antilles carrier A.L.M. Flight 980, was ditched in the sea approximately 30 miles east-northeast of St. Croix, Virgin Islands. The cause of the accident, in which 25 people lost their lives, with a further 37 of the 63 on board being injured, was operational stupidity; the National Transportation Safety Board determined that it was probably due to fuel exhaustion which resulted from continued unsuccessful attempts to land at St. Maarten, until insufficient remained to reach an alternative airport. A contributing factor, they determined, was the reduced visibility in the approach zone because of rain showers – a condition not reported to the flight.

The correspondence between the N.T.S.B. and the F.A.A. opened on 10 September 1970, with a letter from the former's chairman John Reed, to Shaffer, which read in part:

The preparation and bracing of the passengers to accept decelatory forces is, in the Board's opinion, an integral part of configuring the aircraft for a ditching or emergency landing. As such, it should not be left to memory, but should be placed on the emergency checklist for these procedures.

Therefore we recommend that:

The item, "warn passengers", be inserted into the emergency landing or ditching checklists of all carriers, and that it be one of the last items prior to impact, yet sufficiently advanced to insure adequate time to permit the passengers to brace for a crash.

This matter of telling the passengers to prepare themselves for an impending emergency landing seems so basic that one would have thought it would have already been a part of the F.A.A.'s civil air laws written into all the air carriers' operations procedures.

On 28 September 1970 – 18 days after the original letter, and 149 days after the crash – Kenneth Smith, the acting administrator of the F.A.A., replied to the N.T.S.B.'s letter to Shaffer.

We agree that while the emergency procedures' sections of the major air carriers' operations manuals are comprehensive in regard to an anticipated ditching, the preimpact warning to the passengers is not among the items on the checklist.

We will, therefore, initiate a bulletin to our principal operations inspectors requiring them to see that each of their assigned air carriers includes the item "warn passengers" on their preditching checklists.

Here, we get an insight into the way the F.A.A. operated under Shaffer's leadership. Instead of ensuring that in future all airlines must *by law* have the instructions to warn the passengers of an impending impact, the administration apparently agreeing that there was a fault in their regulations, took the less effective step by having their inspectors quietly talk to the airlines' operations managers instead of issuing a demand that all carriers include these instructions in their emergency procedures. This evidently, was a face-saving device, used by the F.A.A. with the one major flaw that there was no legal compulsion on the airlines to maintain this standard at all times.

On 29 October 1970 – 180 days after the accident – John Reed of the National Transportation Safety Board, again wrote to Shaffer after a hearing in San Juan, Puerto Rico, in conjunction with this DC–9 ditching and at which the captain of the aircraft testified that his public address system was inoperative.

The N.T.S.B. recommended that:

The Federal Aviation Administration take the necessary steps to insure that no flight now requiring the briefing of passengers regarding emergency procedures be dispatched without an operable public address system. The system should be functioning so that the flight deck crew can speak to the passengers and a cabin attendant can speak to the passengers from at least one cabin station.

Once again, we see a recommendation concerning basic flight safety which points to a gap in the F.A.A.'s own regulations.

On 5 November 1970 – 7 days after the N.T.S.B.'s letter, and 187 days after the accident – Shaffer replied:

We will consider the Board's recommendation as a petition for rule making and will therefore issue a notice of proposed rule making on public address systems in the very near future.

There is absolutely no logic in this response. As the N.T.S.B. had pointed out, the F.A.A.'s mandatory regulations *already* called for the passengers being warned of an impending impact or ditching. How else were the airlines expected to warn them other than by using the public address system? And again, if one section of the regulations state that the passengers must be warned, why not make it an offence for an aircraft to take off with its public address system out of action? Here we see exactly the same thing: a flaw in the F.A.A.'s regulations being pointed out by this "underling" the National Transportation Safety Board, after they had uncovered it during an accident investigation.

On the same overall subject, a further series of letters was exchanged between the Safety Board and the F.A.A. On 14 September 1970, John Reed wrote to John Shaffer, pointing out that in 1964, the F.A.A. had issued a directive to their field personnel that they put special emphasis on checking the adequacy of the fabric-to-metal type seat-belts. He also pointed out that since that time, deficiencies had been noted in an accident involving United Air Lines at Norfolk, Virginia, in 1967, and an in-flight turbulence incident reported by United Air Lines in a Caravelle in 1965, after which one of the four injured passengers died three days later.

Failure of such seat-belts to lock effectively under loading (continued John Reed) had been attributed to wear of the cam serrations. It is also thought by many that with acceleration conditions, flailing of the loose end of the belt may activate the cam and cause unwanted release of the belt. The demonstrated inadequacy of this seat-belt locking mechanism leads the Board to the conclusion that there is a serious question as to the reliability of this safety device.

A review of accident records over the past ten years reveals no indication of the failure in the metal-to-metal type of seat-belt.

In view of the above, the Board recommends:

That the F.A.A. take the necessary steps to eliminate, within a reasonable time, the use of fabric-to-metal type of seat-belts in aircraft of U.S. Registry and require the metal-to-metal type of seat-belt with a standardized activating device.

Once again, it is interesting to note the dates mentioned in that N.T.S.B. letter. The deficiencies in the fabric-to-metal design were originally uncovered in 1963. They were again reported in 1965 and 1967, and were still being found in 1969 and 1970, all of which resulted in recommendations from the Safety Board, and all of which were turned down by the F.A.A.

On 21 October 1970 – 37 days after the N.T.S.B. had made that recommendation and 172 days after the DC–9 ditching accident – Kenneth Smith, the acting F.A.A. administrator replied:

> As noted in your letter we have emphasized to field inspection personnel and operators the importance of maintaining the seat-belt webbing material and latch serrations in proper condition so that the serrated cam would grip the fabric securely and not permit the webbing to slip through the buckle when latched. Service experience since that time has not indicated the need for replacement of this type of belt.

Here again the F.A.A.'s less than alarmist position becomes evident as Smith says experience had not shown the need for replacement of this type of seat belt and yet, curiously enough, later in the very same letter, says that the F.A.A. is conducting a technical review of such belts.

But the evidence of a design fault was starting to mount against the F.A.A. In that reply, dated 21 October 1970, to an N.T.S.B. letter on the same subject, and following yet a further series of reported cases of these restraining devices unintentionally coming undone, Smith said that: "the service experience . . . had not indicated a need for replacement of this type of belt." However, in yet another letter from Shaffer dated 2 December 1970, there is an admittance that the seat-belt design is faulty. "The F.A.A. amended the . . . specifications . . . recommending that metal-to-metal belt locking mechanisms be used . . . " The operative word is "recommending". The F.A.A. was "recommending" – not "enforcing" – the usage of metal-to-metal type seat-belts.

A major point of contention, made adequately clear in this 1970 correspondence, is that the F.A.A. left matters of major safety importance to the manufacturers to make recommendations to the owners of the aircraft, on the assumption that in this fashion the information would get around. By permitting the plane makers to "recommend" certain changes in the design or installed equipment, this firstly means that the F.A.A. does not have to publicly admit that it has certificated a piece of equipment that was subsequently found to be faulty; and secondly, it allows the maker to have a go at quietly correcting the situation, seemingly off his own bat, and in such a fashion that no one can later blame him for installing faultily designed equipment in the first place. This supposed method of correction is known as the issuance of Service Bulletins.

The other, and more positive way of ensuring that faults in design or equipment be immediately corrected, is for the F.A.A. to publish what is known as an Airworthiness Directive. This is similar to an order, directed at all owners and operators of the faulty equipment as well as the manufacturers, that aircraft which contain such design faults must be immediately modified (or at least by a certain date if the problem is not one of dire emergency), and if this is not done, these machines must remain on the ground. Certainly, the publication of an Airworthiness Directive is a drastic measure – but then it is needed in order to correct drastic safety situations.

The Airworthiness Directive should be issued in every instance where accidents and incidents have uncovered faults in equipment or operations – but this is not done. The only reason one can find for not issuing Airworthiness Directives, is that of saving face. Because Airworthiness Directives are widely distributed, the aviation industry (in which one has to include the F.A.A.) seems to be of the opinion that they tend to publicly show that faulty equipment was permitted to be manufactured, to be operated by the airlines and, furthermore, to be somehow passed as being of a sufficiently high standard by the F.A.A.'s own inspectors.

However, the one great difference between correcting faults by means of Airworthiness Directives and the Service Bulletins that are quietly handed around by the manufacturers, is that the former must be relayed to all operators of the faulty equipment who must in turn, implement the instructions contained therein: whereas Service Bulletins, which are not legal mandatory requirements, have been known over the years to have been lost *en route*, or have been received by the operator of the faulty equipment in amongst sheathes of other papers, and neither their presence nor their importance has been observed. This is exactly what happened in one of these cases of the failure to remedy the seat-belt design that comes undone. Somewhere along the line, the February 1968 Service Letter which *recommended* that metal-to-metal seat-belts only be used, got lost, and this owner never realized that such a *recommendation* had been made. If the F.A.A. had issued a mandatory Airworthiness Directive instead of leaving the matter in the hands of the manufacturer to try and inform all the owners of his aircraft, it would have been a breach of the Federal Air Regulations for (a) that owner to install unsatisfactory seat-belts, (b) to permit the aircraft to take off with unsatisfactory equipment on board, and (c) for him to be sold and have installed, the faultily designed seat-belts in the first place.

In March 1971, a Pan American Boeing 747 encountered severe turbulence while it was climbing away from the airport, and during the shaking, six passengers and a stewardess were seriously injured, with a further sixteen people receiving minor injuries.

Following their investigation, the National Transportation Safety Board wrote to the F.A.A. on 7 April 1971:

Seat-belt Discipline: During the accident, the seat-belt signs were "on" throughout the flight. However, of the 2 hospitalized passengers, one indicated that she did not have her seat-belt fastened, and the other had his seat-belt fastened but it was very loose because he was not able to take up the slack of the belt. Both of these passengers were injured when their heads struck the ceiling on the initial sharp downdraft, but were able to maintain their seated positions during the remaining turbulence encounter.

The Safety Board recommends that:

1. Seat-belt discipline be strictly enforced when the seat-belt signs are "on". Attendants should make a careful visual inspection of all seat-belts before take-off and offer assistance to anyone encountering difficulty with a tight fit. When the seat-belt signs are "on" for prolonged periods, a public address announcement should be made at regular intervals.

747 Overhead Bin Failures: During this encounter with turbulence several of the overhead storage bins in the passenger compartment dropped open, allowing their contents to spill out. It is not known if these reported failures contributed to any injuries of cabin occupants. However, the Safety Board recommends that:

2. Locking mechanisms be inspected and either replaced with locks of a new design or the defective lock mechanisms be returned to serviceable condition by rework or repair.

3. The F.A.A. correct any crashworthiness deficiencies in Boeing 747 overhead storage bins by establishing a deadline date for compliance with any modification requirements.

Economy Seat Head-rest Separations: During this accident, several seat head-rests were reported to have been thrown from their seat units. Examination of the like head-rests in another Pan American Boeing 747 revealed that all such units tested were easily removed by hand without deactivating the lock mechanism. It is not known if these reported failures contributed to injuries, but the Safety Board recommends that:

4. The F.A.A. examine the seats with a view towards improving the

crashworthiness of seat/head-rests and establishing a deadline date for compliance with any modification requirements.

Narrow Aisle Stretchers: Following the abort of the flight and the landing, difficulty was encountered in removing from the aisle, passengers suspected of having back injuries. This was because the aisle widths were too narrow for standard stretchers, resulting in great difficulty in transferring patients from lying positions in the aisle to stretchers.

The Safety Board recommends that:

5. The F.A.A. advise medical facilities serving airports to stock narrow "carrying boards" or narrow stretchers that can easily be used in the space of an air carrier passenger compartment aisle to facilitate the removal of non-ambulatory patients.

As a result of these N.T.S.B. recommendations, Shaffer did not order one single regulatory amendment to alter the situation. Instead, he dealt with each problem in a fashion that suggested changes would be made, but in no case ordered that it be immediately and mandatorily corrected.

In his reply, dated 7 May 1971, he said, for example:

Seat-belt discipline: Federal Air Regulation 121.317(b) requires that "... each passenger shall fasten his seat-belt and keep it fastened while the seat-belt sign is lighted". It is apparent that some passengers do not abide by this rule especially when the seat-belt sign is left lighted for protracted periods. We will issue an operations bulletin to all of our inspectors who have certificate responsibility for air carriers and their training programs, emphasizing the importance of oral announcements and better surveillance to assure compliance with seat-belt fastened commands and security. FAR 121.317(a) requires that seat-belt signs be visible to all passengers.

The other problems were treated in a like fashion. In other words, no mandatory corrective actions were taken and the Safety Board's recommendations for compulsory measures, like so many hundreds of others, were not adopted by the F.A.A. The position before and after Shaffer had written that falsely assuring statement, was unchanged.

More often than not, the F.A.A. tried to cover its own failings by "suggesting" and "recommending" that people take certain actions and do certain things, more than insisting that corrections be rapidly made.

And so Shaffer's failure to implement National Transportation Safety Board recommendations continued unabated. Seven of them made to the F.A.A. following the crash of a Boeing 720 belonging to Western Airlines on 31 March 1971, when part of the tail assembly broke, were completely dismissed.

Yet again, he chose "the easy way out" by having the manufacturers of the Hansa HFB–320 executive jet issue a non-mandatory Service Bulletin in order to try and warn operators of these aircraft of the ever present danger that unless properly locked, the pilot's emergency exit window could blow out in flight. He evidently preferred this Service Bulletin persuasion to the more positive action of legally enforcing a change in the faulty design once it had revealed itself in an accident or potential accident situation.

As has been seen in numerous cases in the past, trying to correct deficiencies in aircraft designs solely by the use of the secret, unpublished Service Bulletins, seems to have increased the aviation danger factor, more than having increased safety.

The National Transportation Safety Board's recommendations that were invariably not adopted by Shaffer, included some potential problems that were uncovered by crash investigators during the course of their work. For instance, on 1 June 1972, John Reed of the N.T.S.B. wrote to Shaffer on the subject of self-illuminating emergency exit door handles. In earlier correspondence, Shaffer had rejected this idea out of hand even though the Board had pointed out numerous cases of passengers having difficulties in locating the handles after night-time accidents, or those in which a fire involved the giving off of enormous amounts of smoke.

Absolutely nothing has been done about this problem, either by Shaffer during his stay as F.A.A. administrator, or by anyone else since.

It is not as though large amounts of money would have to be expended in incorporating most of the recommendations put before the Federal Aviation Administration – it is not as though hard experience had not been adequately demonstrated to Shaffer and his minions, especially one called Rudolph, that when design deficiencies were uncovered, the only positive way of correcting the situation in the fastest possible time was to issue a mandatory order against the manufacturer and all the owners of the aircraft involved, demanding the immediate incorporation of the modifications. The exact definition of his office's powers and responsibilities is set forth in Section 103, of the Federal Aviation Act of 1958, and it reads:

In the exercise and performance of his powers and duties under this Act the Administrator shall consider the following, among other things, as being in the public interest:

(a) The regulation of air commerce in such a manner as to best promote its development and safety and fulfil the requirements of national defense;

(b) The promotion, encouragement, and development of civil aeronautics;

(c) The control of the use of navigable air space of the United States and the regulation of both civil and military operations in such air space in the interest of safety and efficiency of both;

(d) The consolidation of research and development with respect to air navigation facilities, as well as the installation and operation thereof;

(e) The development and operation of a common system of air traffic control and navigation for both military and civil aircraft.

Section 307 (e). The Administrator from time to time may grant exemptions from the requirements of any rule or regulation prescribed under this title if he finds that such action would be in the public interest.

Section 601 (a). The Administrator is empowered and it shall be his duty to promote safety of flight of civil aircraft in air commerce by prescribing and revising from time to time:

(1) Such minimum standards governing the design, materials, workmanship, construction and performance of aircraft, aircraft engines, and propellers as may be required in the interests of safety . . .

(2) Such minimum standards governing appliances as may be required in the interests of safety . . .

Section 601 (b) . . . The Administrator shall exercise and perform his power and duties under this Act in such a manner as will best tend to reduce or eliminate the possibility of, or recurrence of, accidents in air transportation, but shall not deem himself required to give preference to either air transportation or other air commerce in the administration and enforcement of this title.

The reader should measure Mr. Shaffer's performance of his duties against the definition of them in this law.

The last mentioned quotes from the Act give one the suspicion that Shaffer, and indeed many other administrators of the F.A.A., might have misunderstood their function. Even though the Act does permit use of "considered judgement" in how they go about achieving the goal, Section 601 does *not* empower them to experiment with peoples' lives by using the known "hit-or-miss" methods of bringing about design fault changes by "recommendation" more than by "legislation", while, at the same time, keeping the whole matter as quiet as possible so as not to give the badly designed equipment a taint of unreliability, thereby scaring innocent passengers and intending purchasers away from it.

CHAPTER 5

THE fabrication of the millions of parts needed for DC–10 fuselage number 1 rapidly took shape. On 13 January 1970, the forward body section, from the rear of the cockpit to the leading edges of the wings, which had been made by Convair at their San Diego, California plant, was shipped to Douglas at Santa Monica. The same afternoon, it was joined to its already completed nose section. The first part of the aircraft was now 55 feet in length, 19 feet 9 inches in diameter, and weighed around 15,500 lbs. Shortly afterwards, the 33 feet long overwing fuselage section was also delivered by Convair, and it was rapidly followed by the 38 feet rear body portion.

As the different air frame sections arrived at Santa Monica, there was then installed all the wires, pipes, controls and other supplemental bits and pieces before they were trans-shipped to the Long Beach plant for final assembly.

By the first week in April 1970, DC–10 number 1 *looked* like an aeroplane, with the horizontal stabilizers being fitted into place on its tail by one engineering group, while another was attaching the leading edges into the box-like wing structures.

The nose landing gear was in place, but the main undercarriages had been temporarily removed in order to cut some minor relief holes into the underside of the fuselage. In the cockpit, all the frames for the instruments and the cabling were ready for instrumentation. Douglas engineers were also putting in the finished trim inside the cabin, made up of large prefabricated sections already complete with double insulation and face panelling. Lavatories in modular form were also in the final stages of being installed.

By 2 April 1970, only 171 of the thousands of fabricated parts had not arrived at Long Beach plant. An order for valves had been reported missing *en route*, and the tail cone, into which fits the Auxiliary Power Unit that supplies both electricity and compressed air while the machine stands at the airports, had not been completed as some further development work in this area was required.

While DC–10 number 1 was in its last stages of being rigged out in a hangar that was built for DC–8 production (the much larger DC–10 hangars had not been finished), fuselage number 2 was coming together in final assembly across on the other side of the Long Beach facility. Work was also going on in various stages at various places to the remainder of the Douglas products: the DC–8 Series 60s, the DC–9s, and the initial five DC–10s.

Comparative dimensions to scale.

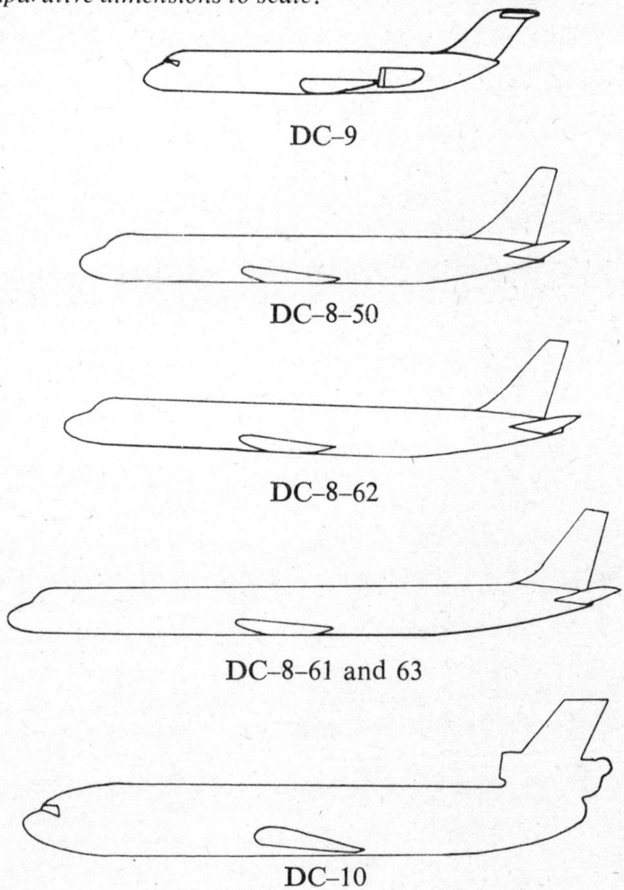

At last, DC–10 number 1 was towed to a new functional check-out building where the upper fin was attached to the vertical part of its tail and flight control cables were rigged. (Engines and passenger seats on the normal production line would at this point have been installed before the machine passed through that section.)

The bodily completed DC–10 rolled off the assembly lines. It was weighed and delivered to the Douglas pre-flight building, where crews were to conduct pneumatic, hydraulic, electrical, fuelling and many other tests before it could take to the air for the first time.

However, it was during this period of the very initial construction of the parts for DC–10 number 1, that our story really starts.

At that time when McDonnell Douglas was in close contact with Convair – a division of General Dynamics Corporation – they were told by the latter that General Dynamics was of the opinion that the DC–10's cargo-door latching and locking design was of such a nature that it was practically certain that one day, somewhere in the world, at least one of these cargo doors would come off in flight during the 20-year period that Douglas was expecting to be the "life" of the machine. If this happened, and they were sure it would, disastrous consequences would occur.

General Dynamics made it very clear to Douglas that they were only the sub-contractor – the company that was building the parts for McDonnell Douglas, from the design plans being supplied to them by McDonnell Douglas. The report went on to say that General Dynamics knew that safety of the design was not their responsibility, however they were of the opinion that something should be done.

General Dynamics engineers considered the doors' locking systems so risky, that besides writing to McDonnell Douglas, explaining that they were sure a door would come off in flight, they also conducted a "failure" analysis of the problem. This showed that *if* the door did come off, it would also cause the collapse of the passenger cabin floor. A copy of this report was *also* sent to McDonnell Douglas.

(It is alleged by plaintiff's Counsel in litigation currently in process in Los Angeles that before this "fault analysis" was circulated amongst the designers and engineers working on the DC–10, someone in the Douglas Aircraft Company, made a copy of this report *but omitted* the analysis consequence that once the door had come off at altitude, the floor would collapse. Douglas now denies this.)

And of course, when it was realized that there was a possibility of the DC–10's floor collapsing, it did not take too long for them to also conclude that this would affect the aircraft's flight control cables that were ducted *through* the floor beams.

All this was ignored by McDonnell Douglas, and the construction of DC–10 number 1 continued, exactly as the company had planned.

CHAPTER 6

IN Washington D.C., to complement John Shaffer, President Richard Nixon put forward a Kenneth Myers Smith as a candidate for the position of deputy administrator of the Federal Aviation Administration. Smith started his duties in this capacity on 11 May 1970.

Mr. Smith (according to an F.A.A. release), brings more than 30 years of aerospace industry experience to his new job, which is second-in-command of the Federal agency responsible for the regulation of flight safety and the operation and maintenance of the air traffic control/air navigation system.

Smith was born in Sacramento, California, on 3 January 1920. He studied aeronautical engineering and air navigation at the St. Mary's and California Polytechnic Universities under the U.S. Navy V5 programme in 1943 and 1944. He completed his commercial pilot's course at the Friedken School of Aeronautics in California in 1945.

His first job was in 1941 as an electrical installer with the Consolidated Vultee Aircraft Corporation in San Diego, California, shortly after he left high school. By January 1952, with a period in between in navy aviation duty during the Second World War, he had risen to the rank of assistant service supervisor, prior to being transferred to Washington D.C. as their local representative, and later becoming chief of flight and service of the firm's guided missile division.

In September 1952, Smith was appointed military sales coordinator with Consolidated Vultee, before moving up the scale to manager of long range planning and director of military relations. He was a corporate vice-president from July 1960.

In May 1962, Smith parted company with his old firm in preference for the position of vice-president of marketing with a division of Bell & Howell Inc., in Pasedena, California. In April 1964, he was vice-president and assistant general manager of the

Aero Commander division of the Rockwell Standard Corporation in Oklahoma City, and was promoted to the position of general manager eight months later.

He became president of an Oklahoma City management consulting firm in April 1966, leaving them in May 1967, to take up the presidency of Windecker Research Inc. He was appointed vice-chairman of their board in June 1969, resigning that post to become the deputy administrator of the F.A.A.

Smith lasted at the F.A.A. an even shorter period of time than Shaffer and it appears that most of his work there was "deputizing" for Shaffer while the latter was out of his office.

CHAPTER 7

BACK at Long Beach, the 29 May 1970, was quite momentous as far as the DC–10 was concerned. It was on that day that Douglas engineers had scheduled a "pressure vessel test" on the naked airframe of prototype number 1.

The machine itself had been built to the stage where it consisted of fuselage skin and wings; there were no cabin fixtures or engines attached. The idea of the test was to find out just how strong the outer fuselage was. Engineers closed and sealed the doors (all the windows were in place), and then pressurized air was pumped inside it until something burst.

The test was progressing satisfactorily until, suddenly, and at a pressure way below that considered to be the minimum, the forward right-hand-side cargo compartment door exploded open. This was the last thing anyone had expected to happen. It was a certainty that the design of the door operation or latching mechanism was the fault, and this was proven to be the case after the engineers had made a thorough inspection.

Douglas had been involved in constructing the DC–10 up to that time, for just on sixteen and a half months. Vast amounts of money had been spent on the project. And with the prototype number 1 now a living shell, with the production lines tooling up not far behind it, that was not the time to learn that the compartment door designs were faulty. But not only did the door come flying open during the tests, the forward section of the cabin floor also collapsed when there was a sudden differential of pressure in the passenger and cargo compartments. Up to that time, the DC–10 design did not cater for any relief venting between these compartments, but very soon after certain modifications were made.

The engineers who inspected prototype number 1 after the pressure tests must have seen, or at the very least must have realized, that with a collapsing of the passenger cabin floor, especially at the front of the aircraft, all the controls, cables, and wires

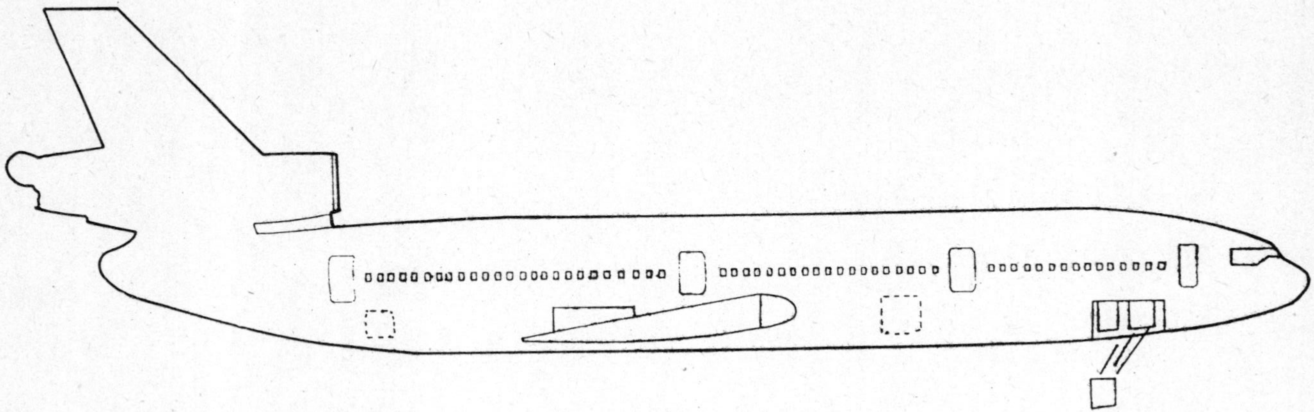

The cargo compartment door that exploded open in the pressure tests.

which were being ducted *through* the floor beams, would have been severed or, at the very least, stretched and jammed. Through the forward passenger cabin floor beams pass the control wires and cables to all of the engines and flight control panels on the wings and tail. To have this section collapse, could even then be seen to be the beginning of a major disaster. If the aircraft were in flight, the pilots would lose all control and the thing would crash to the earth killing everyone on board as well as anyone who happened to be on the ground at what would inevitably be a wide area of wreckage. If this happened over a built-up city, hundreds of people could die, including the maximum of 361 on board, were it a fully-laden flight.

The Douglas engineers realized that the addition of pressure relief vents between the passenger compartment and the forward and middle cargo holds were necessary because they added them. But the only logical reason for not establishing the same extent of venting in the rear section, into the aft cargo hold, was the lack of design strength in the DC–10's passenger cabin floor over that compartment. It had been suggested that if this had been even attempted, the floor could collapse under the weight of a few people standing on it.

In March 1974, the then president of the Douglas Aircraft Company, who was at that time under great pressure and strain because of the Turkish Airlines DC–10 disaster, said:

> On 29 May 1970, while conducting a ground pressurization test during the construction stages of the DC–10 fuselage number 1, the forward cargo door had been improperly latched and the door came open due to cabin pressure loads. The cabin floor sustained local damage in the vicinity of the door. Because of this incident, all cargo doors were redesigned to incorporate a "vent door", linked to the lockpin actuating handle and sized so as to make it impossible to pressurize the aircraft if the door was not properly latched and locked. The system was installed on all DC–10 aircraft prior to delivery.
>
> The F.A.A. requested, and Douglas performed a demonstration of the adequacy of that "vent door" in terms of prevention of pressurization of the aircraft. A compliance report was prepared and submitted to, accepted, and approved by the F.A.A. as part of the type design and certification.
>
> The F.A.A. required 32 separate systems fault analyses on the DC–10 at this stage for certification of the aircraft, which were submitted and are part of the type certification.

In addition to that, Douglas went far beyond and conducted 175 separate failure mode and effect analyses on sub-systems which involved 1,107 separate components, and 3,400 different failure modes. So a great deal of fault analyses and fault testing was accomplished, some by F.A.A. requirement, and much more of our own initiative.

Arvin O. Basnight, the director of the Federal Aviation Administration's office in Los Angeles, was fully aware of this early door failure during these pressure vessel tests, even though no F.A.A. officials were present when it happened. He said that the right-hand forward cargo door exploded open in the test after it had been hand-cranked closed rather than shut by the normal electric motor mechanism. Basnight had admitted that after they had heard about this door bursting open, F.A.A. inspectors . . .

> . . . went into the matter to the extent of making the best judgement we could. We did not require what, in retrospect, might have been required – a so-called fault analysis testing.

With these statements of earnest testing and miles of computer analyses, no one outside a small group in McDonnell Douglas and a similarly small group in the F.A.A.'s Los Angeles office, knew anything about this early door failure until the latter part of March 1974. The National Transportation Safety Board knew nothing about it, even though they were involved, in June 1972, in the investigation of a near major disaster when a cargo door, in similar fashion, flew open with the aircraft up in the air.

Besides, with the benefit of hindsight, we will later be able to determine that even though claims were made about all the magnificent tests being carried out, they quite evidently did not stop at continued repetition of faults that would eventually end in the world's worst air disaster.

CHAPTER 8

WITH the prototype fuselage number 1 completed, the DC–10 moved under its own engine power for the very first time on the morning of 23 July 1970. Soon after, a whole series of ground tests, including progressively faster movements down the adjoining Long Beach Airport runway, brought the designers and engineers to the stage of having it take to the air. The first flight was scheduled to take place on 15 September 1970, but with the forward cargo door repaired, the passenger cabin floor rebuilt and modified in parts, and all the other tests seemingly successfully completed, there was no need to wait for the scheduled date.

On the morning of Saturday, 29 August, the huge DC–10 took to the air for the first time. The flight lasted 3 hours and 26 minutes and the machine reached an altitude of 30,000 feet, and a maximum speed of 345 miles per hour.

Its journey took it up and down the Californian coast before landing at the inland Edwards Air Force Base which was to be its flight test headquarters. On that occasion, Clifford L. Stout, the DC–10 project pilot was at the controls; Harris C. Van Valkenburg the deputy chief engineering pilot, was the first officer; and John D. Chamberlain was the flight engineer and second officer.

After it had landed at Edwards, the pilots said that they had had some troubles but they were described as "minimal". The only major points noted were that the air-speed calibration was made difficult by the loss of three trailing cones used for precise readings, and they also had a problem or two getting the nose landing gear "down" and locked in a free-fall test. But the pilots were unanimous in their praise for the view out of the large front windows throughout their first landing into the Edwards Air Force Base. During their 6° nose-up "flare" immediately prior to touching down, they estimated that they could see the runway as close as about 40 feet in front of them, which meant they had an exceptional range of visibility.

They also reported that the DC–10 stalled in a "classic and

docile" style adding that the nose pitched down slightly as the stall was reached, with good lateral and directional response.

On board the aircraft were 25,000 lbs. weight of instruments that radioed back to the ground engineers at Long Beach, thousands of pieces of information every second. This in turn permitted them to keep fully abreast of what was happening up in the air to both men and machine.

Here was the signal for the release of some fine prose from the McDonnell Douglas publicity-getters:

The new McDonnell Douglas DC–10 tri-jet transport is a 'good neighbour' airplane which will enter and leave airport areas with scarcely more than a polite hum and a woosh. And it won't be trailing long plumes of dark smoke. Not even a whisp.

The DC–10, rather large compared with the first generation jets, has three engines – one under each wing and one which appears to be thrust through the vertical stabilizer. Each of these massive General Electric CF-6 engines produces more than twice the thrust of the jets on conventional four-engined transports. But they generate significantly less noise.

It took years of dedicated development work by General Electric and McDonnell Douglas engineers to attain the low noise levels and smokeless exhaust. Airport communities and environmentalists will agree that their committment was well worth the effort.

The DC–10, produced by the Douglas Aircraft Company division of McDonnell Douglas Corporation, is the first new transport to be certified under the stringent new sound standards established by the Federal Aviation Administration. . .

Prior to this time, the American Air Line Pilots' Association had conducted an evaluation survey into the then unflown plane, and this constituted the first part of their "appraisal".

They were not too critical of the aircraft but naturally, pointed out a number of small items which they considered needed further attention. They indicated that many features of the DC–10 were similar to those of the DC–8 – which were similar to the DC–9 – not too different from the DC–7 – and quite like some that were in the DC–6 and the DC–4.

During the final portion of the Phase 1 era (the A.L.P.A. Report states), the committee noted that a large number of McDonnell Douglas and industry inspired safety improvements have been included in the DC-10. Although we have not been successful in obtaining all of the recommendations we made, the DC–10 will represent a significant step forward safety-wise through the adoption of many new safety concepts. The experience gained in building the DC–8/DC–10 fleet, has been heavily drawn on by McDonnell Douglas in their selection of materials and design practices. The General Electric Company has built many engines and their capability is widely recognized.

One feature in the DC–10 – as well as the DC–9, 8, 7 and 6 – that has been proven to be disadvantageous at certain times, is in the design of the engine's operation mechanism, for the pilots to be able to select reverse thrust while the aircraft was still up in the air. At least two major accidents have been caused by pilots using this "feature" either intentionally (as in the case of the Alitalia DC–8 at Kennedy Airport, New York) or unintentionally when it was shown that if the pilot pulled the throttle back at a fast rate, the engines could automatically go into reverse thrust. This is certainly not what one would call "a good design concept". However, it is one that has been accepted by the pilots without comment.

Some of the other points they raise in their report are well worth citing:

CONTROL FEEL: The size of the DC–10 requires that all flight control surfaces be hydraulically powered. There is no feedback of control surface forces to the pilot. Control feel is provided by a load feel mechanism.

Where flight control synthetic feel systems are utilized which modulate control feel to the pilots at higher speeds, adequate safeguards should be built in to preclude application of excessive control movement in the event of failure or malfunction in the feel system. Warning of failure of such systems should be installed. . .

Since there is no way for pilots to test the malfunctions that could conceivably occur in the load-feel mechanism, we must accept the fact that McDonnell Douglas engineers have taken every foreseeable precaution in designing the system so that there will be no hard-over control inputs that might overstress the aircraft and injure passengers.

This "control feel" and the knowledge of whether the automatically induced feel equipment is working or not, is quite an important point. It is similar to an automobile's steering. The further you turn the steering column, the more tension is felt in the wheel. This instinctively gives the driver an indication of the extent of the turn being effected. Without this "tension feel", steering a car would be that much more difficult and besides, if the tension were to be automatically induced, the driver should know whether the equipment is working properly or not, for obvious reasons.

CIRCUIT BREAKERS PRINT STRIP LIGHTING: Recommendation: Circuit breaker groups by system should be colour coded in the strip lighting or by another identifiable procedure.

Comment: When it becomes necessary to identify a circuit breaker in a certain system it would speed up the identification if the system area would be identified by colour coding.

This is another logical suggestion that had to be made to the Douglas engineers. When a circuit breaker "pops", it does so because a fault has developed in a part of the electrical systems. Should this happen, it could quite easily be imperative that the flight engineer be able to identify the circuit very quickly so as to implement remedial action. If he has to firstly refer to charts and books and reference cards to find out which circuit is the cause, the wasted seconds could permit the problem to be compounded more than lessened.

But that is not all:

Recommendation: The circuit breakers should be lighted in such a manner as to make failed circuits readily discernable and that this be done other than by floodlighting the control board panels.

Comment: In the past there has been inadequate means of determining a 'popped' circuit breaker during reduced lighting conditions. A 'popped' circuit breaker should be identifiable from all locations, including head on. This required some sort of increased visibility in the head of the circuit breaker button.

Remembering that a whole panel of circuit breakers makes it more difficult to identify just one that is causing problems, possibly something along these lines should have been thought of long before the pilots had to point out the difficulties.

Recommendation: Protection of the rain repellant discharge switch should be provided so that it will not be inadvertently pressed when the cockpit door lock switch is wanted and the rain repellant discharge is not wanted.

Comment: The close proximity of the two switches requires protection of the most critical switch, that is the rain repellant discharge switch, to prevent inadvertent operation. The alternative is to move one of the switches.

Another very logical suggestion that should have been considered by the Douglas engineers long before this. With both switches practically the same in appearance and sitting side by side, one can easily imagine the situation of an aircraft coming in for a landing, a cabin crewmember at the cockpit door for some reason, and the wrong switch being operated, showering the pilots' windscreens with rain repellant solution at the very time when they need to see clearly and precisely out of the windows. Any liquid on the windscreen will cause visual distortions which could have a major consequence on the continued safety of the flight.

Recommendation: A distinctive vocal warning system should be used, or a light warning on the communicator panel to be associated with each sound warning.

Comment: This is very much needed to separate the warning signals more distinctly for safety reasons.

With the greater complexity of the newer aircraft, and therefore the greater number of systems and the enhanced warning devices, this is a logical suggestion made by the pilots, but unheeded by the cockpit designers. Already we have bells, horns, clackers, buzzers, beepers, clockers and goodness knows what else, all operating as warning sounds, and it would be quite easy for someone some day to mix up, for example, a horn with a clacker warning signal, and therefore to start remedial action on a system that does not need it, while failing to take action on another that is crying out for it.

Recommendation: All locations where coffee cups and other liquids might spill, should be protected so that electronic and other types of failures will not occur.

Comment: It is necessary to protect against any eventuality that can and will occur. Therefore, it is necessary to design suitable covers to protect all vulnerable equipment.

The Douglas Aircraft Company should have learned its lesson from Boeing of a couple of years previously. It was widely known in the industry that when the engineers were testing the inertial navigation system in the Boeing 747, they were flying high over the Pacific Ocean, letting the equipment do its work, while sitting back and enjoying a cup of well-earned coffee. Something happened and someone put their coffee cup on the floor of the cockpit, and moments later it was accidentally knocked over with the liquid chasing around and finding, of all things, the main electric circuits to the triplicated equipment. All of a sudden the coffee short-circuited the whole lot and the pilots were completely lost.

Pilots are only human. The air in the high-flying aircraft is very dry. They need to keep supplying the liquids being slowly drained from the body, especially on long flights. Would it not be logical to build in to the design *in the first place*, cup holders adjacent to each of the cockpit crew's seats, so that they have *somewhere* to place the cups in between sips? Is that not basic logic? Then if so, why was it not done?

If you have ever been in an aircraft cockpit and seen the myriad of knobs and switches all around the pilots' habitat, you will possibly have wondered how on earth any of them have been able to negotiate the actions of getting in to or out of the seats without, at some time or another, bashing their heads on the overhead switch fingers, or knocking them out of their set positions. Then again, even if the overhead switches are lever locked, this still does not mean that you are not likely to implant one of these sharp protrubances in your head while negotiating the contortions of getting in to or out of the seat. This is another point not corrected in the DC–10, and there are still further more that the passengers can see and they directly affect his safety.

Recommendation: The cabin door windows should have a de-fogging feature and should be in such locations and of such a size as to permit the attendant to see out when she is in a seated position.

Comment: The cabin door windows have no de-fogging feature and are in such a location that it is impossible for the attendant to see out. We realize that the slide packs on the doors are of such a size that this is a problem. Nonetheless, the item is extremely important.

It is, of course, very important, if the airlines claim that the cabin staff are the ones responsible for getting the passengers out in the case of an emergency occurring. If this is so, it is imperative that the crew member at each door can see outside the machine so that they know when it has come to a stop, and so that they may gain warning of any problem that may be existent outside the particular exit prior to opening it and in such a fashion causing an unnecessary line to form to it ready to escape.

What if there was a fire just outside the exit? To open the door would let the fire come straight inside. How would anyone know if there was a fire outside if one could not see out?

On another point, because of the DC–10 design with the passenger oxygen generation systems and cabin air regurgitators being built into the backs of the seats, airlines that purchase these machines have to also buy seats that have been specifically designed for them. Firstly, passengers should be warned that it is very dangerous to place hand luggage *under* the seats. In decelatory forces in an accident, the briefcase under the seat could rush forwards, pushing the passenger's legs forward with it and then jamming them under the seat in front. This is an extremely dangerous hazard, yet no one is told not to do it.

Some airlines however, actually *recommend* that hand baggage be stored under the seats. How dangerous can you get?

But besides causing a possible hazard, the Air Line Pilots' Association also points out that with the DC–10, the hand baggage is also likely to damage (a) the wrapping in which the underseat life-vest is stored, and even (b) the life vest itself. So what is the logic of permitting baggage to be stored under the seats and then placing the life vests right at the back so as to try and avoid damage, but in a position where the passengers cannot reach them when needed?

Ah, but now we come to another problem in the DC–10 design.

Recommendation: The over-head carry-on luggage compartments should have adequate latching devices, should be of an adequate number for all passengers, and should be stressed to carry any weight placed on them.

Comment: The latching devices appear satisfactory. There are insufficient luggage compartments for all passengers and they are only stressed to carry a certain maximum weight load. There is just no way that the attendant is able to ascertain what the weight load in an individual compartment is.

Here is a definite danger feature in the DC–10. It is no wonder that Douglas aircraft passengers have had in the past, and will have in the future, sore heads incurred in accident situations, both minor and major, when these understressed luggage racks come pounding down on top of them

A danger exists in *all* large capacity aircraft. After most emergency evacuations from them, you cannot help noticing the problems that have occurred. The exit slides down which the people on board escape, are practically useless if there is even a slight breeze blowing outside at the time they are extended. Possibly the industry expects accidents in these machines to be confined only to times of calm meteorological conditions? In any sort of a wind, the evacuation process is severely held up. This danger feature is something the whole industry has known about for at least the past 4 years but absolutely nothing has been done about it.

Recommendation: All cabin doors should open and close with the same mechanical movement.

Comment: The two front main cabin door operating mechanisms in the DC-10, operate differently from the remaining doors.

Here is proof of McDonnell Douglas design clumsiness. Here is ineptness of design; you do not expect companies which boast about safety features they claim are in their products, such as McDonnell Douglas, to include such unnecessary complications as different doors, different methods of operation. This is so basic, that it is from this point that one should look more critically at the DC–10, and all its other relevant design features.

Furthermore, what was the F.A.A. doing to permit such a ridiculous situation to exist in the first place? The F.A.A. – and the other civil air departments – certificated this machine as being "acceptable".

Even worse, on some of the doors there is an "Exit" notice on one side, with a grab handle next to it. But the operating handle is way over on the other side of the door.

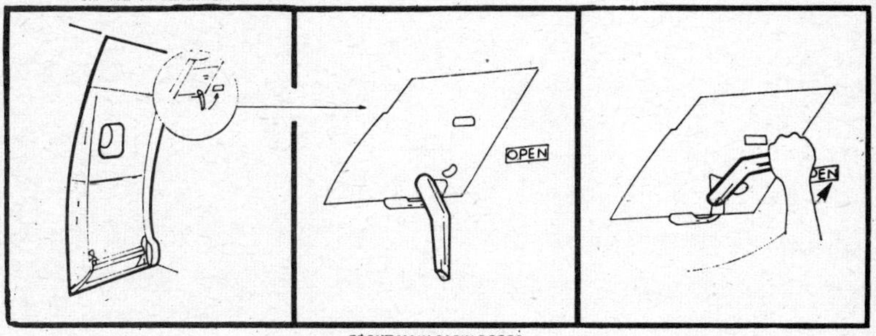

FRONT MAIN CABIN DOORS

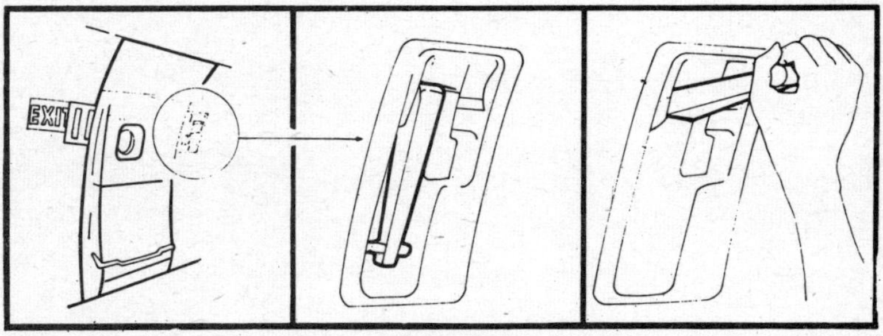

OTHER DOORS

Can you not imagine a group of passengers, following an accident at night, feeling their way towards these doors, possibly catching a glimmer of the luminous "Exit" notice and feeling around in that area for the method of getting the door open only to come across the grab handle and, then being unsuccessful in trying to turn *it*, leaving the area under the impression that the door will not open? Can you not imagine the confusion this will cause in an already confused situation? Further complicating the situation are well documented reports of crash incidents (see Godson, *Runway*) where emergency lighting failed to operate.

Recommendation: It is recommended that a satisfactory means be developed for evacuation through the openable flight-deck windows for emergency use.

Comment: Inertia reels for the crew members out of the windows, would be very desirable and they have been installed.

Two of the "crews" emergency instructions' briefs that have been seen have large CAUTIONS in them, warning the crew that because of the composition of the escape rope (it is a nylon type material) and the length of the drop from the cockpit windows to the ground, flight crew members who try to make a rapid departure via this method will end up more than likely with badly friction-burned hands, thus adding *them* to the injured lists and precluding their assistance any further with the evacuation. So where they learned of inertia reel installations in all the DC–10s seems to be quite a mystery.

The next section of the comments in the pilots' document moves to the subject of Auxiliary Power Units, the fourth engine built into the section under the aircraft's tail that supplies both compressed air and electricity while the machine is static at an airport. This Auxiliary Power Unit is purely a generator and in no way assists with the thrust needed for flight.

Recommendation: The APU should be designed and installed to operate safely. No other concern, such as noise should be permitted to influence the installation by adding or eliminating components which may reduce the safety of the basic design.

Comment: On some previous aircraft containing APU installations the problem of noise influenced modifications for noise insulation purposes. These modifications later contributed to a fatal accident resulting from fire in this portion of an aircraft. The lessons of previous accidents should be evaluated in the design of this installation.

But evidently they have not been. Another point the pilots have completely overlooked at this stage, is that even though there is a fourth engine buried in the tail of the DC-10 and even though that engine contains a fire and gives off large amounts of heat, the structure and materials surrounding the Auxiliary Power Units could be on fire, and neither the pilots nor anyone else on board the machine could do much about it.

And that leads us on to the next points.

Recommendation: Compartments adjacent to the APU be provided with fire protection installations.

Comment: Previous experience has shown the need for this protection.

Recommendation: No hydraulic or fluid lines other than those necessary for its operation should be installed adjacent to, or in the

APU compartments, including the air inlet and exhaust sections. No combustible materials should be installed in or adjacent to the APU compartment. The APU should be located in an area which is as remote as practicable from other control systems and functional equipment.

Comment: In the event of a fire in the APU compartment it would be undesirable to risk possible fire penetration of combustible materials which would intensify the fire.

Also on the subject of dangerous areas as far as fires are concerned:

Recommendation: Every possible effort should be made by the manufacturer to ensure adequate protection of the hydraulic components located in the landing-gear wheel-well areas.

Comment: In-flight wheel explosions and/or brake fires have caused considerable damage to the systems in the past, including fatal accidents when the components are not adequately protected.

Surely junior designers would realize the dangers of having extremely combustible materials situated close to or surrounding very hot items such as the wheels and braking systems when they are raised into their stowing areas soon after a take-off? The similar flaw in the French built Caravelle showed the world what could happen. It was designed with flexible pipes containing highly flammable hydraulic fluid, that ran right next to the wheels and undercarriage brakes in the wheel-well bays. Swissair can tell you a story or two about this design feature in their Caravelles. This feature contributed to the crash of a Caravelle near Zurich which involved the death of 80 people.

Now the emergency equipment:

Recommendation: Emergency operation of the main doors should be possible in case of power failure or track deformation.

Comment: The doors are pneumatically operated when normal electric power fails. There still might be deformation of the door tracks which would render inoperative both the electric and pneumatic actuation systems. McDonnell Douglas considers this is no more likely than jamming of conventional doors.

This comment is an indication that McDonnell Douglas have not made *all possible* safety design improvements in this aircraft. Surely each new design should eradicate problems that have made themselves evident in the older machines? Besides, the British manufacturer, Hawker Siddeley Aircraft Ltd., also used the idea of doors running up on overhead tracks in their nearly impossible to sell Tridents. The only problem was these doors jammed very easily if the person opening them was not cool, calm and collected and standing exactly in front of the operating handle. They claim these Tridents have now been modified. And the other difference is that the Trident holds up to 130 people – the DC–10 up to 355. The Trident has an average of 6 exits (4 doors and 2 windows) – the DC–10 has only 8 (all doors).

Recommendation: The emergency or alternative brake system should utilize the normal pedals and operate with similar forces and feel. Anti-skid protection should be included in the alternate or back-up system if at all possible. Complete redundancy should be provided.

Comment: Accumulators in each hydraulic system provide a source of power for emergency braking in the event of multiple hydraulic system failures. Application of the brakes is through the rudder pedals with normal anti-skid protection. Douglas feels there is adequate accumulator power available to stop the aircraft safely with two hydraulic systems inoperative. In a critical system such as brakes, there should be sufficient redundancy so that a single component failure cannot cause total loss of braking capability.

Again, no action by Douglas on the Air Line Pilots' Association report. The next plan was for the pilots to critically appraise the DC–10 after they had been flying it for about a year. And in the latter report, as we shall see, they will also include collected comments from the airline crews that worked on board the aircraft.

CHAPTER 9

By 12 November 1970, the earlier problem that had been experienced with the securing of the cargo doors had, according to Douglas, been remedied.

These doors were operated by an electric motor. There was also a provision for them to be opened or closed manually by using a drill type ratchet into a special socket on the outside of the door, that was directly connected to the motor's armature. Prior to this time, in order to close and lock it, all the operator did was to hold down an electrical spring-loaded switch in a small panel in the fuselage just to the left-hand side of the door. By holding the switch in the "down" position, the electric motor would wind the cargo door (which opened outwards and upwards) down into the "closed" position, and once it was flush with the sills, a gearing mechanism inside the frame would hold it closed while the motor would then wind around a series of grips over the locking bars located in the bottom sill. The idea was that once the motor had stopped working, the door was (presumably) closed. However, there was no way at all that the operator could see or check that it had actually completed the cycle and was locked.

After the pressure vessel test on 29 May 1970, when the forward cargo door suddenly exploded open, this had prompted the design engineers to come up with a modification that would "correct" the fault. A series of lock pins (marked "A" on the following diagrams, which also have been very much simplified so as to make them easier to understand) move in behind a circular armiture once the lock grips, marked "B", were closed, ensuring that they would not come open. The lock pin bar "C", was positioned manually by the door's operator, and it was connected to a series of push-rod linkages "D", and a bell-crank "E", with a handle "F", which was located on the outside of the door.

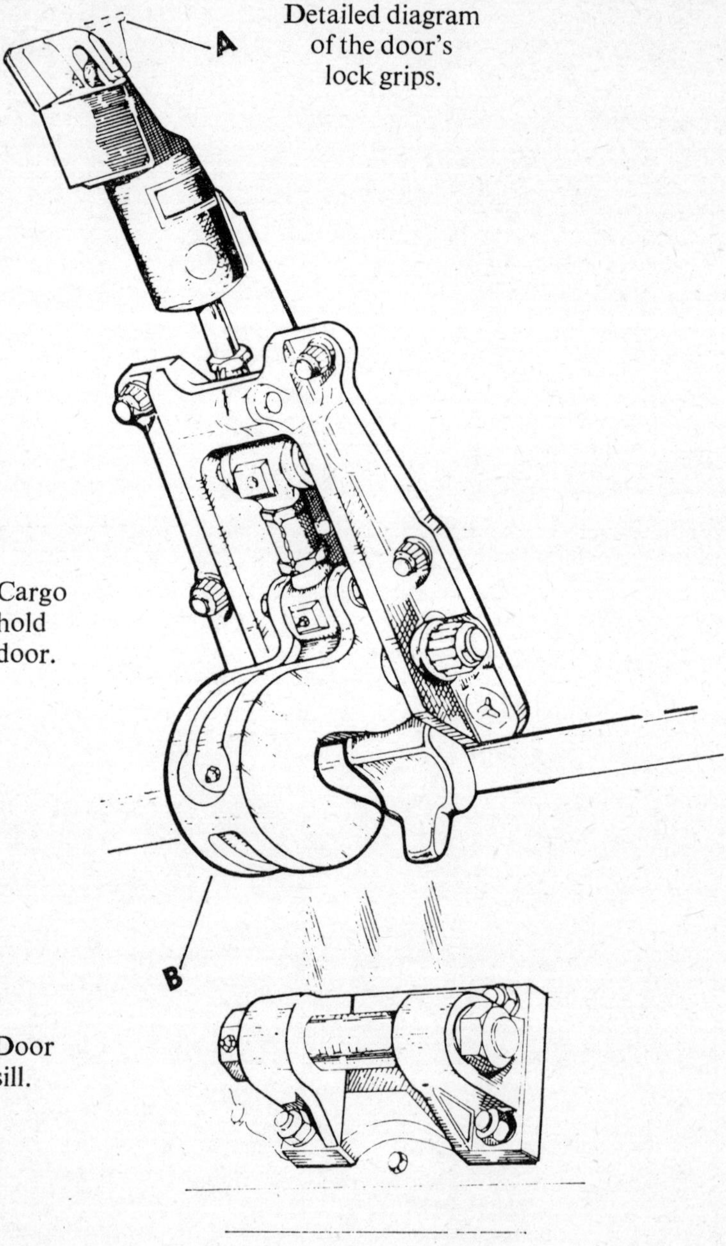

Detailed diagram of the door's lock grips.

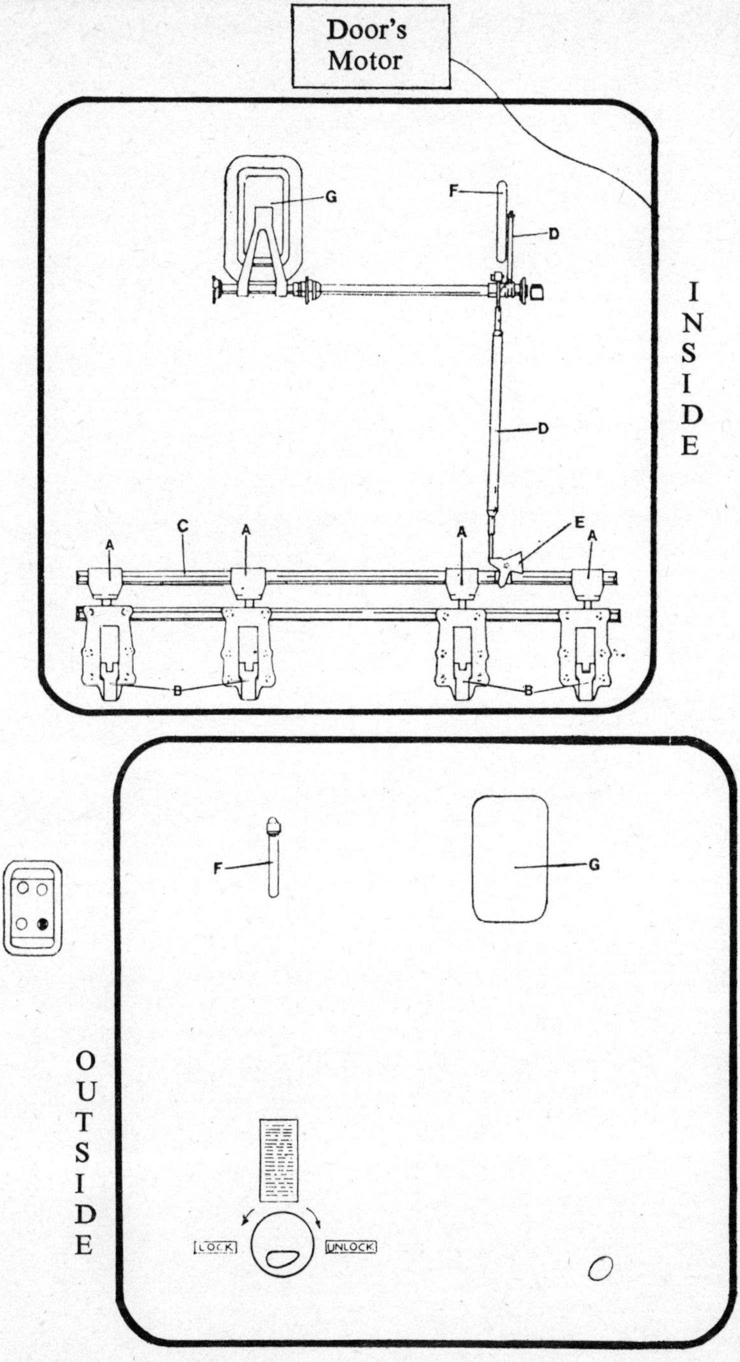
Door's Motor
G
F
D
D
C
A
A
A
E
A
B
B
INSIDE
F
G
OUTSIDE
LOCK
UNLOCK

Then, to make sure that everyone realized that the door was properly secured, they added a small "vent door" "G", into the main door itself, and again connected it by a rod and bell-crank to the operator's handle. The theory behind this thinking was that once the door had been electrically wound closed and the lock grips were positioned over the lock bars, the operator could then easily pull the locking handle down, which in turn would move the lock pins behind the lock armiture thus ensuring that the grips were properly closed and at the same time making sure that they could not spring back open, while the handle movement would also close the small "vent door" and thereby give the operator a visual indication that the door was properly closed and locked.

This, Douglas and the F.A.A. considered, was a completely fool-proof system because if the door was not properly locked, the "vent door" would remain open, initially giving the operators a visual warning, but at the same time, the "door closed" light on the cockpit control panel would remain "on", and if no one noticed either of these things, the flight crew would be unable to pressurize the aircraft after taking off because they would be losing the build-up of air inside the machine, through the opened "vent door". Magnificent – a completely fool-proof idea – *so they thought.*

At around this time, when the modifications of the doors were being contemplated on the prototype and early production models, Nicolas Schipper, a senior design and construction specialist working with the R.L.D. – the Dutch civil aviation authority – wrote to the F.A.A. and reasserted his earlier stated fears that dire consequences would occur if pressurization was lost in any of the different cargo or passenger compartments in the DC–10. It was some years earlier, around 1969, that Schipper had first made his presence felt in this regard soon after the announcement

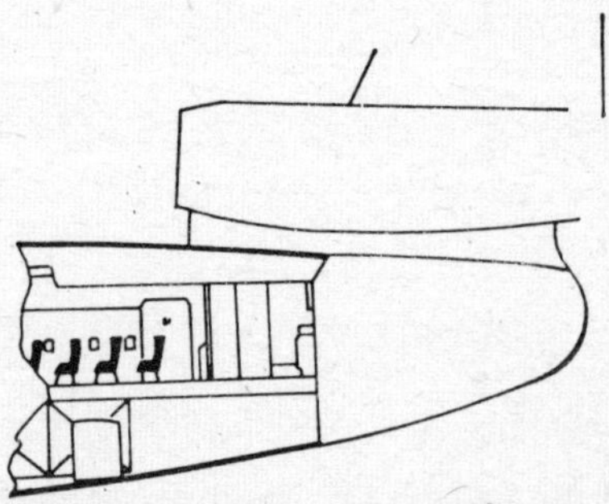

that jumbo-sized jets were being planned. He first stated his fears at an I.C.A.O. technical conference.

Now again, he reaffirmed his earlier point, in a letter to the F.A.A., that if pressurization was lost in any compartment in the DC–10, the floor would collapse and the aircraft would be out of control. He made this point quite clear in a letter he sent to a friend in September 1972. In it he remarked that he felt it was unfortunate that the F.A.A. was slow to make new regulations, since three years earlier, Schipper had explicitly warned them at an I.C.A.O. meeting and, later in Washington, of the probable results of decompression accidents in Boeing 747s and DC–10s. He predicted that collapse down or up in the case of the DC–10 would block the controls. Nothing up to the time of his writing this letter, he remarked, had been done. The F.A.A. made no changes as a result of Schipper's letter. However, it was because of the partial collapse of the floor on 29 May 1970, during the pressure vessel tests, that the designers had incorporated pressurization relief vents between the passenger cabin and the forward and middle cargo compartments. But because of a construction strength difficulty, they were unable to modify the same area of floor above the aft cargo compartment.

The relief vents in the floor meant that if pressurization was lost in either the passenger or forward or centre cargo holds, the air pressure inside all three of these places would rapidly equalize through the relief vents, thus not causing a pressure/weight increase on the section of the cabin floor between these areas.

Of course, with Schipper apparently realizing that an utterly dangerous situation existed in the floor strength and design of the DC–10 aircraft, either some fast talking on the part of McDonnell Douglas, or the hushing-up of the matter by the Dutch civil aviation department, or the cancellation of the DC–10 order by K.L.M. would have been on the cards. K.L.M. did not cancel. The Netherlands government department, of which Schipper was an employee, did not refuse to certificate the DC–10. And so it must be presumed that this whole design/floor strength business was ignored by either McDonnell Douglas or the Netherlands R.L.D.

For such a matter as this to be made public would, of course, be the spelling of disaster were an accident to occur to a K.L.M. DC–10 in which even the most minutest of factors was seen to have been either the design or the floor strength. Mountains of money would change hands from airline and government departments to aggrieved passengers and their relatives. And one must also assume that both K.L.M. and the Dutch R.L.D. are prepared for such consequences, in the same way that both McDonnell Douglas and the Federal Aviation Administration, who were both of the same knowledge after the door blew out on the 29 May 1970, must also have been prepared.

Civil aviation government departments that are aware of major defects in either an aircraft design, its operation, or its maintenance, must be ready for a volley of writs.

But even more interesting is the fact that K.L.M. was in allegiance with three other European carriers, Swissair, Scandinavian Airlines System, and Union de Transports Aerien. These four airlines all in unison, ordered DC–10s. Now surely, one is entitled to ask, if the Dutch civil aviation department was of the knowledge that there were design faults of a potentially dangerous nature in a certain aircraft, and if it were indicated to them that their national carrier was showing an interest in buying the machine (together with its cohorts), surely it is the legal and moral responsibility of the government department concerned to warn and advise their sister bodies in Switzerland, Scandinavia and France of the dangers that exist? And if these warnings were given and not heeded, surely, one is entitled to ask, would this not indicate that the Swiss, the Scandinavians and the French were as equally as culpable as the Dutch?

Interesting, is it not, that in late 1969, there was the opportunity for the responsible French government authority to know, or at least suspect, that design faults existed in the DC–10? If they were, this in no way stopped U.T.A.'s machines being freely certificated to carry passengers.

And what about Swissair (the mostly-American-owned Swiss airline) that said in a letter: ". . . our company readily accepts any reasonable suggestion and shuns no effort to always improve both the quality and safety of its services." Was Swissair of the knowledge that they were buying a machine with major design faults in it? Did Schipper advise the Swiss Federal Office of Air of his fears for the safety of innocent people flying in the DC–10?

Both Swissair and the Swiss Federal Office of Air knew that aircraft under their control and already in use contained the fabric-to-metal type seat-belts, which, we have already discussed, unintentionally come undone. Nothing was done to correct this fault.

Besides, the Swiss together with the Scandinavians and French, are members of I.C.A.O. Did they miss hearing Schipper's lecture?

CHAPTER 10

THE DC–10's testing programme continued unabated, but the Corporation's financial position was worsening. One of the important questions of the day was: how many DC–10s would Douglas have to sell in order to "break even"? As at the 31 December 1970, they had reported orders for 121 of the planes, with options on another 117.

However, James F. McDonnell their chief executive officer, later said that the break-even point for the DC–10 was 438 aircraft: "plus or minus 5%". He also stated that the project represented an investment of just on $1,500 million of private capital.

Meanwhile the test programme was continuing just ahead of schedule. By the time 300 test flights had been logged – representing over 500 hours of airborne time – they had arrived at a point one third of the way through it. The "flutter" (shaking) tests had been completed, the aircraft taken up to its ceiling altitude of 42,000 feet and to speeds of 503 miles per hour. Take-offs had been made at a gross all-up-weight of 380,000 lbs. some 32,300 lbs. heavier than the design limits.

All four of the test aircraft had been in the air simultaneously and up to ten flights had been made in a single day. Several of them had exceeded 6 hours in duration, with the longest lasting for 7 hours and 20 minutes.

Hydraulic, electric, pneumatic and fuel systems were all operating satisfactorily, and results from the initial evaluation of the automatic flight control systems were described as: "encouraging".

On 29 January 1971, F.A.A. officials from their Los Angeles office began flying the DC–10, having been provided with approximately 90% of the data estimated to be needed for the aircraft's certification.

As the tests continued, everyone working on the project could see that they were two months in advance of the schedule and delivery of the first two aircraft, it was announced, would be made on 29 July 1971. American Airlines and United Air Lines, the two carriers who initially placed orders, were each to receive one of the wide-bodied tri-jets for crew training purposes, in preparation for inauguration of commercial services later in the year.

CHAPTER 11

THURSDAY, 29 July 1971, was a momentous day for the McDonnell Douglas Corporation. It was awarded a Certificate of Airworthiness for the DC–10, and once that part of the ceremony was completed, Jackson McGowen, the president of the Douglas Aircraft Company, then officially handed over the first two aircraft, one to United Air Lines and the other to American Airlines.

Up to that time, Douglas had received 127 orders for their machine, but due to financial reasons and the dropping off of the number of passengers, both American and United had cancelled some options that they had held, reducing the sales figure to 96.

Once the ceremonies were over, American Airlines, as the first to place an order, was the first to fly their DC–10 away, closely followed by the United Air Lines crew, after a false start or two, in theirs.

Around the same time, it was announced that there would be five DC–10s in service – under the American Airlines banner – by the end of the year, with twenty more serving their routes by the end of 1972.

As the machines moved around the country, thousands of stories appeared in the different local newspapers, eulogizing its two main qualities: that of being much quieter than earlier aircraft, and the fact that its engines did not emit clouds of smoke during the former and latter parts of a flight. These points, of course, were adequately drummed into the press by McDonnell Douglas publicity officers, realizing that they were about to lose this battle once the Lockheed TriStar entered passenger services.

For instance, the *Newburyport News* (Newburyport, Massachussetts) said in part:

> A B.O.A.C. airliner took off over our heads at the end of the runway at Boston Logan Airport, its Rolls Royce engines spewing black streamers, the roar pounding our livers.

"There goes the loudest and dirtiest engines in the business", an airport type said . . .

At 182 feet long, the DC–10 is more than 50 feet shorter than the 747, has a swept wingspan of 155 feet 4 inches. It's her girth that makes for her load-carrying ability, seating eight people abreast with a configuration that puts no person more than one person removed from an aisle.

So what McDonnell Douglas has designed here is an airplane that will fly more people as far, for less money than existing SSTs with cleaner engines and less noise.

Granted it will not fly them to Paris in under three hours. But it will do it in well under eight.

And what is it that is planned for the U.S. SST by the way of engines? These Rolls Royces that dirty up the skies and pound the earth . . .

We supsect that other airlines will be forming a line to the right to buy DC–10s. What we hope is that other aircraft producers will follow with the same kind of environmentally orientated development of the new breed forecast by the DC–10.

Certainly, Rolls Royce engines *were* exactly as he said – but so were Pratt & Whitney's. And the reason for thinking that McDonnell Douglas were ever-present in the background while this story was being written, is because it was well known at that time that the Rolls Royce RB 211 engines on the DC–10's competitor, the Lockheed TriStar, were even quieter than the General Electric CF–6s which Douglas had chosen, and moreso gave off even less pollution. The writer's assumption about the SST (Supersonic Transport) came from an equally as spurious source.

As the DC–10s continued to move around the countryside, their fame for quietness and cleanliness grew appreciably. Civic leaders and members of local aircraft noise action groups were given days outings at the airports while the DC–10 made a series of landings and take-offs, joining in the normal traffic streams. Everyone was completely impressed, including the *South Bay Daily Breeze* (sic), another local paper, which went to the extent of publishing a faked "photograph" of a supposed DC–10 soon after lifting off, with, carefully printed underneath it, the words: "Staff Photo". Amazingly enough, this fakery was even reproduced in one of McDonnell Douglas's publicity handouts.

CHAPTER 12

THE American Air Line Pilots' Association had completed its second look into the operational and design problems of the DC–10 that had come to light during the early test flights.

Even though there was a large list of earlier located troubles still "outstanding" – having not been corrected by the McDonnell Douglas Corporation – these were now added to, with some more that had been uncovered during the actual operating of the aircraft. It is notable that even though there were a large number of incidents reported to Douglas with reference to difficulties in closing and locking the cargo compartment doors, the A.L.P.A. evaluation did not consider this point at all.

During the first year, the pilots had asked all their members to report incidents connected with the aircraft's design and operations – and in they flowed. On top of this, McDonnell Douglas gave a group of specialists access to their fuselage number 2, from the 11 to 13 August, so that they could test fly it and then submit their report. This group also found faults and the stewardess's complaints, in particular, tried to give reason for their slap-happy, unbusinesslike approach to waiting and waitressing on American airlines which, at its worst, has to be seen to be believed. The apparent universal American attitude of being rude and abrupt is not acceptable outside their land.

Nearly all aspects of food service on the DC–10s came under fire as a highly dangerous affair, possibly designed in such a fashion by Douglas to accommodate the maximum number of passengers in order to make their machines inviting to purchase. But while being more economical in many ways, the danger factor was, in consequence, substantially increased.

From the angle of passenger service, and certain operations of same, the DC–10 is a digression from normally accepted safety standards. But this must have been overlooked, like so many other things, when the F.A.A. gave the aircraft its certificate of airworthiness. And as there also has been no change in the interior cabin set-ups for foreign airlines, it seems quite evident that the F.A.A. did a disservice to all other civil air government departments that simply "rubber stamped" its certificate.

The Air Line Pilots' Association report on the DC–10 concludes with the following piece:

> The A.L.P.A. Evaluation Committee finds that the DC–10 has many good structural and operating qualities designed into it. However, because of construction time available, cost, or lack of agreement, some of our recommendations have not been included in the final product. We believe that a number of our suggestions will be given consideration in later DC–10 models and certainly should be considered in follow-on airliner types.
>
> We recognize the close design tolerances required in order to manufacture a competitive, economically successful airline aircraft. The only way to do this and still have a good safety record is through eternal vigilance. We call on all segments of the industry to provide vigilance and to continue their work towards design improvements to the DC–10.

It would seem, however, that no one bothered to carry it out.

CHAPTER 13

On 3 November 1971, at Cleveland, Ohio, a Beechcraft 99 was taking off on scheduled commuter Flight 101. The co-pilot gives us a précis of the story:

> At first I thought we hit an obstruction on Runway 23 Left. The next thing that I saw was the right front cargo hold door swinging up and down and finally coming to rest in the up position.
>
> I was flying the aircraft at the time and had been cleared for a take-off. At the time of the door coming open, we had just lifted off and were at an altitude of 10 to 15 feet above the runway surface. The baggage door opened, a mailbag then flew out and smashed into the propeller. The pilot-in-command, Captain Krout, then told me that he would take over the controls and abort the take-off. The abort was accomplished as an "everyday" occurrence. The pilot brought the Beech 99 back onto the runway and safely stopped. Captain Krout should be praised for saving our aircraft. Highly professional.

Co-pilot Thomas A. Hirsch, 26, worked for Wright Airlines, and was seated in the right-hand flying seat when this happened.

At 2230 hours, both pilots performed their pre-flight checks in readiness for the journey from Lakefront Airport. Fuel loading was checked and was correct. The cargo nets used to tie down the freight were inspected and so they said, found to be in order. Back in the cockpit, Hirsch radioed the weather station and asked for meteorological details for their proposed flight route to Chicago.

At 2250, the little airplane was waiting at the end of the runway for permission to take-off. Within moments of it lifting up from the concrete, the right forward baggage compartment door came open and the aircraft started swinging around to the right.

The National Transportation Safety Board found that an

electric malfunction existed in the baggage door warning system. On the inside of the baggage door there was a switch, which, when the door was properly closed and locked, extinguished a light on the pilots' control panel. If the door was open, the light should remain on and act as a warning that all was not well. However, if there was a fault in this microswitch, then the warning light could remain out at all times, giving the pilots a false indication that the door was closed and locked. The other problem that could arise was that if the light bulbs themselves were "burned out", there was no way of testing for this condition other than someone getting out of the aircraft and going around all the doors, opening them to see whether the appropriate lights came "on".

The investigators found that the lights for both of the forward freight doors were "burned out". They replaced the bulbs and then realized that the one for the front left door would remain "on" even though the door had been properly closed and locked. Further tracing revealed that there was a fault in the switch itself inside the door which did not cut "off" the power to the warning light when the door was closed. A closer examination of the nose cargo compartment showed that the webbing, used to contain the freight, had not been installed.

In a review of the Beechcraft Flight Manual, Step 23 – "Before Starting Engines", required the pilot to make a test of the lights in the warning system. Research revealed that Wright Airlines had been using a typed-out "check list" of things to be done before an aircraft takes off, and Item 15 on this instructs the pilots to check to see that all the warning lights were extinguished.

The Federal Air Regulations, CRF 23.787, states: "(b) there must be a means to prevent the contents of any cargo compartment from becoming a hazard by shifting".

Beech aircraft had been notorious over the years for faults of this nature developing in its designs, and the Beechcraft 99 had been outstanding in this regard. The 99 was certificated and approved by the F.A.A. under their "delegated option authority" on 8 July 1966, to be a machine that was acceptable for the carrying of cargo and passengers. Under this scheme, Beechcraft issued their own certificate of airworthiness to their own design and their own manufactured product. This, the F.A.A. deemed, was quite acceptable, so long as their engineers may, if they choose, inspect and test fly the machine on a "spot check" basis. However, in this particular case they elected to do absolutely nothing at all.

As a result of these types of faults being cited as the causes of accidents to these types of aircraft, the Beech Company was flooded with writs, and in their attempts to correct this situation, it is known that they started by settling out of Court for large sums of money. Mind you, these actions were not solely for problems in the Beech 99, but in the whole range of machines built and certificated as being of an acceptable standard to carry passengers. In one case in June 1971, in Santa Ana, California, a $21.7 million judgement was achieved against Beechcraft, which is one of the highest awards ever.

But, one asks the morality of such a situation existing in the first place. If the administrator of the F.A.A. is charged by Congress with the all-seeing, all-acting responsibility for the maintenance of aviation flight safety, surely *he* should also be involved in this legal situation?

Even though at this time, the reader is wondering why we are discussing such small-time subjects as doors coming open in 9 and 12 seater aircraft, one can purely suggest that the information be remembered for later reference as our story develops.

CHAPTER 14

By the middle of October 1971, the Lockheed Corporation announced that their TriStar test flight programme was also running ahead of schedule. Rolls Royce had upgraded their engines to a power output of 42,000 lbs. of static thrust, and these newer and even quieter powerplants were now being used in the tests.

Robert Shuberth, the chief engineer of Lockheed's advanced design and technology department, also added that the fourth TriStar would take to the air that October, with the fifth in November 1971. He went on to say that in automatic landing tests, the system performed so well that a maximum variation of only 20 feet from the runway centerline had been recorded.

Over in Japan, a strange situation was in existence. In 1970, certain Japanese air routes – notably the Tokyo to Osaka run – were becoming rapidly overcrowded with aircraft. The Japanese political parties were locked in debate because of the growing and desperate air-lanes situation. They, like many other bodies around the world, also came to the conclusion that it would be eased by replacing smaller jet aircraft with the new wide-bodied machines, but the government would then have to decide which makes of aircraft they would permit the airlines to operate.

At that time, Japan Air Lines, the country's foremost air carrier was about to take delivery of the first of its order of Boeing 747 Jumbo Jets. One of the secondary group, All Nippon Airways, had also been looking around for an appropriate wide-bodied machine to back up its application for a larger number of services on certain routes – including Tokyo to Osaka. They had – evidently on the quiet, because McDonnell Douglas hand-outs of that time reveal no existence of it – placed an order for 6 DC–10s through the huge Japanese finance house cum consortium, Mitsui.

The order was kept quiet probably because a large number of local groups located around the perimeters of the major airports, and especially Osaka, were taking action to stop the noise and pollution caused by the increase in air traffic, and were especially adamant at the size and passenger carrying capacity of the wide-bodied jets.

Publicly, All Nippon Airways had placed some orders for aircraft, but these were for the smaller Boeing 727 Series 200, and Boeing 737 machines. For some strange reason, even though it was agreed later to be fact, not one ounce of publicity regarding All Nippon Airways' order for 6 DC–10s, made through the finance division of the huge Mitsui cartel, leaked out.

October 1971 was also the time when John C. Brizendine was promoted to the position of executive vice-president of the Douglas Aircraft Company, immediately under Jackson McGowen.

The appointment followed more than three years as the vice-president – general manager of the DC–10 programme. In that position, he was responsible for directing the engineering, flight development and production of the DC–10, beginning one month after the corporation announced its decision to build the tri-jet.

Brizendine was born on 3 August 1925, at Independence, Missouri. During World War Two, he served in the Navy as a pilot of multi-engined, land-based bombers. After the war he graduated from the University of Kansas, receiving a bachelor of science degree in aeronautical engineering in 1949, and a master's degree in 1950. In that same year, he joined the Douglas Aircraft Company and was assigned a position in the flight test division. He subsequently served as the test project engineer on the F–4D Skyray, a high-performance interceptor produced for the U.S. Navy, and in 1956 held a similar position during the experimental development of the supersonic F–5D Skylancer.

When flight-testing of the four-engined commercial DC–8 began in 1958, he was assigned as a test project engineer for that programme, subsequently being appointed manager for the DC–8.

In April 1963, one week after the company announced the decision to proceed with the DC–9, Brizendine was made assistant general manager of Douglas, and manager of the DC–9 programme, with prime responsibility for engineering, flight development and manufacture of the short-to-medium haul range twin-jet.

On 1 December 1965, he was promoted to vice-president – engineering, of Douglas and he held that position until his appointment as head of the DC–10 programme in May 1968.

At the end of October 1971, the advertisements for the new aircraft started to appear:

THE NEW DC–10:
McDONNELL DOUGLAS BUILDS IT,
AIRLINES ARE NOW FLYING IT,
YOU'LL LOVE IT.

Our famous DC airliners have met every generation's needs for fast comfortable and dependable air travel every year since 1935 – Now we're delivering another great line of DCs – the exciting new DC–10 luxury tri-jets – With the DC–10, airlines can now offer passengers true wide-cabin spaciousness and luxury on flights to many airports bypassed by larger planes. And its quiet, smokeless engines make it a 'good neighbour' wherever it goes – DC–10s are the only new-generation jet-liners that can profitably fly inter-city routes as short as 173 nautical miles, as well as intercontinental non-stop flights as far as 5,270 nautical miles. They will continue the reputation of operating efficiency and passenger-pleasing extras that make McDonnell Douglas DCs the choice of leading airlines the world over – Ask your travel agent or airline reservations' office to book you aboard a luxurious DC–10. You'll love it.

These forward looking airlines have chosen the DC–10: American Airlines, United Air Lines, Northwest Airlines, Trans International Airlines, KLM-Royal Dutch Airlines, Scandinavian Airlines System (SAS), Swissair, UTA-French Airlines, Overseas National Airways, National Airlines, Air Afrique, Alitalia, Continental Airlines, Air New Zealand, Sabena Belgian World Airlines, Lufthansa German Airlines, Atlantis, Finnair, Delta Air Lines and Western Airlines.

Still no mention of the All Nippon Airways-Mitsui order for 6 of these machines.

CHAPTER 15

On 10 December 1971, a paper was presented to the Air Transport Association's Engineering and Maintenance Conference, held in Washington D.C., by L. B. Zambon, the vice-president – engineering of American Airlines.

The paper itself was long and technical, but the conference was told that with aircraft prior to the DC–10 and the Lockheed TriStar, the manufacturers used to come to the airlines and say something like: "We are designing an aircraft to fill this part of the commercial airliner market spectrum. What inputs would you like us to make?"

However, with the DC–10 and TriStar, the manufacturer – airline approach was different, being instigated by the carrier: "We want an airplane meeting the following performance, size and mission requirements, and we want to consider some specific maintenance goals in its design. What do you propose?"

The A.T.A. Conference heard how American Airlines "put the screws" on the Douglas Aircraft Company. They formed a number of working groups:

> Each system working group as identified, included the Douglas design engineer, reliability engineer, and maintenability engineer along with the American Airlines engineer specializing in that system with the proviso that other airlines committed to the DC–10 could, and we expect would, assign someone to each of the 10 groups. The working group activities included design reviews, engineering subtask (design study) reviews, failure mode and effect analyses, fault location procedures, and scheduled maintenance requirement reviews: as well as orientation visits by Douglas design engineers to American Airlines maintenance and overhaul facilities. In addition, American Airlines participated with Douglas in review and selection of vendor proposals for significant aircraft systems.
>
> As other airlines became committed to the DC–10, they

participated in the activities just described, joining the airline-manufacturer partnership.

So now we see that airlines that placed early orders for DC–10 aircraft were as deeply involved in basic design considerations which ultimately resulted in the design flaws inherent in the DC–10. The leader of the whole lot seems to be American Airlines.

But the conference was also told that:

> . . . the DC–10 structures testing program, including component and major section strength and fatigue tests, is the most comprehensive of any similar program to date.

And again:

> The structure (of the DC–10) has been designed and tested to continue to carry fail-safe loads after sustaining more extensive principal structural element failure than required by the Federal Aviation Regulations. The designer's objective here, which goes beyond the requirement for safe structure, is to provide a maintenable structure in which failures will progress slowly and be detected easily, so that structural repairs can be made before they develop into major economic maintenance problems.

Certainly the testing may be the "most complete", but was it good enough? And was it good enough, especially in the light of the "door exploding open" incident 19 months before this lecture was given, when they should have realized there were design deficiencies in the door and the floor? Surely American Airlines staff members who, according to themselves, were working so closely with Douglas, should have made *some* suggestions about how to correct this door/floor/flight control cable mess? With the airline admitting daily co-operation with the designers, surely someone there, too, must be held responsible if a design fault is later found to persist. Some design faults also seem to be tied-in with the economics.

> The DC–10 power plant installation features fan cowls, fan thrust reversers and core engine cowls which are integral with the airframe . . . The maintenance advantages . . . are obvious. The nose cowl, although not integral with the aircraft, is attached to the engine by 12 fasteners and may be removed and reinstalled on a replacement engine within 30 minutes. These features . . . represent a multi-million dollar saving in spares inventory costs.

National Airlines, as will be seen later on, were involved in two near calamities, one of which killed one passenger, injured three others, and practically caused a disaster that would have left at least 128 dead. Here American Airlines admit their involvement in the creation of this "feature", which even though found by the N.T.S.B. not to have been the cause of the incident has clearly been shown up as faulty.

The second one was caused, according to the National Transportation Safety Board, by: ". . . the failure of company maintenance personnel to complete the installation of mounting bolts on the inboard panel of the No. 1 engine core cowl. This failure resulted from non-compliance with established company maintenance procedures."

It will be noticed that the words "core cowl" are mentioned both in the N.T.S.B.'s adjudication of the causes of the National Airlines accident on 8 July 1974, *and* the American Airlines statement made at the A.T.A.'s Engineering and Maintenance Conference, by their vice-president – engineering, Zambon.

The most interesting point at this stage, is that American Airlines is the one making all these ballyhoo statements – the same American Airlines that will be seen to be involved in an accident that will set the scene for the 346 deaths just north of Paris, two and a half years after.

CHAPTER 16

On 23 February 1972, a press statement was released by Mr. Freddie Laker, the managing director of the British charter company, Laker Airways, to the effect that they had signed a contract with McDonnell Douglas for the purchase of two DC–10s. It added that the machines were originally earmarked for All Nippon Airways, via the Japanese Mitsui finance and business cartel. Mitsui in turn, had originally picked up some of the options on purchase that had been let slide by either American Airlines or United Air Lines. These gave Laker two early positions on the production line, numbers 47 and 50.

But Laker's announcement was premature. The previous November, after All Nippon had cancelled the order leaving Mitsui holding the bag as far as the arrangements with Douglas were concerned, Mitsui contacted Laker's to determine if the airline would be interested in any of the 6 DC–10s then being assembled for All Nippon Airways. The company was interested and engaged in negotiations with both Mitsui and Douglas.

However, it was not until 23 June 1972, that Mitsui finally signed a sales agreement with Laker's and, on the same day, advised Douglas that they had at last, assigned two of the machines they had ordered for All Nippon over to the British charter carrier. This left Mitsui with the problem of getting rid of the other four.

Naturally, this mess resulted in at least one intriguing question: if All Nippon Airways could suddenly decide that they were no longer in love with the aircraft they had agreed to purchase, why was Mitsui trapped?

The answer seems to once again turn to some strange (to us) Japanese customs. It appears that when two Japanese companies do business between themselves, no written contracts as such change hands, but they simply come to a "mutual understanding". However, in the case of a Japanese company dealing with a western concern such as the McDonnell Douglas Corporation, the arrangements are worked out, written down, signed and sealed. Now in further Japanese tradition, so it is understood, when All Nippon had second thoughts about the DC–10, all they did was to telephone Mitsui, advising them that they were no longer interested. But Mitsui, being in possession of a signed and sealed legal document with McDonnell Douglas, simply could not phone Long Beach and do the same thing.

Even though this might sound a rather strange way to do business as far as us westerners are concerned, it is not untypical in Japan. Mitsui would have to accept the fact that All Nippon had changed its mind – they would be bloody annoyed about it – but *c'est la vie*. However, *next time* they were approached by All Nippon Airways to finance one of their projects, the business arrangements need not necessarily be "so favourable".

This seems to be the explanation of how the airline could change its mind, yet the finance cartel then had the problem of disposing elsewhere of the order.

Further investigation into this "love/hate" affair between All Nippon Airways and the Douglas DC–10, seems to hinge around a change in the airline's president. It is understood that soon after the new company president took office in late 1971, one of the first things he did was to cancel All Nippon's order for six DC–10s with Mitsui.

Why he decided to cancel, is still a mystery. Neither is it known why All Nippon then turned around and placed an order for six of the DC–10's direct competitor, the Lockheed TriStar.

Of course this cancellation meant that Mitsui, who seem to be in nearly every other type of business other than air transportation, had the problem of getting rid of these six machines, and this would mean that the aircraft, after their manufacture, were to be standing around at the McDonnell Douglas factory at Long Beach awaiting Mitsui's negotiations of sales.

CHAPTER 17

ON 2 May 1972, Continental Airlines DC–10, registered number N68041, departed from Tucson International Airport, Tucson, Arizona, at 1152 hours, with nine persons on board, at the start of a pilot's training flight.

Just before midday, the aircraft was at an altitude of 5,000 feet and circling around the Ryan radio beacon. The training called for one circuit of the mast prior to commencing an instrument approach to Tucson Airport.

As the machine completed the last turn of its swing around the beacon, the crew felt a mild vibration and seconds later, the co-pilot-instructor reported no oil or oil pressure in the number 2 engine – the one located in the DC–10's tail. Within moments of him bringing this to the attention of the engineer, the number 2 engine low pressure and master warning lights came "on"; the "thrust reverser unlock" light flickered "on" and "off", and a "thumping sensation" was felt in the cockpit. The crew immediately shut down the engine and the aircraft came straight in for a touch down.

After landing, an inspection revealed that the whole of the low-pressure turbine assembly, the turbine rear frame, and the reverse thrust assembly, all of which were supposed to be at the back of the engine, were simply not there. The investigators instituted a ground search in the vicinity of the Ryan radio beacon where most of about 100 pieces of metal were recovered.

As a result of this accident, the National Transportation Safety Board wrote to John Shaffer the administrator of the F.A.A., pointing out the dangers and making the following recommendations:

1. Require that the provisions of General Electric Service Bulletin (CF6–6) 72–177, dated the 28 March 1972, which recommended adding a sleeve tube to strengthen the high-pressure turbine pressure tube, be incorporated in all affected CF6–6D engines at an early date.

2. Require 'C' sump borescope inspection and engine oil consump-

tion monitoring, similar to the procedures recommended in General Electric Alert Service Bulletin (CF6–6) A72–273, and require repetition and continuation according to experience, with attendant appropriate action, until an improved 'C' sump assembly is provided by General Electric.

This letter was sent to Shaffer on 7 June 1972. From the details, it will be seen that Continental Airlines had evidently *not* elected to incorporate the two mentioned Service Bulletins – which is their own prerogative as Service Bulletins are not mandatory upon the industry. But besides not being enforceable, they are nice, quiet ways of trying to effect design fault changes.

Shaffer's reply, dated 30 June 1972, was in part, as follows:

This problem is being studied extensively by the manufacturer and the Federal Aviation Administration. The preliminary indication that the loosening of the stiffening rings on the pressure tube in the high pressure turbine region was the primary cause of the problem has since been discounted as the result of the most recent investigation of the occurrence. Nevertheless, the modification to strengthen the high-pressure air tube is being incorporated in all early production engines at the first opportunity. . . .

It is not believed that issuance of an Airworthiness Directive would contribute to increased airworthiness of the DC–10 aircraft as compliance with this Service Bulletin is being obtained through voluntary adoption by the carriers involved and is being closely monitored by F.A.A. inspectors.

Shaffer said that the cause of the accident was something other than that determined by the National Transportation Safety Board, but offers no information as to (a) what then had caused the accident, or (b) who had supposedly carried out this later study. In contradicting the Board, Shaffer was disagreeing not only with the N.T.S.B. but also those who assisted in their investigation, General Electric, Continental Airlines, United Air Lines, and others.

In the N.T.S.B. report, released on 18 October 1972, the probable cause of this accident had not been changed.

Then, if the carriers were supposedly complying with the Service Bulletins as Shaffer had said, why did the N.T.S.B. have to point out that these Bulletins should be made mandatory? The innocent passenger might ask himself: "*Anything* that affects aviation safety should be made mandatory in the first place – should it not?"

It is also interesting to note that Shaffer, as far as the correspondence files are concerned, never substantiated his claims that the N.T.S.B.-determined causes of this accident were wrong.

CHAPTER 18

It was on 19 May 1972, when Captain Richard T. Zettel was making final preparations to his aircraft at the Albuquerque International Airport, New Mexico, while the 8 passengers were emplaning and strapping themselves into their seats.

At 1320 hours, everything was ready to start Ross Aviation's scheduled air taxi flight to Los Alamos, New Mexico. The airplane was a small Beechcraft 65–B80 Queen Air, registered number N841NS.

The passengers had been checked-in at the Ross counter in the terminal building and the cargo, consisting of eight parcels weighing a total of 121 pounds, was delivered to the machine by Wayne Lillard, a Ross Aviation employee. Lillard stated that the pilot placed two parcels weighing a total of 35 pounds into the rear cargo compartment. The two men then loaded the front cargo hold with the remaining six parcels, and the last to be put in place was a box of Control Data Inc., electronic circuit boards, after which Lillard said he saw the pilot close the cargo door, but he could not say whether he saw him lock it.

At 1324:40 hours, the tower controller cleared the Beech Queen Air for take-off.

Robert E. Gill, the assistant airport manager, and Paul G. Flood, a local construction engineer, were standing near the intersections of runway 17, 12 and 21, and they observed the Queen Air approach them, then lift off the concrete and climb into the air. Gill estimated that the aircraft's altitude was between 50 and 100 feet, whereas Flood said it was between 150 and 200 feet when it passed them. They both agreed that the left propeller was stopped and the landing gear and flaps were in the "up" position. The machine continued to climb very slowly for a few hundred feet when it started a shallow left turn. Both Gill and Flood said that the turn was wide and continued intermittently until the aircraft was facing the direction from which it had taken off. During this time, they estimated that it lost 50 feet of altitude.

At 1328:35, Captain Zettel radioed the tower asking for a clearance to return and re-land, and he was immediately given the O.K. However, no further transmissions were received from the aircraft.

After being momentarily straight and level, the plane was observed to assume a slight nose-up attitude and then roll to the left at a very rapid rate. The nose then suddenly dropped down during the roll. The aircraft crashed into the ground, 7,000 feet from where Gill and Flood were standing, right wing first, with the machine upside down. It disintegrated, and the wreckage was scattered over an area of 60 by 280 feet as it exploded into flames. All nine persons on board died.

N841NS had been fitted at the Beechcraft factory with a forward baggage compartment door that was hinged at the top, (in similar fashion to those on the DC–10), and was opened by turning a D-shaped handle. The latching mechanism was three sliding bayonets, one on either side of the door and the third on the bottom, which were held in the latched position by an over-center cam. An electric micro-switch was mounted at one of the bayonet points in the door frame and this in turn was connected to the left engine starter switch. In this fashion, theory has it that the door must be in the fully locked position and the micro-switch actuated by the pressure of the locking bayonet arm, before the pilot could start the left engine.

Shortly after the accident, at 1341 hours, a U.S. Air Force B–57, landed on runway 17 and reported "debris" on and adjacent to it. A runway sweeping machine was immediately sent to the area and the operator got out of his cab and picked up a box that had been smashed apart and had evidently contained electronic circuit boards. A thorough search of the entire runway and its surrounds revealed a huge collection of these boards, together with cardboard, plastic foam, airline shipping tags, the tip of a propeller, some aluminium sheeting, a tin can, and a quantity of wrapping paper.

The investigators arrived at the scene and impounded the fault reports and maintenance sheets for N841NS, which revealed the following:

3 May 1972: Forward baggage door handle broken. Action: replace handle.

6 November 1971: Forward baggage compartment door bent. Action: paint.

7 November 1971: Safety switch in forward baggage door does not operate. Action: circuits changed and parts ordered.

This item was then carried over to the "deferred list". The word "inoperable" was added to the page.

Just before the accident, the aircraft was inspected in preparation for a major overhaul, and the person who looked it over revealed to the investigators that he had found that the left-hand engine *could* be started with the forward baggage door ajar. He checked further, and saw that even though the safety micro-switch was mounted in the door frame, there were no wires running to it. Neither could any be found that would have appeared to have been part of the system.

Between June and December 1968, there were eight pilot "write-ups" about either faults in the baggage door or its safety system. From January to May 1969, seven more discrepancies or malfunctions were noted. On both 18 December 1968, and 11 January 1969, the baggage door had been hit by the left propeller, and there were recurrences on 2 April 1969, and 20 April 1969, before the records went "missing" for the period November 1971, to February 1972.

The National Transportation Safety Board determines that the probable cause of this accident was the inadvertant opening of the forward cargo compartment door and the subsequent discharge of cargo, which caused damage to the left propeller and additional drag, at a critical phase of flight. The Safety Board believes that had the door-unsafe-indicating system been operational or had the security of the forward cargo compartment door been ensured, the accident would have been avoided.

A few days after the crash, on 7 June 1972, the chairman of the N.T.S.B. wrote to John Shaffer of the F.A.A. and outlined what they had discovered at that time in their investigations. He also pointed out that this was not the first time that a Beechcraft had been involved in a cargo door coming open with the aircraft in flight.

There was the accident on 3 November 1971, at Cleveland, Ohio, when the nose door suddenly opened and a 60 lbs. weight mail-bag fell out, smashing itself into the right propeller.

The letter to Shaffer noted the earlier accident, and another list of explicit recommendations from the N.T.S.B. landed on his desk.

CHAPTER 19

In summary, Service Bulletins are non-compulsory pieces of paper. They are not made public, and therefore they are not widely known about. They are not explicit enough with regards to the "risk factor". They are a completely "hit-or-miss" way of trying to cajole owners into modifying their aircraft. On the other hand, Airworthiness Directives are positive. They are widely publicized. They are legal documents that must be immediately obeyed or, at the least, complied with in the period of time specified, if one is given. Therefore, where matters of human safety are concerned, the *only* way of guaranteeing that design faults will be corrected is by the issuance of Airworthiness Directives

The first Service Bulletin concerning the DC–10s cargo doors that enters our story is number 52–27 issued on 30 May 1972. Fifty-two is a code number for Douglas aircraft doors – twenty-seven, means that this is the twenty-seventh Service Bulletin McDonnell Douglas had issued for the DC–10 in this specific area.

The aircraft affected by this Bulletin were manufacturing numbers 1 through to 45. American Airlines were operating numbers 3, 5, 7, 9, 12, 13, 20, 21, 22, 23, 24, 30, 31 and 37. Continental Airlines had numbers 34, 40, 41, 43 and 44. National Airlines were in possession of 14, 16, 18, 19 and 38. And United Air Lines had 4, 6, 8, 10, 11, 15, 17, 25, 26, 27, 32, 35, 39, 42 and 45. It will be noted that aircraft fuselage numbers 1, 2, 28, 29, 33 and 36 were not included in the lists as needing modifications. Number 1 was to be later delivered to American Airlines; number 2 was McDonnell Douglas' own machine; number 28 was earmarked for Northwest Orient Airlines and likewise number 36; while 29 and 33 were still at Long Beach having been ordered by Mitsui, the huge Japanese cartel, for All Nippon Airways, but at the last minute, and for unknown reasons, All Nippon cancelled their order.

Affected aircraft other than those listed (so says McDonnell Douglas) will be modified prior to delivery or included in a subsequent revision to this Service Bulletin.

Reason: Three operators have reported failure of the electrical latch actuators to latch/unlatch the cargo doors. Latch actuator failure is attributed to an excessive voltage drop reducing the output torque to the actuator. This condition may prevent electrical latching/unlatching of the hooks. Failure to provide adequate torque necessitates manual latching/unlatching of the doors, resulting in flight delays. Increasing the wire thickness between the circuit breakers and the actuators will reduce the voltage drop and provide adequate torque to operate the latch hooks under all anticipated conditions.

Description: This modification revises cargo door latch actuator wiring as follows:

(1) Replaces wiring between the cargo door latch motor circuit breakers in the centre accessory compartment and the door latch relays on the forward, centre and aft cargo compartment equipment panels, with wiring of a heavier thickness.

(2) Replaces wiring between the aft cargo door latch relay on the centre cargo compartment equipment panel and the terminal strip on the aft cargo door frame, with wiring of a heavier thickness.

Compliance: Recommended.

Approval: This Service Bulletin is F.A.A. approved.

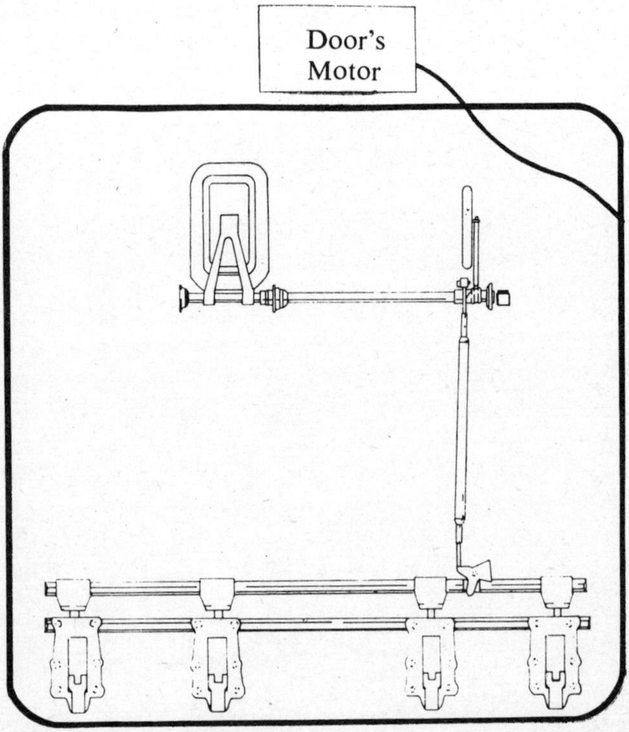

What all this means is that when there is a large "drag" on the electricity supply and the aircraft's voltage drops, because of the amounts of power required by the operating motors in order to open and close the cargo compartment doors, if the power is low, the motors will get insufficient to complete the job of locking or fully opening them. McDonnell Douglas were then "suggesting" that the airlines put in a stronger or thicker wire so that more current would be available to these motors and, therefore, the door opening and closing sequence would not be affected if the overall power in the aircraft was low.

Hidden behind this Service Bulletin to the outsider, was the fact that the door opening/closing motors, especially when the doors were being closed, could stop working, (a) before the locking was completed, and (b) without the operator realizing that the motor had stopped because of the low power and that the door was *not* fully locked.

To close a cargo compartment door on a McDonnell Douglas DC–10 requires an operator, standing on a ladder or adjustable height platform, to be able to reach a switch inside a small door-operation-control-panel situated next to and about halfway up the left-hand side of the door. The switch itself is a spring-loaded device which returns to the central "off" position when not kept "up" (to open) or "down" (to close the door). The operator then holds the switch in the "down" position and the motor, situated just inside and in the centre of the upper sill, starts winding the door closed. Once it had been wound flush with the sills, the operator must still keep the switch "down" for a further four to seven seconds during which the gearing system inside the door holds it closed and then starts winding around the 4 lock-grips over the lock-bars located on the lower door sill. If the operator does not hold the switch "down" for a sufficient length of time, the lock-grips will not be completely around the lock-bars, the door will not be properly secured, and yet there was no way that the operator may *visually* know whether the operation had been successfully completed.

The next step in the closing process is for the operator to pull "down" a handle, located in the outer upper left-hand section of the door itself. The theory was that if the lock-grips were in place around the lock-bars, the handle would move a "lockpin-bar" into a position (a) ensuring that they were in place, and (b) stopping them from unintentionally coming open.

Prior to this Service Bulletin, DC–10s were being held up at airports because the drive motors were not getting sufficient electric power to complete the locking process. The operators could not see what was happening. And indecision reigned as to whether it was not locked because of the power supply, or because of some other unknown fault that could have developed within the hidden mechanism inside. There was no way at this stage of our story, that the operator or anyone else for that matter, could actually see what had gone wrong. The locking of these cargo doors was a completely "chance" affair. All they knew was that for one reason or another, something must have been wrong because they could not easily stow the locking handle in the third part of the operation.

Up to this stage, the DC–10 had been in service for ten months with the four American carriers, and during that time there had been a large number of hold-ups, the causes of which were eventually and supposedly traced to the electric power supply.

CHAPTER 20

JUST under 200 people were on board American Airlines Flight 96, scheduled to depart Los Angeles, California, at 1350 hours, on its way to La Guardia, New York, with intermediate stops at Detroit, Michigan, and Buffalo, New York, when the ground service engineers were having difficulty in shutting the aft cargo compartment door. It took them eighteen minutes to close and lock it, while the crew and passengers in the DC–10 – registered number N105AA – sat and waited. The date was 11 June 1972.

This was not the first time that American Airlines personnel had experienced this type of trouble. Later, when F.A.A. inspectors went through the company's Maintenance Information Summaries, they uncovered ten similar entries, all concerned with closing and locking this door.

Normally, the airlines process this information within the following 30 days so that any recurrent problems, such as this one, would have been noticed and acted upon. But for some strange reason it was missed out, and no one realized the difficulties being encountered by all of the DC–10 operators – other than the Douglas Aircraft Company that is, and they were not telling.

The other possibility is that because there was a huge number of reports of operational difficulties, the door problems were prognosticized as being of a relatively minor nature, whereas the flight control and engine faults were being given full consideration. The airlines could very easily "cover-up" the fact to their passengers that rather stupid door problems were occuring. But it would be difficult to hide major flight difficulties or engine disintegrations, from their customers' knowledge.

So in order to maintain face in the light of their own advertising campaigns, it was evident that the troubles that could be seen, experienced, or even read about in the newspapers by their passengers, should be the ones on which all maintenance efforts must be concentrated.

CHAPTER 21

REASONABLE weather conditions existed in the vicinity of Detroit Metropolitan-Wayne County Airport on 12 June 1972. There was a blanket cloud covering at about 4,500 feet above the ground, and underneath it visibility was restricted to around one and a half miles because of smoke and fog. The wind was a light six knots from the southeast, and the temperature 61° F.

American Airlines Flight 96 was running just under one hour late because of a hold-up at its starting point. It was a Douglas DC–10–10 – registered number N103AA – on a scheduled passenger flight from Los Angeles, California, to La Guardia, New York, with intermediate stops at Detroit, Michigan and Buffalo, New York. It was the same flight, only one day later, to that involved in the eighteen minutes hold-up because of the aft cargo door closing problems.

It was scheduled in at Detroit at 1730 hours but landed at 1831, and was at the passenger loading gate at 1836. The hop from Los Angeles had been uneventful, and once parked, the 38 passengers leaving the flight at Detroit, disembarked. The DC–10 was "topped up" with 1,426 gallons of kerosene. Forty new passengers then joined the sixteen who were continuing on to Buffalo, and as everything was being readied, an American Airlines foreman, R. Holyworth, conducted the usual pre-flight exterior walk-around check in order to see that all was in order on the outside of the machine.

At this stage, everything seemed normal even though the ramp service agent who had closed the aft cargo compartment door reported that he had difficulty in locking it. He had shut the door electrically and listened for the motor to stop running before attempting to pull down the locking handle. As it would not come down with the application of normal pressure, he somehow got his knee to it and forced it to lock, then called over an American Airlines mechanic who was told the story, quickly checked it and pronounced it as being all right. The cockpit crew knew nothing of the difficulty as their control panel light had gone "out", showing, so they thought, that the door was secured.

The drama of a developing accident situation usually starts well and truly behind the scenes away from the passengers' gaze. And so, with Flight 96 ready to continue its journey, we shall combine all the behind-the-scenes conversations chronologically, using different typesettings and making slight adjustments here and there so as to maintain continuity of conversations going on at the same time.

Here is what happened as the flight crew and air traffic controllers experienced it.

The times given are actual local times at Detroit Metropolitan Airport.

Conversations on the radio are shown in this typeface.

Conversations in the DC–10's Cockpit are in this typeface.

Conversations over other radio frequencies to which the DC-10 was not tuned are shown thus.

Conversations over telephones or inter-link head-sets, thus.

1903:35	**Co-pilot**	**Metro Clearance Delivery from American Ninety-six. Information Delta** (what they were to do as soon as they had taken off) **and airways Buffalo** (and their clearance details of both height and direction for the flight to Buffalo).
	Clearance	**American Ninety-six. Be right with ya. Just a moment.**
	Co-pilot	**Roger, Ninety-six.**
1904:15	**Clearance**	**American Ninety-six. Cleared to Buffalo Airport as filed. Right heading** (after taking off) **to zero six zero** (degrees). **Vector Jay five five four** (turn on to the heading towards that air way), **maintain four thousand** (feet altitude), **expect two five at ten after departure** (expect to be cleared to a height of 25,000 feet, 10 minutes after leaving the runway), **departure eighteen ninety-five** (the departure controller is on radio frequency one eight nine point five MHz), **squawk zero two zero zero** (dial up and transmit on your transponder frequency code 0200 so that the ground radar may identify your aircraft and its height).
	Co-pilot	**American Ninety-six. Cleared to er, Buffalo Airport. After take-off turn right, that's as**

		filed, after take-off turn right to, er, heading, er, zero six zero, for radar vectors Jay five five four. Maintain five thousand – that's maintain four thousand, but expect flight level two five zero, ten minutes after departure. One eighteen niner five, and zero two hundred transponder.
	Clearance	**O.K. American Ninety-six. Clearance correct. If I gave ya eighteen ninety-five** (as the departure controller's radio frequency), **it should be one one eight point four for departure.**
	Co-pilot	**I was probably wrong. Er, one one eight point four. Thank you, sir.**
	Clearance	**One eighteen four. So long.**
1905:00	**Co-pilot**	**Good night.**
		(The conversations all around continued normally, including the co-pilot's request to Departure Control/Ground, for his clearance to start up his engines.)
1910:25	**Co-pilot**	**Ground. American Ninety-six.** (O.K. for) **push-back from Gate 10?**
	Ground	**American Ninety-six. Roger. Call** (me) **when** (you are) **ready to taxi.**
1910:36	**Co-pilot**	**Wilco.**
1913:35	**Co-pilot**	**Ground. American Ninety-six** (ready to) **taxi.**
	Ground	**American Ninety-six** (taxi out) **to Runway 3 Right.**
	Co-pilot	**Three Right.**
1914:00	*Cleveland M*	*Windsor?*
	Departure	*American Ninety-six* (at) *Metro* (politan Airport, Detroit) *is, er, Windsor, or whoever you are, he's gonna be off* (the) *Metro* (runway) *at two zero* (at twenty minutes past the hour).
	Cleveland M	*OK. Ah, . . . ah, niner thousand* (feet is the initial height clearance we can give him).
	Departure	*Niner thousand* (feet).
1916:40	*Departure*	(Regarding) *American Ninety-six. Ah, any chance of* (a) *higher* (altitude) *for him? Is that ten thousand* (feet) *over the airport?*
	Cleveland R	*American Ninety-six can have, er, two . . . one zero.*
	Departure	*Say again.*
	Cleveland R	*Two one zero* (21,000 feet)*, er, for American Ninety-six.*
	Departure	*Thank you.*
1917:15	**Co-pilot**	**American Ninety-six. Going to** (change frequency to) **one twenty-one one.**
	Ground	**Roger. Good day.**

	Co-pilot	**Goodday, sir.**
1918:00	**Co-pilot**	**Metro. American Ninety-six is ready** (in the holding position just off the end of runway) **Three Right.**
	Local Cont.	**American Ninety-six. Taxi into position** (on the runway) **and hold.**
	Co-pilot	**Position and hold.**
1918:47	*Pilot*	*Are you ready, er, Paige? – Uh, Paige?* (The pilot is offering the co-pilot the chance to effect the take-off).
	Co-pilot	*Oh, thank you.*
	Pilot	*Wanna try one?*
	Co-pilot	*Alright, sir. Thank you.*
	Pilot	*I'll bring it up to a stop here* (holding it on the runway).
	Co-pilot	*O.K.*
	Pilot	*It's still 'position and hold' for us on the air?*
	Engineer	*O.K. We've got anti-skid to go* (on the check-list).
1919:15	**Local Cont.**	**American Ninety-six. Maintain runway heading** (after lifting off). **Contact** (the) **departure** (controller on your radio). **Cleared for take-off.**
1919:19	**Pilot**	**American Ninety-six is cleared for take-off.**
	Engineer	*Checklist completed.*
	Local Cont.	**Roger. Maintain runway heading** (after taking off).
	Pilot	**To maintain runway heading.**
1919:26	*Pilot*	*O.K. You can take it with the rudders* (take over control of the rudders with your feet).
1919:35	*Local Cont.*	*Departure. American Ninety-six is on the runway.*
	Departure	*OK.*
		(Sound of engine noise level increases.)
	Pilot	*He said maintain heading* (after taking off).
	Co-pilot	*Yeah.*
	Pilot	*Yeah.*
	Pilot	*OK. You've got it* (the control of the aircraft).
	Co-pilot	*OK.*
	Engineer	*Ninety-eight point six* (engine pressure ratio speed either being read from dials or the planning chart).
	Pilot	*Get your hands on the wheel* (the flight stick).
	?	(Unintelligible).
	Co-pilot	*I got cha.*
1919:48	*Pilot*	*Vee One* (the aircraft was now up to 135 mph).
	Pilot	*Rotate* (meaning 'lift the nose off the runway for take-off').
	Co-pilot	*Your forward main gear up?*

	Pilot	*OK. Then your forward* (landing) *gear has a long way to go.*
1920:10	**Local Cont.**	**American Ninety-six. Maintain runway heading. Contact departure** (controller on radio frequency) **one eighteen four.**
	Pilot	**One eighteen four. Maintain runway heading. American Ninety-six.**
1920:32	**Pilot**	**And American Ninety-six.** (We are now on) **your frequency. We're off Detroit Metro** (we have just taken off from Detroit Metropolitan Airport).
	Departure	**American Ninety-six. Detroit departure radar contact. Climb and maintain six thousand** (feet).
1920:42	**Pilot**	**OK. You want us to climb** (to) **and maintain six thousand** (feet) **American, er, Ninety-six.**
1920:50	**Departure**	**Roger. American Ninety-six. Now turn right** (to) **heading zero six zero** (degrees).
1921:00	**Pilot**	(Turn) **right to** (heading) **zero six zero** (degrees). **American Ninety-six.**
	Pilot	*OK. You got 'em?*
	Co-pilot	*Yeah.*
	Co-pilot	*Flaps up.* (Sound of flaps lever being moved.)
1921:23	**Departure**	**American Ninety-six. Turn right, heading zero nine zero** (degrees).
	Pilot	(Turning) **right to zero nine zero** (degrees.) **American Ninety-six.**
	Pilot	*Zero nine zero* (degrees). *You got it?*
1921:30	*Cleveland R*	*American Ninety-six. Cleveland.*
	Cleveland R	*American Ninety-six. Cleveland.*
	Cleveland R	*Departure. Windsor, departure.*
	Departure	*Go ahead.*
	Cleveland R	*Is that American Ninety-six west of LaSalle, or east of Metro?*
	Departure	*Ah, just northeast of Metro there – about 8 miles.*
	Cleveland R	*OK. Radar.*
	Departure	*Thank you T.J.*
	Co-pilot	*Slats up.* (Sound of a click as slats handle is moved.)
	Pilot	*Whoever designed that thing. I'll tell ya, he – ooh.*
	Co-pilot	*They're really something to behold. Huh?*
	Engineer	*Gear handle to* – (Unintelligible) – (Sound of a click as flaps handle is moved.)
1922:15	**Departure**	**American Ninety-six. Turn right** (on to a) **heading** (of) **one one zero** (degrees). **John, er,**

		Jay five five four (the ground beacon signal and its direction) **when ya intercept** (it).
	Pilot	**OK. Take over on Vector ninety-four when we intercept** (its signal). **American Ninety-six. And, er, six thousand** (feet) **still our** (cleared) **altitude?**
1922:33	**Departure**	**OK. I'll have something higher for ya in just a moment, and that'd be Vector Jay five five four, to resume normal navigation on** – (pause) – **er, Jay five five four, to American Ninety-six. Climb. Climb and maintain, er, flight level two one zero** (21,000 feet altitude).
1922:45		(Sound of a warning horn.)
	Pilot	**OK. Er, climb** (to) **two one zero** (21,000 feet). **American Ninety-six. We are out of, er, fifty-five hundred now.** (We have just passed up through 5,500 feet altitude just now.)
1922:54	**Departure**	**And would you verify – were you issued** (with a flight plan for) **Jay five five four? I misunderstood you, I believe.**
	Co-pilot	*Yeah. Flight plan route.*
1922:58	**Pilot**	**That is correct. Er, take over on five five four and inter** (cept), **and flight plan route. American Ninety-six.**
	Departure	(Makes two clicks with microphone key.)
1923:07	**Departure**	**American Ninety-six. Call Cleveland Centre on** (radio) **frequency one two six point four. Good day.**
1923:07	*Pilot*	*Call for 'em Paige, and I'll try and set 'em up for you. OK?*
	Co-pilot	*Thank you.*
1923:13	**Pilot**	**Er. Repeat that frequency. I'm sorry.**
	Departure	**One two six point four. Twenty-six four. Good day now.**
1923:18	**Pilot**	**Good day, sir.**
	Co-pilot	(I can do this) *but I'll be late*, (changing radio frequencies).
	Pilot	*OK. I expect you to do it when I'm a few minutes late.*
	Co-pilot	*Thank you.*
1923:35	**Pilot**	**Good evening Cleveland Centre. American flight Ninety-six is out of, er, seven thousand** (feet altitude) **now for two one zero** (21,000 feet).
1923:45	*Engineer*	*I'm giving you three seven to get ten.*
1923:47	**Cleveland R**	**American Ninety-six. Squawk code one one zero zero and indent.** (Turn your transponder

		on to code 1100 and push the code transmit button on it so that we may identify you on radar.) **Maintain flight level two three zero – and report reaching.** (Climb to 23,000 feet and report by radio when you get there.)
1923:58	**Pilot**	**OK. One one zero zero and ident. And report reaching two one zero. American Ninety-six.**
	Co-pilot	*Two three zero.*
1924:05	**Pilot**	**Ah, correction. Was that two three zero?**
	Cleveland R	**Two three zero. American Ninety-six. Report reaching.**
1924:10	**Pilot**	**OK. American Ninety-six.**
1924:28	*Pilot*	*There goes a big one up there.*
1924:32		(Sound of a high amplitude noise).
	?	*Oh, shit.*
1924:35	*Pilot*	*What the hell was it – I wonder.*
	?	(A whistle made by one of the crew.)
1924:35		(Sound of the number 2 (in the tail) engine's fire warning horn and at the same time the sound of the altitude warning horn.)
1924:39	*Pilot*	*We'll pass the fire warning.*
	Co-pilot	*Which one?*
1924:42	*Engineer*	*We've hit something.*
	Co-pilot	*We've lost* – (pause) – *lost an engine here.*
1924:46	*Pilot*	*Er, which one is it?*
	Engineer	(The number) *two* (the engine in the tail).
	Pilot	*Is pitch hold grabbing on?*
	Engineer	(Engine) *number one* (on the left wing) *is still good.*
	Engineer	*And, er, captain . . . we'll have to . . . to check this out.*
	Co-pilot	*OK. We apparently* – (pause) – *master warning. This board's got* (an indication of) *an engine fire over here. Yeah, we got the engines one and three* (engines number one and three, those under the wings are still working).
	Co-pilot	*Do we have hydraulics?*
	Pilot	*No.*
	Pilot	*I have full rudder here* (meaning that the rudder pedals have swung into a full tail left position and jammed).
	Pilot	*Yeah.*
	Engineer	*Hydraulic pressure's OK.*
	Stewardess	*Is everything alright up here?*
	Co-pilot	*You go back to the cabin.*
1925:16	**Pilot**	**No. Ah, Centre, this is American Ninety-six. We got an emergency.**

1925:22	Cleveland R	American Ninety-six. Roger. Returning back to Metro?
	Pilot	*Yeah.*
1925:25	Pilot	Ah, negative. I want to get in to an airport that's in the open. Where's one open? (He wanted an airport that did not have any houses and a built-up area all around it.)
1925:29	Cleveland R	American Ninety-six. Start (a) right turn. Heading'll be one seven zero (degrees). Maintain one zero thousand (10,000 feet altitude). Go ahead.
	Pilot	Right turn to one zero thousand?
1925:37	Cleveland R	Turn right (to) heading one eight zero (degrees). Maintain ten thousand (10,000 feet altitude). Go ahead.
1925:48	Cleveland R	American Ninety-six – Cleveland.
1925:51	Co-pilot	We got one seven zero (degrees for the new) heading, sir. And, er, maintaining twelve thousand (feet altitude).
1925:54	Cleveland R	Nine six. Roger. Type of emergency?
	Co-pilot	Yeah. Yes, sir.
	Pilot	We have control problems. We have no rudder – got a full jam. We've had something happen, I don't know what it is.
1926:06	Cleveland R	American Nine Six. Understand. Cleared to maintain, er, niner thousand (feet); (barometric) altimeter (setting is) two nine eight seven (inches of mercury); be, er, radar vectored (led by radar) back towards the ILS course (the instrument landing system signal for) runway three (at Detroit Metropolitan Airport). Do you want the (emergency) equipment to be standing by?
1926:22	Co-pilot	OK, sir. Er, er, say again the heading and we'll let down slowly to nine thousand (feet altitude).
	Cleveland R	Heading'll be two zero zero (degrees). Two zero zero (degrees) for American Nine Six. And do you wish the (emergency) equipment to be standing by?
	Co-pilot	Affirmative.
	Cleveland R	Understand. Full (emergency) equipment standing by at Metro (politan Airport).
	Co-pilot	Yes, sir.
	Cleveland R	Roger.
1926:40	*Cleveland R*	*Hello – East Arrivals – Stripe.*
	Arrivals	*Yes, Sir.*
	Cleveland R	*Yes, Sir. We have an emergency. American*

		Ninety-six. He is a heavy DC–10 coming into Stripe right now. Er, I don't know what type of emergency it is – (pause) *– Er, I'm just picking this up from the guy that's working him right now. I think he said he thinks he's hit something. He'd like the* (emergency) *equipment standing by.*
	Arrivals	*OK. Thank you M.C.*
	Cleveland R	*Presently he is twelve miles north of Stripe. I'm descending a North Central* (Airlines aircraft) *that's at Stripe, to get the American* (Airlines aircraft) *in first.*
	Arrivals	*OK.*
	Cleveland R	**And American Nine Six, you say you believe you hit something?**
1927:00	**Co-pilot**	**Er, I don't know sir. Just stand by one. Stand by until we assess the situation. We've got here, er, definite problems.**
	Cleveland R	*I think he says he's hit something. He'd like the* (emergency) *equipment standing by.*
	Arrivals	*OK. Thank you M.C.*
1927:21	**Pilot**	**OK, now. We've got, er, problems. I've got a hole in the cabin. I think we've lost number two engine** (the one in the tail). **We've got a jammed rudder, full left rudder, and we need to, er, get down and make an approach. I guess Detroit Metro** (politan Airport) **would be the best and, er, can you vector** (lead us by radar) **around?**
1927:42	**Cleveland R**	**American Nine Six. Roger. Turn further right now. Heading'll be two zero zero** (degrees).
	Pilot	**OK. Two zero zero** (degrees). **American Ninety-six.**
	Cleveland R	**Roger. And that the left, er, rudder is jammed.**
1928:00	**Pilot**	**Er, it's possible. We don't know what the problem is. We've got a hole in the side of the airplane and we've got a full left rudder here. But we're under control and we're heading, er, two three zero** (degrees) **at the present time, letting down slowly to nine thousand** (feet altitude).
	Cleveland R	**Roger. Two three zero** (degrees), **and continue around the, er, turn to a heading of two seven five** (degrees).
	Pilot	**Continue** (the right turn) **around** (to) **two seven five** (degrees), **sir.**
	Cleveland R	**Roger. Can you make a standard rate of descent?**
	Pilot	**Er, negative. Er, we gotta go a little slower.**

	Cleveland R	**Understand. Slow descent. And how about turns?**
1928:15	*Arrivals*	*Carlton Metro.*
	Cleveland R	*Yes.*
	Arrivals	**North Central Nine Eighty-five. Descend to five thousand** (feet altitude).
	Arrivals	*American Ninety-six. Ten miles north of Stripe radar. You can put 'em on about a . . .*
	Cleveland R	*He's on a two seventy five* (degrees) *heading right now. Just a second.*
	Arrivals	*That's fine – (pause) – and did you ever find out exactly what his emergency was?*
	Cleveland R	*Er, he says he's not sure. He's got no left rudder. He thinks he has some kind of explosion or he hit something. He's not sure yet. They are still talking to 'em, tryin' to figure out what's the matter.*
	Arrivals	*Oh boy.*
	Pilot	**Er,** (the) **turn we can give you, er, is close to fifteen degrees** (banking of the aircraft) **maximum.**
	Cleveland R	**OK. That'll be fine.**
	Cleveland R	*OK. I don't know how high he is, I'll get back to ya.*
	Arrivals	*He's twelve two* (12,200 feet).
	Cleveland R	*Twelve two.*
	Cleveland R	*How long can he go? He'd like a gradual descent.*
	Arrivals	*Descend him to five* (5,000 feet).
	Cleveland R	*Five thousand. Going to it. Mark it.*
1928:25	**Pilot**	**I have no rudder control whatsoever, so our turns are gonna have to be very slow and cautious.**
1928:31	**Cleveland R**	**Understand.**
1928:33	*Co-pilot*	*OK. We've got full control on this though, however, so I guess we're slow enough so we can probably use differential directions with engines. Thank goodness it's* (engines number) *one and two we've got* – (pause) – *er, one and three.*
1928:54	**Cleveland R**	**American Nine Six. Continue descent to five thousand feet. Say the altitude now.**
1928:58	**Co-pilot**	(We are at) **twelve thousand** (feet altitude and we are letting down) **to five** (thousand feet).
1929:10	**Arrivals**	**North Central Nine Eighty-five. Descend and maintain, er, three thousand five hundred** (feet altitude).
	Cleveland R	*Hello, er, East Arrivals.*
	Nth Central	**OK. Nine Eighty-five. We'll continue**

		(our) **descent to three thousand five hundred** (feet).
	Arrivals	**Yeah.**
	Cleveland R	*OK. He's got no number two engine. No rudder control. And this is for American Ninety-six.*
	Arrivals	*Yeah.*
	Cleveland R	*And that's all we have right now.*
	Arrivals	*OK.*
	Cleveland R	*OK.*
	Cleveland R	*Carlton. He's no rudder control at all. A hole in the cockpit. Number two engine out. He's from Metro. Just took off.*
	Carlton	*Hello, er, East Arrivals.*
	Arrivals	*Yeah.*
	Carlton	*OK. He's got no number two engine. No rudder controls, and this is for American Ninety-six.*
	Arrivals	*Yeah.*
	Cleveland R	**American Nine Six. If you are able to, squawk code, er, zero two zero zero** (so that the radar stations may identify you) **on your transponder please.**
	Co-pilot	**OK.**
	Carlton	*And, er, that's all we have right now*
	Arrivals	*OK.*
	Carlton	*OK.*
1929:57	**Cleveland R**	**American Nine Six. Altitude now?**
1929:58	**Co-pilot**	**Er, eleven thousand two hundred** (feet).
	Cleveland R	**OK. And about** (a let-down rate of) **er, two hundred feet a minute then?**
	Co-pilot	**Yes, sir.**
	Cleveland R	**Roger.**
	Cleveland R	*Want to get Litchfield to hold their guy's* (their aircraft) *at Milan* (radio beacon) *for a while. I'm coming in high with 'em* (American Ninety-six). (American is) *comin' this way for the ILS* (Instrument Landing System radar signal) *course.*
	Cleveland R	**And American Nine Six. You wanna make a left turn now, to a heading of er, two four zero** (degrees). **Two four zero – once again.**
	Co-pilot	**Two four zero** (degrees).
	Co-pilot	**Two four zero, on the heading?**
	Cleveland R	**Yes, sir. Two four zero, so it can give you, er, a little more room for your descent.**
	Co-pilot	**Yeah. Give me plenty of room to start a long approach** (to the runway).

	Cleveland R	**Roger.**
	?	**Two seventy-five heading?**
	Cleveland R	**Two forty. Coming down the ILS course, he wants about twenty miles to get in.**
	?	— (unintelligible) —
	Cleveland R	*No. You do it for me.*
	Arrivals	*Hello.*
	Carlton	*East Arrivals. Stripe again.*
	Arrivals	*Yeah.*
	Carlton	*American Ninety-six'd like to – he's on a two forty heading now. This is at his request. He wants to join the ILS and then come back up* (he wants to join the ILS radio signal going outwards from the airport and then turn and come back down to it).
	Arrivals	*OK. Fine. I'll push him down there south of Windsor.*
1931:12	*Pilot*	***OK. Give me about fifteen*** (degrees extention) ***on the flaps now. Watch it carefully.***
	Carlton	*He's only out of eleven* (thousand) *five* (hundred feet altitude) *now.*
	Arrivals	*OK. Thank you.*
	Carlton	*Two hundred feet a minute on the descent.*
	Arrivals	*M.C.*
1932:06	*Engineer*	*We'll be landing* (with our aircraft weight at) *about two hundred and nintey-two thousand* (pounds).
	Cleveland R	**And American Nine Six,** (you) **have traffic** (another aircraft) **out to the two o'clock position** (in front of you and slightly to your right) **three miles** (distance) **and east bound and off your right wing now. Three miles** (distance and a) **parallelling course** (at the) **present time.**
	Pilot	**Er, we didn't get that. We're working on the airplane here.**
	Cleveland R	**American Nine Six. Call** (me when you are) **leaving one zero thousand** (10,000 feet altitude).
	Pilot	**Er, say again our reference. Er, our altitude, sir.**
	Cleveland R	**Nine Six. Call** (me when your altitude is) **out of ten thousand** (feet).
1932:45	**Co-pilot**	**We're out of** (under an altitude of) **eight** (thousand) **seven hundred** (feet) **for, er, five thousand** (feet), **right?**
1932:48	**Cleveland R**	**American Nine Six. Roger. Turn back right – now – heading'll be two eight zero** (degrees).
	Co-pilot	**Two eighty** (degrees).

	Cleveland R	**Roger.**
1933:00	**Co-pilot**	**I'm guessing that we're gonna have to have a long turn on this final** (approach) **for the ILS** (radio beacon). **I have no control on rudder and steering directions.**
	Cleveland R	**Roger, I'm planning about** (a) **twenty miles** (turn). **'ll that be enough?**
	Co-pilot	**Er, I hope so. Thank you. I'll keep you advised.**
	Cleveland R	**Roger.**
	Pilot	**My rudder pedal is jammed all the way to the left. I have no trim tab system or stabilizer trim** (on the control surfaces on the tail).
	Arrivals	*You wanna give me American Ninety-six now?* (Do you want to hand over its ground control to me?)
1933:40	*Cleveland R*	*East Arrival – Stripe.*
	Arrivals	*Go ahead.*
	Cleveland R	*OK. I got two* (aircraft) *holding at Stripe. Right now, you wanna take the North Central* (Airlines aircraft)?
	Arrivals	*Yeah. North Central Nine Eighty-seven is* (on the radio frequency to Cleveland) *radar, and do ya want to give me American Ninety-six now?*
	Cleveland R	*American Ninety-six, er. Just a second. —* (pause) *— Two eighty* (degrees) *is the heading; eight thousand feet* (his present altitude); (his) *rudder is jammed in the left position on American Ninety-six. No stabilizer and no trim. He wants a long approach for the ILS* (landing) *course.*
	Arrivals	*OK. We'll turn him on* (to a planned landing course for a distance of) *twenty five* (miles) *out, then.*
	Cleveland R	*Say again?*
	Arrivals	*We'll turn him on about twenty-five out. That heading looks real good right now. Descend him to three thousand* (feet) *though.*
	Cleveland R	*Go to three thousand* (feet altitude) *with him now?*
	Arrivals	*Yeah. Three thousand.*
1934:28	**Cleveland R**	**American Ninety-six. Now cleared to** (let down and) **maintain three thousand** (feet altitude).
1934:34	**Co-pilot**	**Three thousand.**
	Cleveland R	*Arrivals. Are you ready for American Ninety-six?*

	Arrivals	*Yeah. Put'em on.*
	Cleveland R	*OK.*
	Cleveland R	**American Ninety-six. Contact Metro Approach now** (on radio frequency) **one two, er, one moment.**
	Co-pilot	**Normal frequency, or special frequency?**
	Cleveland R	*Normal frequency or special frequency for them?*
	Arrivals	*Er, normal frequency's alright.*
1933:13	**Cleveland R**	**American Ninety-six. Metro** (politan Airport) **Approach** (Controller on radio frequency) **one two five one five. One twenty-five, fifteen. Good night.**
1935:18	**Co-pilot**	**Thank you. One twenty-five one five.**
	Litchfield	*Hey, er, Windsor.*
	Cleveland R	*Yeah.*
	Litchfield	*You guy's working that American Ninety-six? This is, er, Litchfield* (a nearby radar station).
	Cleveland R	*Yeah. He's going back to Metro. Had an emergency.*
	Litchfield	*Yeah. Is that him about fifteen* (miles) *due south of Carleton?*
	Cleveland R	*Yeah.* (He's) *just leaving eight thousand* (feet altitude) *now.*
	Litchfield	*OK. Ok, fine then.*
	Cleveland R	*Rocky? Detroit High.*
1936:00	*Detroit*	*Detroit High.*
	Cleveland R	*Yeah. American Ninety-six* (has) *gone back to Metro* (politan Airport).
	Arrivals	*Carleton. Arrivals* (at) *Metro.*
	Cleveland R	*Yeah.*
	Arrivals	*Yeah. We still haven't talked to this American Ninety-six yet.*
	Cleveland R	*OK. I gave him* (the radio frequency of) *one twenty-five fifteen. Check it.*
	Arrivals	*OK.*
1936:08	*Co-pilot*	*We've got a nice rate of descent. Even if we have to touch down this way, we're doing well.*
1936:30	**Arrivals**	**American Ninety-six. Detroit.**
1936:32	**Co-pilot**	**Loud and clear, sir. And we're through five thousand five hundred** (feet) **for three thousand** (feet).
1936:45	**Arrivals**	**American Ninety-six. Turn right, heading three six zero** (degrees); **descend and maintain three thousand** (feet); **vector** (work your way towards) **the ILS for** (runway) **three Left final approach course;** (pressure) **altimeter** (setting is) **two niner eight five** (inches of mercury); **braking**

		(on the runway surface) **clear for all types of aircraft.**
1936:55	**Co-pilot**	**Er, thank you. Heading, er, three** – (garbled) –
1937:05	**Co-pilot**	**We'll probably have no brakes, er, so we'll try to reverse** (thrust the engines) – (pause) – **we're heading north and we're outta forty-eight hundred** (feet altitude) **for, er, three thousand** (feet).
	Arrivals	**American Ninety-six. Roger.**
	Arrivals	**American Ninety-six. Turn right, heading zero two zero** (degrees). **Your position is thirteen miles from the marker** (the radio beacon that is six miles out from and directly in line with the start of the runway). **You're cleared for** (using the) **ILS** (for runway) **three Left approach.**
1937:40	**Co-pilot**	**Zero two zero** (degrees), **and cleared for the ILS three Left approach** – (garbled) – **American Ninety-six.**
	Arrivals	**American Ninety-six. Do you have the localizer alright?** (Are you receiving the Instrument Landing System radio signal alright?)
	Co-pilot	**Yes, sir. We're turning in** (towards it) **slowly here now.**
	Arrivals	**OK. You're going a little bit through the west side of it. Turn right, heading zero five zero** (degrees).
1938:25	**Co-pilot**	**Ah yes, sir. We're recovering now to zero five zero.**
1938:40	*Pilot*	*Well, gimme the gear.* (Lower the landing gear, the undercarriages.)
1939:10	**Arrivals**	**American Ninety-six. You're nine miles from the marker now.**
	Co-pilot	**Thank you.**
	Pilot	**Er, we'd better have the emergency equipment stand by, er, at present.**
	Arrivals	**American Ninety-six. All the equipment is standing by now.**
1939:39	*Pilot*	*OK. Here we're coming into the ILS. I'm gonna start slowing her down. Give me twenty-two* (degrees extention) *on the flaps.*
1940:20	**Arrivals**	**American Ninety-six. Five miles from the outer marker.**
	Co-pilot	**American Ninety-six.**
1940:59	*Pilot*	*Alright. We got the green lights?* (There are some green lights on the pilot's controls showing that the landing gear is "down" and "locked".)

1941:10	**Arrivals**	**American Ninety-six. You're two and a half miles from the marker. Contact the tower** (controller on radio frequency) **one two one point one. Good night.**
1941:15	**Co-pilot**	**Good night, sir.**
1941:40	**Co-pilot**	**American Ninety-six is approaching the outer marker, inbound.**
	Local Cont.	**American Ninety-six. Roger. Continue your approach.**
	Co-pilot	**Ninety-six.**
	Local Cont.	**American Ninety-six. The wind is** (from a direction of) **one two zero** (degrees), **at five** (knots).
1942:25	**Local Cont.**	**American Ninety-six is cleared to land.**
1942:42	**Co-pilot**	**American Ninety-six. Cleared to land.**
1942:42	*Pilot*	*Give me thirty-five* (degrees extention) *on the flaps.*
1943:38	*Pilot*	*I have no rudder* (operation) *to straighten it out when it hits* (lands).
1943:53	*Co-pilot*	*OK. Now engines off at your discretion.*
1944:25	*Pilot*	– (garbled) – *keep 'em in reverse* (the engines in reverse thrust). *I can hold the brakes. Ahhhhhh.*
1944:55	*Pilot*	*SHUT'EM* (the engines) *DOWN.*

Besides the whole incident being one shrouded in extreme luck, the DC–10 when it landed 1,900 feet down the Detroit runway, immediately started to veer to the right. The captain applied reverse thrust to the two operating engines, but the machine continued to run off the right-hand side of the runway. As it veered further to the right, the co-pilot applied full reverse thrust to the left engine and brought the right one out of reverse and into its normal "idle". This provided some directional control and the airplane started moving back towards the concrete. It was finally halted, 8,800 feet from the start of the strip, and the pilot pushed the "emergency evacuation button" which heralded the beginning of the people on board leaving the crippled machine in a hurry.

In order to relive the airborne drama as the passengers and cabin crew experienced it, together with further details obtained from the flight deck personnel, here is a chronological combination of the events, obtained from the staff a short while after the accident.

(Co-pilot Paige-Whitney) On 12 June 1972, we took our load sheet flight release and flight plan to Gate 10 and boarded the DC–10 for Flight 96 to Buffalo from Detroit. The checklist was completed, and after a four or five minute delay because of open cargo doors, we started the engines and were pushed back from the Gate.

(Captain McCormick) We departed the Gate at Detroit at 1911 hours. During taxiing out to the runway, I asked First-officer Paige-Whitney to check the book for legality of take-off on Runway 3 Right. He verified that we were legal. The "Before Take-off Check List" was run and we were cleared into position on the runway. I decided to allow Paige make the take-off as I wanted to practice American Airlines' training system of allowing the first-officer to fly, and I would monitor.

(Co-pilot Paige-Whitney) The take-off was normal and smooth. The autopilot was put "on" at 6,000 feet altitude during our climb, and at 10,000 feet we rotated the vertical speed control and the autopilot to 1,000 feet a minute "up", to enable the aircraft to accelerate from 250 to 340 knots. At approximately 11,500 feet altitude, I remember we broke out on top of the clouds and thinking "what a beautiful day".

(Captain McCormick) I had just looked at the digital read-out on my altimeter at 11,740 feet and an indicated air speed of 260 knots, when it seemed to me that the rudder pedals "just exploded" to a full-left-rudder position. I had my feet on the pedals at the time. My right leg was thrown back against my seat with extreme force. Dirt and debris hit me in the face so severely, it almost felt like a firecracker had gone off right in front of my nose. I was practically blinded by dust in my eyes. When I could get my eyes open again, the first thing I looked for was the windshield; I just knew it had to be gone. I could see it, but I couldn't believe it. I started to reach out and touch it, but was distracted by the flashing red autopilot light, so I went down to push the "lights out" that were flashing on the instrument panel. At this time, I noticed I had a "red flag" (a warning signal) on my airspeed indicator. I thought, "the radar dome is gone". Paige had never taken his hands off the flight wheel, so I told him to let me have it; I wanted to "feel out" the controls. He immediately took his hands off the wheel when I took over. He said: "I think it's going to fly".

(Co-pilot Paige-Whitney) Then there was this loud "bang" from the left rudder pedal snapping full forward at a fantastic speed; the number two engine throttle control hit the "stop" with also a lot of noise. The number one and number three throttle controls also moved back to within a half-an-inch of the number two's position. The autopilot had disengaged itself, and the airplane then rolled slowly to the right. The rapid rearward movement of the throttles probably caused the momentary "jerk" in the airplane. The captain pulled the headset back onto his head (it had been knocked backwards when the rudder pedal flew back), and said: "What happened? We had a mid-air collision?" I had been looking outside and had not even seen the shadow of another aircraft, so my thoughts were immediately on the number two throttle movement and the rudder "explosion", and I contemplated a number two engine disintegration as being a possible source of the control problems, rather than a mid-air collision. The airplane was responding to control movements at our then 220 knots speed. The captain had taken control, talked on the public address system to the passengers and was checking everything in the cockpit. The girls then called for help in removing one of them from what they said was a "hole in the floor". The captain said: "Yes", to the engineer's request to go back and help.

(Captain McCormick) My headset was blown off my head and was on the back of my seat. I replaced it and radioed the Cleveland Centre and told them that we had an emergency on board and would have to get to an airport immediately. I knew I had a pretty sick airplane on my hands. In the meantime, the nose of the machine was dropping noticeably. At the time of the explosive decompression, we were just breaking out on top of the clouds into clear skies above and sunshine. I did not want to go back into the overcast on instruments again until I could determine the integrity of the instruments and controls. I looked over at the engine throttles and they were retarded. The number two engine control was clear back against the "stop"; the number one and number three controls were about half an inch to an inch in front of the number two. I asked Paige if he had pulled the controls back and he said: "No". I pushed the thrust levers forward and found that the number two engine did not respond; it just stayed in in-flight idle. I then tried moving the number two lever back and forth, then I could feel that it was no longer connected to the cables. It just stayed in in-flight idle. I noticed that the engine instruments seemed to be normal. Previous to all this, it kept going through my mind, "fly the airplane".

(Engineer Burke) Soon after the decompression and as we were in a descent back to Detroit, I received a call on the interphone from one of the cabin attendants to the effect that part of the cabin floor had collapsed and two stewardesses were trapped in the area and also that there was a hole in the fuselage. I relayed the message to the captain and asked if it was OK to go back and try and help. After getting out of my seat, I went to grab my hat, but all three of our hats were missing from their positions on the hangars on the cockpit door. While I was looking in the coat compartment for my hat, another phone call came from the cabin. This time the attendant said that the two stewardesses were safely out of the hole, and everything was under control. I went back to my seat to prepare for an emergency landing.

(Stewardess Sandra McConnell) The "No Smoking" signs were turned "off" almost immediately after take-off, and I remained in my seat next to the right-hand emergency exit at the rear for about ten minutes, waiting for the captain to "level off". I was talking to Bea Copeland (who was in the seat next to the rear left exit) and we had both unfastened our shoulder harnesses and seat-belts. Suddenly I heard a very loud explosion and I was thrown from my seat onto the divider partition that was behind the bar. I believe I was right next to the partition and not in front or behind it. I could see the hole in the lounge floor and I had the sensation that I was slipping down into it. The floor around me kept crumbling down and I got myself up and back to the lavatory. I felt that if I walked on the floor it would completely collapse.

(Stewardess Bea Copeland) After the "No Smoking" signs went off, I unfastened my seat-belt. Knowing that we did not have many passengers to serve, I stayed and talked to Sandy who was also in her seat, waiting for the plane to level off. Suddenly, I felt myself being lifted from my seat. The next thing I remember, I was on the floor. The bar was lying beside my head. A piece of the ceiling was on my head and something was trapping my foot. I could see down into the cargo compartment. I tried to move the object that was on top of my head. When it came back into position again, I yelled for help. Seeing that no one could help me, and remembering that I was on an airplane, I tried again to remove this object off my head. This time I succeeded so I next began to try and free my shoe. When I couldn't get my shoe out, I tried to get my foot out of the shoe – and succeeded. I pulled myself up, and climbed out of the area, over the bulkhead, and in to one of the rear lounge seats. Cyd and Carol were there to help me. Janice met me with a portable oxygen mask. I told her I didn't need it and so we walked forwards to the next door up on the right-hand side where Janice called the engineer and told him we were out of danger.

(Stewardess Carol McGhee) The head stewardess had decided not to open the rear lounge for cocktail service. Since my working position is in the lounge, I stayed in my seat next to the front right-hand side exit for a longer period than normal. At the time of the incident, I was still strapped in my seat. I heard a sound like a "frump", and saw the escape hatch from the downstairs galley fly up from the aisle floor and hit a passenger on the head. Turning towards the cockpit, I saw the door burst open and the crew's hats flying out. At that same time, a "sandy-like" atmosphere came gushing out of the door. Simultaneously, the cabin filled with a "cool fog" – it was different to what was coming out of the cockpit. I reached above for the oxygen mask and found none. I then realized I had no difficulty in breathing. I went to the cockpit and asked if they were alright. Their answer was: "Yes". I then slammed the door and proceeded to take care of the passenger who had been hit on the cheek by the underfloor galley hatch cover. She, and three other passengers were hysterical. After calming them and applying a thermo swab to the wound, I then proceeded to find the rest of the stewardesses.

(Stewardess Colleen Maley) After the "No Smoking" signs went off, Marlene Saydah (who was stationed next to Exit number 3 Right) came up and sat on the seat next to me, and we continued a conversation that we had previously begun. Approximately one minute later, the air filled with a white "fog-type" substance and we were jerked out of our seats and on to our feet. My first thought was to get an oxygen mask, but none of the masks had fallen down, so I assumed we had not climbed high enough to need supplemental oxygen. I started to survey the cabin and noticed a first-class male passenger standing and lighting a cigarette. I immediately reseated him and asked him to put his cigarette out. I then noticed that a woman in the first-class compartment had a minor cut on her face. I talked to her for a while, trying to calm her down, and then prepared a pink towel, packed with ice, and instructed her to hold the compress on her face.

(Stewardess Marlene Saydah) When the "No Smoking" signs went off, I went to the Service Centre to chat to Colleen who was in her seat. We had been talking only a few minutes when there was a noise and a jolt, and smoke came billowing through the area in which we were. I put my hand over my mouth and nose, looked up to see if my oxygen mask was there, and when it wasn't, I ran to the forward economy class cabin to see if any of the masks had fallen. They hadn't, and I started to shout at the people to put their heads down because a piece of the ceiling had fallen on the floor and I was afraid that more pieces would come down and injure them. I then tried to tell everyone that it was all over and that everything would be alright. I went up to the first-class compartment to check up there, and then Cyd asked us all to go to the Service Centre for a briefing.

(Stewardess Carol Stevens) The "No Smoking" signs went off – I kept my seat-belt fastened. I was waiting until we had gained more altitude before getting up. I don't recall whether the "seat-belt" signs had gone off, it was probably five to seven minutes after the take-off when the incident occurred. There seemed to be a "varrooom" type of noise, accompanied by a light fog and a rushing of air from the front of the aircraft to the rear. When this fog cleared, I could see the rear section of the economy compartment and the lounge. The ceiling panels were hanging down. The floor was caved in. The bar unit had fallen to the left side of the cabin. And then I saw Bea Copeland on the floor. She seemed to be on her side facing forward, with her head towards the centre of the aircraft. She seemed to be trapped and was crying out for help. I could not see the other girl assigned to the seat next to her. There didn't appear to be any immediate problem with any of the passengers in the rear economy-class section. I looked around and towards the forward economy-class compartment. Everybody seemed OK there. I could see a ceiling panel that had almost come completely loose and was dangling at a vertical angle between about seat rows 14 and 15. I then thought about calling the cockpit to ask for help in getting Bea out of the hole, however I knew they would be extremely busy, so I tried to contact Cyd Smith to tell her the girls were trapped down the back. I received no answer. I was hesitant to get out of my seat since I assumed that there was a decompression and was waiting to see if there was going to be an emergency descent. Since I was unable to reach Cyd by phone, I looked around again, and forward towards the front of the plane and noticed that the other girls were up. I then got out of my seat and ran forward. Cyd was at the forward side of the dangling ceiling panel, and I told her we needed help because Bea and Sandy were trapped. A male passenger offered to help and he followed Cyd and I back to the economy-class lounge. By that time, Bea had pulled herself up out of the hole and Cyd was directing her. On the left-hand side, Bea climbed across the divider panel and onto the side lounge seats. I grabbed her hand and pulled her into the economy section. Nearby, we had an elderly Italian couple who understood almost no English, and were extremely frightened, so I tried to comfort them. We then moved the passengers seated in the rear economy-class section, forward.

(Stewardess Cydya Smith) The "No Smoking" signs went off approximately two to three minutes after taking off and I got out of my seat and proceeded to the Service Centre where I began to make coffee and set up for the liquor service. The next few minutes are somewhat of a blur; however I shall try to recount the facts to the best of my ability. All of a sudden the door of the lift to the downstairs galley burst open – if I had been standing just a shade to the left, it would have smashed me in the face – and heavy smoke, or a smoky substance, started billowing out. I was thrown off balance, but managed to grab a handle right at the edge of the Service Centre, which prevented me from falling. As I was in this position, I saw a ceiling panel in the middle of the forward economy-class compartment fall down, and this was the sound I heard. I felt a sensation of weightlessness and then complete silence as though one was in a vacuum. Almost immediately, I looked forward to seat row 8 to see if the oxygen masks had come down – none had – so with that danger out of the way, I moved forward towards the cockpit. I remember seeing the captain's hat on the floor and I felt an urgent need to get up there. Once there, I asked if everything was alright. The captain said: "No", and Paige, the first-officer didn't turn to smile at me and I realized then that I had to hold up my end of the bargain and have something valid and necessary to relate to him when once again I would go back to the cockpit. I left the flight-deck and proceeded through the cabin, assuring the passengers that all was fine. After realizing that decompression had not occurred, I grabbed the public address microphone and instructed the passengers to stay in their seats; do not smoke; and please remain calm. Carol grabbed me and said that Bea was stuck in a hole down the back, so I went after her. As we reached the economy-class lounge, I was devastated by the amount of damage, but I could see Bea and we pulled her out. Bea then informed me that Sandy was trapped on the other side, and so I ran over, calling Sandy's name continuously. Finally, almost out of nowhere, Sandy appeared, and between the two of us and God, we got her over a seat into the lounge area in much the same way as we had rescued Bea. As we walked forward, I instructed the girls to move the passengers away from the rear economy-class cabin, and I then went up to the cockpit. I informed the captain that there were no serious injuries; that the two trapped stewardesses were shaken but safe, and that there was a hole in the fuselage in the very aft of the left-hand side. He then informed me that we were to prepare for an emergency landing. I asked him if we would be evacuating and he said he didn't know at that time. I left the cockpit, gathered the girls together and informed them that we would be preparing for an emergency landing. I made a public address announcement, and the girls went around collecting shoes and other loose articles from the passengers and stowed them away, and then we proceeded to show the passengers the "brace" position. The girls informed me that all was secure, and then they went back through the cabin making sure that the passengers understood what they were to do and briefing them on the respective doors through which they would escape. I went forward to the cockpit and told the captain that the "check-list" was complete, and asked him if he would be giving the "brace" command when the time came.

(Captain Bryce McCormick) One of the stewardesses came forward and I remember her saying: "Do you guys have a problem up here?" It really struck us funny at the time. We roared with laughter and answered: "Yes". We had a problem; but it did relieve the situation. She then asked if we really intended to use the evacuation slides on the ground. I told her that at that time, I hoped not. If we got the landing gear down and we finished up on the runway, we may have to use them. I said that if I wanted her to use them, I will activate the evacuation horn. She then left. About this time, I decided I had better have another talk to the passengers to put them at ease – if that was at all possible. I apologized to them for the inconvenience, and told them that we had control of the aircraft. I also said that just as soon as we got on the ground, American Airlines would provide further transportation for them to their destinations. I tried to speak in as calm a voice as possible. We started our landing approach. I slowed the airplane to 184 miles per hour and was able to keep it at a six to seven hundred feet a minute descent. I thought: "I wonder if I can get it any slower." So I pulled the throttles back a little and with the control column back all the way, the sink rate went up to 1,800 feet a minute descent. This was the first time I heard the co-pilot sound a little bit excited. He really started calling it out. I applied power to the engines again and settled for a 184 miles per hour landing speed. The entire landing approach was carried out with the aircraft yawing 5° to 10° nose to the right in order to stay aligned with the runway. The tension was pretty heavy in the cockpit at this time, so I thought I would try to relieve some of it. I turned to Paige and smiled and said: "Maybe we should have split the flaps?" He turned and looked at me as though I had flipped my wig, and then caught the jest of it and smiled back. He made some remark that I didn't hear, but I think we all relaxed a bit. We crossed the runway's threshold at 100 feet altitude on the radar altimeter, so I squeezed on a little more power to the engines to help me flare the aircraft's nose "up", and we touched down slightly facing to the right. The airplane was in such a flat attitude due to the high-speed approach that I seemed to be a lot lower than usual. The touchdown was smooth and I felt like I had given all on board a good landing. But then, all hell broke loose. The automatic spoilers on the wings lifted themselves up and I pulled the throttles back and went for the reverse thrust levers. I applied full left aileron and was waiting for the reversers to come in. We started running towards the right-hand side of the runway and I felt for the first time that I had lost control of the machine. I yelled for Paige to give me all the reverse power he could on the number one engine (the one under the left wing). He took over the operation of the reverse levers. We ran off the runway almost immediately. Our path was between the adjacent taxiway and the runway. Every time we would run over a cross-taxiway or runway, I would pull back on the stick trying to keep some of the load off the nose-wheels. The ground was rough, but the landing gear handled it very well. As we started coming back onto the runway, I started using a little right foot brake, but it didn't seem to be effective. I guess that was because we were out on the wet grass. We got back onto the concrete with approximately 1,000 feet of runway left. After we came to a stop, I decided that due to the rough ground run we might have ruptured a fuel tank or broken a fuel line. So I elected to activate the evacuation warning horn system and get the people off. As we pulled the horn handle, I saw the fire trucks coming to a stop around the nose of the aircraft. We started the engine shut-down procedures right away and I could hear the stewardesses getting the people out. By the time we had shut everything down and I started to take my seat-belt off, I looked out of the window on my side and there the passengers were, all out on the opposite side of the runway just standing there, looking up at us. The last stewardess on board came up to the cockpit door and said: "All the passengers are off, captain. Good bye. See you later", and down the slide she went. I told the co-pilot and flight engineer to complete the shutdown list and pull the flight recorder circuit breaker. I went back to the rear of the cabin to make sure that there were no passengers left on board. As I went past the first door, one of the firemen was yelling up at me to get out. I called back to him saying that I still had things to do. I went down the left-hand aisle and was checking for passengers, and when I got to the number 3 Left exit, I stepped out onto the wing to make sure there was no fire and to see if the number two engine in the tail was intact. I could then see the damaged horizontal stabilizer and the door to the aft baggage compartment having been torn off. I stepped back into the cabin and continued my walk to check the rear lavatories. I couldn't get near to them due to the floor damage. By this time, the co-pilot was coming down the right side aisle with the flight engineer immediately following him. We could see then that we were in no danger and so we took our time in getting out. We had trouble finding our hats as they had been blown through into the cabin. The stewardesses had picked them up and had thrown them into one of the forward lavatories.

Flight 96 was so close to becoming a major disaster, yet no one realized this until the accident investigators from the National Transportation Safety Board arrived on the scene the following morning. In fact the *only* thing that saved them was the lack of heavy weight on the rear passenger compartment floor in the vicinity of the lounge; the section immediately above the aft cargo compartment. Luck was with them in two ways. Firstly, the lack of heavy passenger seats in this area, simply because it was a lounge and, secondly, the lack of any people being in the area solely because the head hostess had decided that she would not open the bar on that particular flight.

The investigators soon discovered scuff marks on the aft cargo compartment door's securing mechanism, which showed that it had not been fully locked. They checked with the airport staff, and the two people concerned said that they both were solidly of

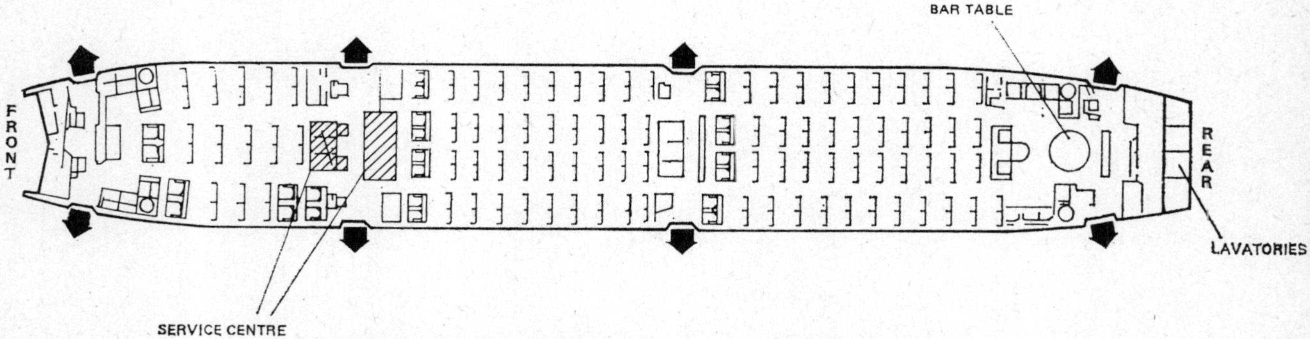

DC-10 configuration for American Airlines, flight 96.

the opinion that the door was properly closed, even though they did admit that pressure from a knee was necessary to get the handle into the "locked" position. They then quizzed the flight crew and it was confirmed that the "open door" light was "out", indicating for all intents and purposes that it was closed and properly locked. *But it was not.*

During the climb away from Detroit Airport, and as the internal air pressure started to build up inside – there would have been approximately six tons of pressure on the door itself – the partially positioned lock-grips suddenly sprang open.

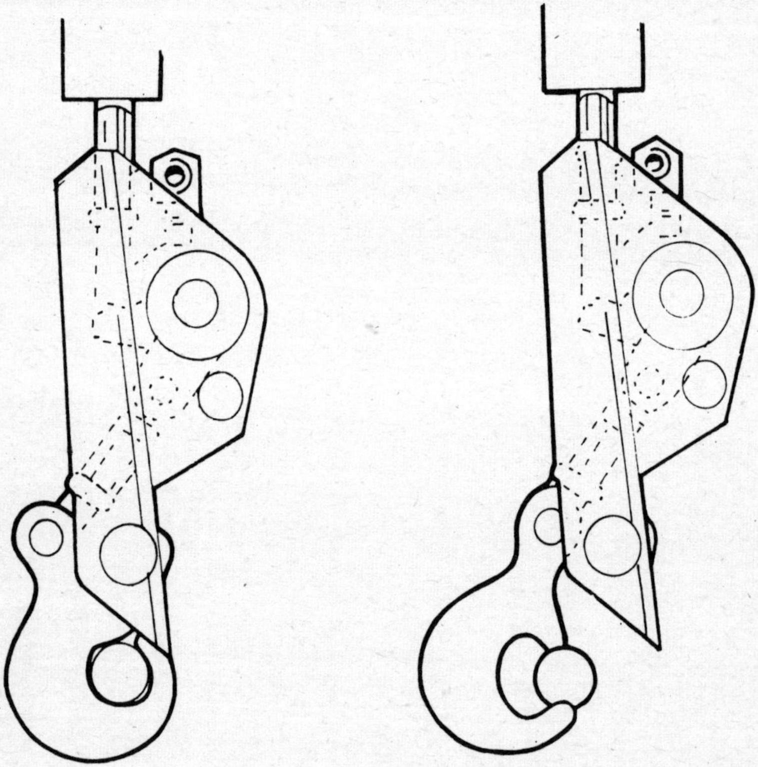

FULLY POSITIONED PARTIALLY POSITIONED

This, the investigators discovered, combined with the pressure inside the cargo compartment, made the door swing outward and upward where it smashed against the side of the fuselage immediately above its normal "open" position, and was then wrenched

Position of the aft cargo compartment door.

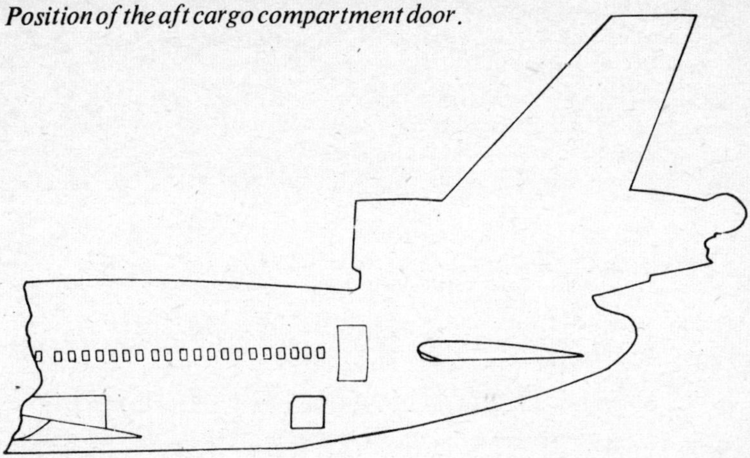

The trajectory course of the door.

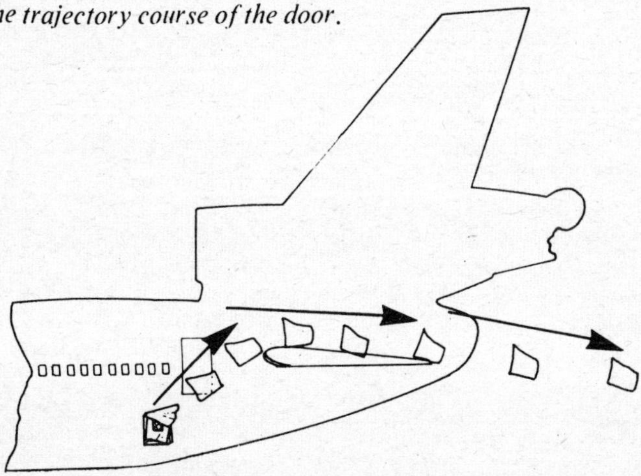

off by the jet stream of air through which the plane was flying.

The pressurization in the aft cargo compartment was now suddenly released (A). As there was no effective venting system between this compartment and the passenger cabin above, the full interior pressure load was now placed on the ceiling and walls of the hold and on the section of the passenger cabin floor immediately above it (B), and naturally, this caused it to collapse downwards (C), releasing the pressurization that was being built up in the passenger compartment (D), and trapping the lounge bar in the now open door space (E). (Diagram on following page.)

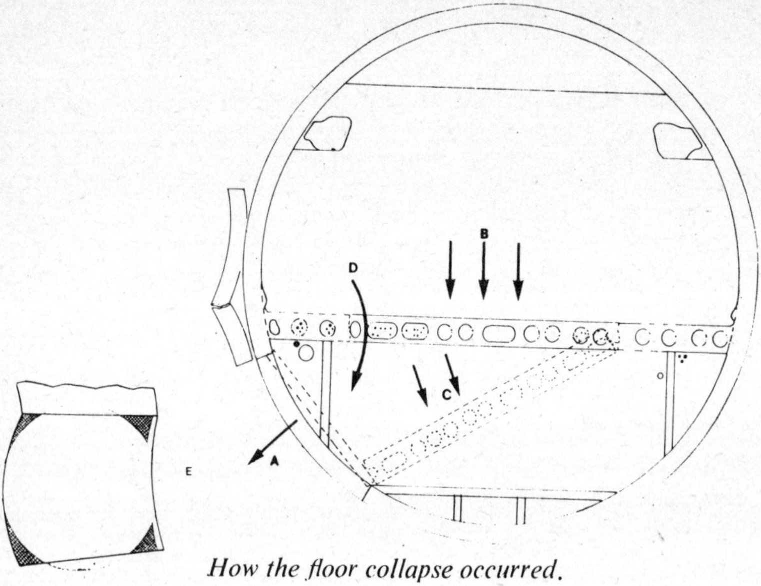

How the floor collapse occurred.

But – and this is where the pilots' problems really started – built *through* the beams that hold the passenger floor in its position, were the control cables and wiring that led from the cockpit to the aircraft's tail. When the cabin floor collapsed, all of these were either snapped completely or were stretched and jammed making most of the tail's control surfaces suddenly inoperative, and reducing the engine's thrust to "flight idle".

The cables which were completely broken were those running along the left-hand side of the fuselage in between the floor beams. They were: (A) the left-hand horizontal stabilizer suitcase handle controls; (B) the horizontal stabilizer position indicator; (C) the rudder controls; (D) the emergency shut-down control for the number 2 engine; (E) the throttle cable for the number 2 engine; (F) the fuel shut-off control cable for the number 2 engine; and (G) the left-hand elevator control.

Other systems also running down the left-hand side of the aircraft but NOT through the weakened section of the floor beams, were unaffected by this collapse, and were: (H) the aircraft's pneumatic system; (I) the number two hydraulic system; (J) the fuel lines to the rear engine; and (K) the number one hydraulic system.

On the right-hand side of the machine, the three main control cables to the tail were stretched and jammed. They were: (L) the

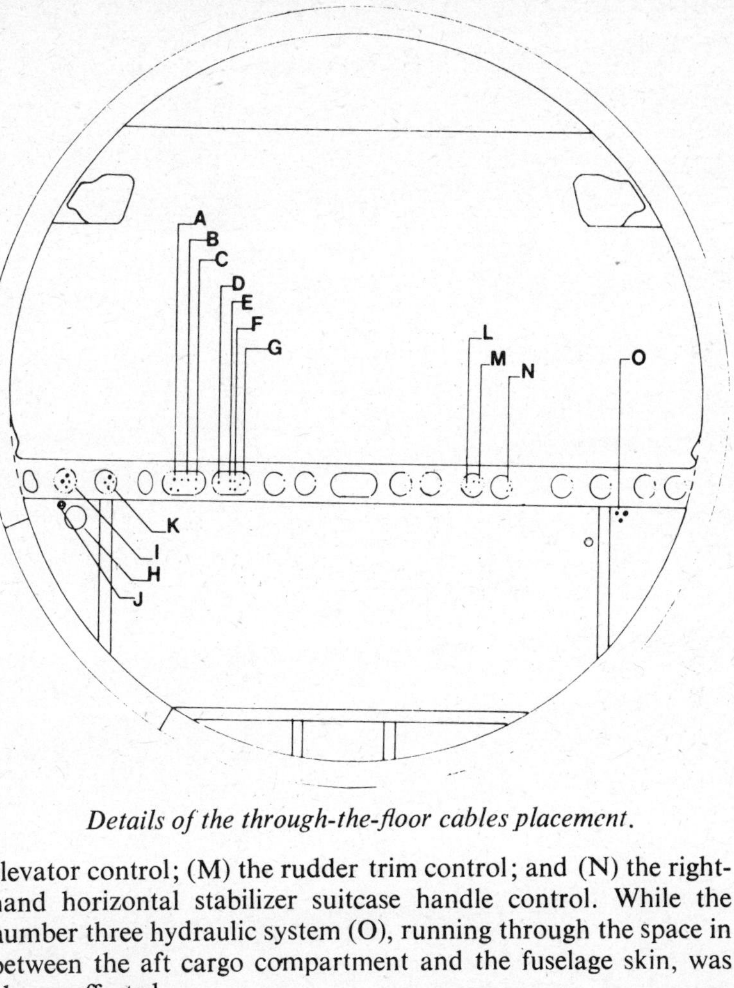

Details of the through-the-floor cables placement.

elevator control; (M) the rudder trim control; and (N) the right-hand horizontal stabilizer suitcase handle control. While the number three hydraulic system (O), running through the space in between the aft cargo compartment and the fuselage skin, was also unaffected.

The pilots lost all control of the engine built into the tail. The only way they could stop it was by turning the "boost-fuel" pumps "off". All the normal methods of controlling this engine were severed, including the fire controls and the fire extinguishant operation cables. Had there been uncontrolled flames in that engine, the pilots could *have done absolutely nothing about it*.

Again it was lucky that the very rear section of the passenger cabin floor collapsed. If this had happened somewhere near the nose of the aircraft, it is quite possible that *all* of the engines and *all* of the control surfaces would have been affected. Because the collapse had occurred down the very back, only the tail units were inoperable.

Further amounts of luck were also on the flight crew's side as, because of the lack of heavy objects on that part of the floor that collapsed, a partial failure only occurred which left them *just sufficient* breathing space to get the machine down.

As we have seen, this was not the first time that the DC–10 had been involved in a major decompression combined with a collapse of the floor. Now, the whole world should have known and realized the dangerous design situation built into this airplane and, with full F.A.A. knowledge, the fact that these faults were permitted to exist.

CHAPTER 22

On 14 June 1972, Richard S. Sliff, the head of the aircraft engineering division of the F.A.A.'s Los Angeles office, was not pleased. He had discovered that the Douglas Aircraft Company were withholding information – details that were growing in importance with every passing hour. They were not sending on the reports forwarded to them by the airlines that were operating their DC–10 aircraft, detailing specific difficulties reported by the ground and flight staffs, that had been experienced in closing and locking these cargo doors. Over one hundred such documents had reached the Douglas Company, and unlike normal circumstances, none of them had been passed on to the F.A.A.

When an airline places an order with a manufacturer, one of the conditions of sale is that the operating company will keep the manufacturer informed of any problems – minor or otherwise – that crop up during the operations. In the first year the DC–10s were in service, a large number of these reports had come in to the Douglas Aircraft Company dealing with problems that had been encountered in closing and locking the aft cargo compartment doors. The Douglas Company was "reticent" at passing these on to the F.A.A. for reasons "not determined". When, (and if) these documents are received at the F.A.A.'s offices, details are cross-filed and should a number of similar reports come in, F.A.A. inspectors should be assigned to look further into the matter not only with the aircraft's manufacturer, but also with its operators.

Douglas had received a large number of complaints regarding difficulties in closing the doors; the seating of the door seals; the loss of pressurization through the door; and so on. But it was not until Sliff telephoned them and asked for further information about these specific faults that the Douglas Company released the information.

While this activity was taking place in Los Angeles, James Rudolph, the director of Flight Standards Service in Washington

D.C., was composing a telegram to take the form of a general notification to the appropriate F.A.A. inspectors, to have them pass on urgent information to the airlines flying the DC–10.

GENOT RWA 2/123 SVC B FF
ALRGNS 1/200 WE/100 ALACDO ALIFO AC 1/200
NOTICE 8340.146 SUBJECT/MCDONNELL DOUGLAS DC–10 BAGGAGE/CARGO
COMPARTMENT DOORS. CNL 15 SEP 72.
PART 1 OF 2.

A DC–10 operator experienced explosive decompression in flight apparently due to failure of a rear baggage compartment door locking mechanism. The incident is under active investigation. As a regulatory measure and to obtain information to aid in the investigation, assigned maintenance inspectors are to:

1. Request DC–10 operators to conduct a one-time inspection of each DC–10 baggage/cargo compartment door before further flight. Inspection is to consist of, (a) check of door fit and rigging, (b) check of door operation and lock mechanism, (c) check the door warning system, (d) door checks are to be accomplished by qualified maintenance personnel located both inside and outside the aircraft, (e) visually check for evidence of structural failure or deformation in door and door frame.

PART 2 of 2.

Door inspections/checks and repairs and adjustments are to be accomplished in accordance with all applicable Douglas instructions. Discrepancies found during door inspections must be corrected before further flight.

2. Report inspection findings and discrepancies and nature of corrective actions on each aircraft immediately by telephone and confirm by dispatch to WE/100 and FS/300. Reports may include inspectors comments and recommendations. Title report DC–10 Baggage Compartment Doors.

RIS 8340.OT.
RUDOLPH FS/1.

Anyone reading this telegram *now*, would certainly be of the opinion that the Federal Aviation Administration was fully aware of the most dangerous situation that existed, and even at *that* time, the full concentration on what was determined to have caused the near disaster, centred solely on the badly designed locking mechanism chosen for these doors. The F.A.A. took no positive steps sending around this Rudolph telegram – the fact is that it was a weak excuse as it was not mandatory.

But the National Transportation Safety Board's investigation into the American Airlines accident was still young.

According to the General Accounting Office of the United

States Government, Sanford McDonnell, the president and chief executive of the McDonnell Douglas Corporation, a wholly owned subsidiary of which is the Douglas Aircraft Company, made two contributions to the re-elect Nixon campaign. McDonnell gave two smallish donations – one of $3,000 to the finance committee, and the other of $1,111.33 to the media committee. The receipt is dated 14 June 1972.

A further delving into these documents shows that contributions to Nixon's re-election campaign from McDonnell Douglas executives, totalled approximately $34,000 prior to 7 April 1972, and after that date, a further $41,000 came from the company's officials.

Sanford McDonnell refused to be drawn into a discussion on this matter. However, his uncle, James S. McDonnell, the chairman of the McDonnell Douglas board, did issue the following statement.

> The contributions were made by me personally and by others of our executives who gave as good citizens to both major parties to help keep the two-party system going for the good of America.
>
> These personal contributions were made in exact accordance with the law and had absolutely no bearing whatsoever on any McDonnell Douglas business. Further, the McDonnell Douglas Corporation did not give a dollar to any political party.

These donations were made, according to the company, well before the near-disaster to a McDonnell Douglas aircraft occurred, but the receipt was issued on the very same day that the F.A.A. discovered that Douglas had been withholding information from them which was of a vital safety nature. At the same time, the Administration was considering the formulation and issuance of an Airworthiness Directive, mandatory by law on the DC–10, to correct serious design faults. Finally, the receipt is dated one day before Jackson McGowen, the president of the Douglas Aircraft Company, entered into a "gentlemen's agreement" with John Hixon Shaffer, the Nixon appointed administrator of the F.A.A., an agreement which, in hindsight, could be seen as contributing to covering up a most dangerous situation, being the design faults in an aircraft that was capable of carrying up to 361 persons.

CHAPTER 23

WITH the accident investigators working away at Detroit Metropolitan – Wayne County Airport, John Shaffer turned up at his office in the huge, white, government building that houses the headquarters of the F.A.A. in Independence Avenue, Washington D.C., on the morning of 15 June 1972. As soon as he had read the latest reports received into the near major catastrophe, he telephoned George Wardy, the president of American Airlines, to see what he thought went wrong with the airplane. He followed this with further telephone calls to others who could have known or were involved in the possible causes, and as he said later, he eventually telephoned Jackson McGowen the then president of Douglas. Shaffer went on:

> We discussed how we were going to fix it. And as a result of that discussion, when it was determined that the thing that was needed was a simple plate added into the latching mechanism of the cargo door, and when we learned how simple the fix was in reality, and how quick we could get the parts, and how easy it would be for Douglas personnel who were familiar with the airplane to campaign the fleet – at the time there were 35 airplanes in the system – we agreed that we could do it by Service Bulletins, principally because the Service Bulletin gives you the instructions, it gives you the parts, and it gives you the trained personnel to accomplish the work. We were under the impression; or rather I was under the impression, that we could virtually do the job overnight.
>
> I agreed with Mr. McGowen that we would do the job with a Service Bulletin. It was not really a request of his. It was because of my confidence in our system; confidence in him and his company; and I guess faith and trust in people to follow through with what they had agreed to do.
>
> I don't believe that at any time were the merits of a Service Bulletin versus an Airworthiness Directive the subject of a topic of conversation or debate. And again I think that any-

one that is familiar with an Air Force background would know what a Service Bulletin can accomplish, and how we won the War in Europe with Service Bulletins. I survived, I think, because of Service Bulletins. So I believe in them, and I think you can get the job done with them.

How Shaffer came to the opinion that a Service Bulletin was adequate simply because it: ". . . gives you the instructions, it gives you the parts, and it gives you the trained personnel to accomplish the work . . .", is a mystery. Did he forget that Service Bulletins were non-mandatory?

Each Service Bulletin contains the following piece – and this is important:

Operators with aircraft listed under Paragraph 1.A.(1), should submit a "no charge" purchase order for service change kits in the quantities shown.

The purchase order must specify the Service Bulletin number and only parts listed herein.

Operators will be advised of delivery schedule upon receipt of the purchase order. Direct purchase order to:

Douglas Aircraft Company,
P.O. Box 1771,
Long Beach, California – 90801,
Attn: Manager, Warranty, 7–21,

A "no charge" purchase order for kits under the provisions of this bulletin must be received not later than 90 days after the issue date noted on Page 1. Any purchase order received after the 90-day period will be processed on a charge basis.

It would be interesting to know *how*, with the need for forms to be filled in, posted, and a time limitation on the "special offer", Shaffer was of the opinion that: ". . . Service Bulletins . . . give you the instructions, give you the parts, and give you the trained personnel to accomplish the work . . ."? Moreover, the Service Bulletin also tells the airlines how long it will take *their personnel* to affect the modifications. It says nothing about Douglas engineers being sent with the modification kits in order that they be rapidly installed. Neither does it even indicate that the kits will be quickly handed around.

But then Shaffer's recollections of his dealings with McGowen differ from those of other people. For instance, Arvin O. Basnight, the chief of the F.A.A.'s Los Angeles office which was in the centre of this intrigue, recalled the instance like this:

I received a telephone call from Mr. Jackson McGowen, president of the Douglas Aircraft Company, who indicated that late on Thursday, 15 June, he had received a call from Mr. Shaffer asking what the company had found out about the problems of the cargo door that had caused American Airlines to have an explosive decompression.

Mr. McGowen said he reviewed with the administrator the facts that had developed which included the need to "beef up" the electrical wiring and related factors, that had been developed by the Douglas Company working with the F.A.A.

He indicated that Mr. Shaffer expressed pleasure in defining a reasonable corrective action and had told Mr. McGowen that the corrective measures could be undertaken as the product of a "gentlemen's agreement" thereby not requiring the issuance of an F.A.A. Airworthiness Directive.

Shaffer had said that the cure for the problem was ". . . a simple plate, added in the latching mechanism of the cargo door . . ." Basnight said that McGowen had discussed the electric wiring with Shaffer. The Service Bulletin to be issued four days later, referred to the placement of placards and the drilling of a small viewing window, and says nothing about: ". . . a simple plate added in the latching mechanism of the cargo door . . ."

Shaffer said: ". . . I agreed with Mr. McGowen that we could do the job with a Service Bulletin . . ." Basnight said that McGowen had told him that the Service Bulletin idea was 100% Shaffer's: ". . . the product of a 'gentlemen's agreement' . . ."

If Shaffer was telling the truth when he said that: ". . . a simple plate . . .", would remedy the problems, the first Service Bulletin "recommending" the addition of such a "simple plate" in the DC–10 aft cargo door operating mechanisms, was not issued until 3 July 1972, which is inconsistent with the remainder of his statement and especially the part: ". . . we were under the impression, or rather I was under the impression, that we could virtually do the job overnight . . ."

And finally, Shaffer said that he made this telephone call to McGowen on 13 June 1972 – that is, the day after the near disaster. Basnight said his records indicate that the telephone call was made on the evening of the 15th.

CHAPTER 24

At 0850 hours on 16 June 1972, the telephone rang in the office of Arvin O. Basnight in Los Angeles, California. It was Jackson McGowen, the president of Douglas on the line. McGowen told Basnight that late the previous evening he had received a call from Shaffer in Washington, asking what Douglas had found out about the problem with the cargo door that had caused American Airlines to suffer an explosive decompression. McGowen said that he had reviewed with Shaffer the facts that had developed, which included the need to replace the electric wiring that had originally been installed to the cargo door motors on DC–10 aircraft, with a stronger, thicker wire.

Once their conversation had finished, Basnight telephoned Richard Sliff, the head of the aircraft engineering division in Los Angeles, and told him what had been said as already they were preparing the wordage of an Airworthiness Directive. Basnight added that Sliff advised him that several steps seemed necessary to prevent further explosive decompressions of DC–10 cargo doors, and they were considering what the F.A.A. would class as: "an appropriate Airworthiness Directive".

Sliff then came around to Basnight's office and while they were chatting, they telephoned James F. Rudolph, the director of the flight standards service of the F.A.A. in Washington D.C., and reviewed with him their plans, and the background data available in Los Angeles. They then informed him that the Airworthiness Board was in session, and asked his guidance in continuing this effort.

Rudolph suggested that they keep going with what they were doing and wanted to be supplied with a copy of the intended Directive, which Sliff had already telexed to Washington in its rough format.

That afternoon, Basnight received a telephone call from Ken Smith the then deputy administrator of the F.A.A., and they discussed the same background information. He was then asked by

Smith why the Los Angeles Maintenance Reliability reports on the DC–10 had not made known the problems of the locking of the cargo doors so that the F.A.A. was as well informed as to their proper function as the Douglas Company.

Basnight replied that the F.A.A.'s reporting system had not disclosed this type of data; the reasons for which he would look into.

Smith queried particularly why the Douglas Company's attitude might be one of not revealing their records regarding the reported difficulties by the airlines in closing these doors. He indicated that he concurred with the Los Angeles office's judgement in issuing an Airworthiness Directive. Smith agreed that one must be issued and he said he would talk to Shaffer about it.

The Airworthiness Board continued to meet and defined the earlier drafted Directive, and Basnight continued to await advice from Washington with regards to the issuance of same.

At about 1700 hours, with Basnight still waiting for the telephone call, he decided to phone Rudolph to find out what was happening, and was told that they had had difficulty in trying to locate Shaffer who was evidently *again* away from his office. Smith joined Rudolph in Washington during the phone call and they said they were planning to have a "long-distance conference" with the Douglas Company together with the 4 airlines that were flying the DC–10s, to ensure that the objectives of effective operation of the cargo doors sought by the F.A.A.'s proposed draft of the Airworthiness Directive, would be accomplished.

Basnight remained in his office awaiting instructions from Washington. In desperation, he again telephoned Rudolph as in Los Angeles they had learned from Douglas that the joint telephone conference had occurred and that steps were being taken to accomplish the objectives of the proposed Airworthiness Directive.

One significant point was that the Directive was "upgraded" by the Board, based on data the F.A.A.'s engineering personnel had worked out with Douglas, to add an additional provision that would require the drilling of a hole in the door near the lock-pins in order to allow a visual inspection of the locking mechanism to be made by the operator after he had closed the door. Basnight told Rudolph about this added modification that had been considered necessary, and asked him whether it would be possible to include this in the message telexed by Smith to the Douglas Company and the airlines involved.

Rudolph then pointed out that it was after office hours in

Washington. The message containing basically the Los Angeles office's draft phraseology had been released by Smith before he went home, and since Smith had signed and authorised it, Rudolph said that he could not make any amendments to it, and suggested waiting until Monday (the next working day) when the matter could be taken up directly with Smith. Rudolph did however, agree with this newly proposed amendment.

Early in the morning of Monday, 19 June 1972, Basnight received a telephone call from Joseph A. Farrarese the head of the Flight Operations Division, Flight Standards Service of the F.A.A. in Washington D.C., who said that with him in the room was a Mr. Slaughter (sic), the head of the Engineering and Manufacturing Division, and that they were calling to advise Basnight that the teletyped message signed by Smith had been received at the F.A.A. offices in the three regions where the DC–10s were in primary use, and that he was acting on instructions and he asked Basnight to see that all but one copy of this message was now destroyed.

Basnight told Farrarese that he had not seen that particular telex, but would conform with his instructions.

Shortly after they hung up, Basnight was handed the message in a sealed envelope, in triplicate, by the Los Angeles duty officer. He discussed the matter of its destruction with Sliff and another F.A.A. executive, before telephoning Farrarese to tell him that they had received the message and were destroying all but one copy. He then reviewed with him Smith's inquiry regarding the Maintenance Reliability reports concerning DC–10 cargo doors. Basnight told Farrarese that he considered there was a significant safety problem with these doors involving both maintenance and reliability, even though in earlier aircraft designs they had not been considered critical as far as the machine's safety was concerned.

Afterwards, Basnight gave the original copy of the message that was to be destroyed to Sliff for his aircraft engineering records and ripped up the other two copies. The message was dated 16 June 1972, and was signed by Smith. It was addressed to the presidents of American Airlines, National Airlines, Continental Airlines and United Air Lines, and read:

In accordance with our telephone discussion of this date, I understand that your company will accomplish the following on all your DC–10 aircraft. This program is effective immediately upon receipt of this telegram because of the possibility of incomplete latching of lower cargo compartment doors, which causes inadvertant opening under pressurization.

It then went on to list a number of minor modifications which the F.A.A., or rather certain people in their office, considered sufficient to try and avoid a major catastrophe occurring.

No one could explain why the message was suddenly retracted and ordered destroyed soon after it had been issued to this select group.

Douglas needed to keep selling and producing DC–10s in order to stay financially viable and in order to recoup the development costs already spent – whether or not major design faults had been uncovered or even existed in the machines. To publicly reveal the faults by way of issuing extensive orders, mandatory or otherwise, could, so they must have felt, turn prospective customers away from the DC–10, possibly to its direct competitor the Lockheed 1011 TriStar, which of course, would not be affected in the slightest when this urgent and long list of necessary modifications to the Douglas machine finally became known.

Basnight was told to destroy all copies of the telegrammed message except for one, by Farrarese in Washington. Farrarese was handed down these instructions from a Norris J. Ansell, the assistant executive secretary of the F.A.A. also in Washington, via his secretary who arrived in her office on Monday morning 19 June 1972, to be confronted with this hand-written note:

> Jan. Call FS (Flight Standards) on these two cables and tell them they must comply with the following three requirements: (a) no copies to be made, (b) no further distribution, and (c) "For Official Use Only".
>
> Jan. For your info, and Denise. These were sent out Friday evening by the communications centre on the 1st floor.

CHAPTER 25

OVER the weekend, McDonnell Douglas had also sent a telegram to the four American operators of the DC–10, stressing the need for the installation of small viewing windows in the cargo doors so that the persons whose jobs it was to close and lock them could see that they were, in fact, locked.

This telegram constitutes Alert Service Bulletin A52–35. One incident has occurred where the aft lower bulk cargo door opened in flight and separated from the aircraft.

It is recommended a lock mechanism viewing window be installed in the forward, centre and aft lower cargo doors. Installation of the window will permit visual inspection to determine the doors are latched and locked prior to flight.

On Monday, 19 June 1972, McDonnell Douglas followed up this telegram with the actual Service Bulletin:

The following rework is typical for all cargo doors:

1. Open cargo door.

2. Locate and install viewing window in cargo door per drawing ADA 7793.

3. Locate and install the following placards per drawing ADA 7793:

A. Install placard, P/N ADA 7795 (Forward and centre doors), and placard P/N ADA 7794 (Aft door), below exterior control panel door. Both placards state:
CAUTION: VERIFY LATCH LOCKPINS ENGAGED.

B. Install placard, P/N ADA 7796 (all doors), on exterior surface of door adjacent to the vent door handle. The placard states:
CAUTION: DO NOT FORCE HANDLE. IF HANDLE WILL NOT OPERATE RESTORE POWER TO DOOR AND HOLD DOOR CONTROL SWITCH TO DOOR CLOSE POSITION FOR ADDITIONAL THREE SECONDS. MOVE HANDLE TO CLOSE POSITION.

C. Install placard, P/N ADA 7792 (all doors), on exterior surface of door above viewing window. The placard is illustrated and states:

LOCK PINS ENGAGED. SYSTEM SAFE.
LOCK PINS NOT ENGAGED. SYSTEM UNSAFE.

4. Functionally check door system and operation by performing steps 1 through 6 Douglas TWX C1–SVC–DC–10–COM–77 dated 13 June 1972. Prior to each flight cargo restraint curtain at aft bulk cargo door, if installed, must be properly secured in place at top, sides and bottom before closing door.

R. N. Frankel,
Manager Customer Services.

APPROVAL: This Service Bulletin is F.A.A. Approved.

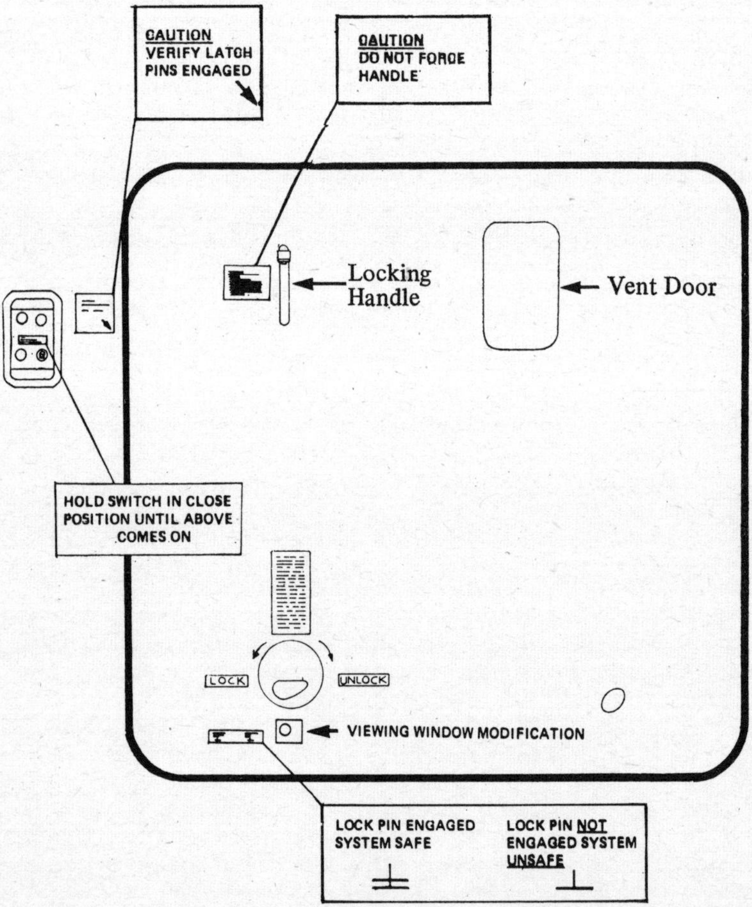

One would imagine that the F.A.A.'s own regulations concerning minimum standards acceptable for safe flight, would be F.A.A. approved. Is there an excuse?

Page 222 of the American Federal Air Regulations: Part 25, Section 784, reads as follows:

(e) There must be a provision for a direct visual inspection of the locking mechanism by crewmembers to determine whether external doors, for which the initial opening movement is outward (including passenger, crew, service and cargo doors), are fully locked. In addition, there must be a visual means to signal to appropriate crewmembers when normally used external doors are closed and fully locked.

(f) Cargo and service doors not suitable for use as an exit in an emergency need only meet paragraph (e) of this section and be safeguarded against opening in flight as a result of mechanical failure or failure of a single structural element.

And what does the F.A.A. have to say about DC–10 aircraft up to and including 19 June 1972, not complying with their Airworthiness Certification Regulation 25.784 (e)? The reply was that the DC–10 lower cargo doors were "considered" to be in compliance with the regulation because: ". . . they had a small vent door in them . . ."

As can be seen from the drawing, the locking mechanism for which the regulation requires a means of: ". . . direct visual inspection . . .", is down at the bottom of the door itself. How does the F.A.A. expect anyone to believe it is serious when it says it considers the vent door at the top, being a means for: ". . . a direct visual inspection of the locking mechanism . . ."?

Furthermore, the Certificate of Airworthiness issued for DC–10s prior to the "recommended" installation of this viewing window, contained the following statement:

This airworthiness certificate is issued pursuant to the Federal Aviation Act of 1958 and certifies that, as of the date of issuance, the aircraft to which it is issued has been inspected and found to conform to the type certificate and therefore, to be in condition for safe operation, and has been shown to meet the requirements of the applicable comprehensive and detailed airworthiness code as provided by Annex 8 to the Convention on International Civil Aviation, except as noted herein:

Exceptions:

NONE.

This would mean that the Certificate purportedly states that the F.A.A. has determined that the DC–10 is in compliance with the 1958 Act. But it certainly was not in the case of Section 25.784 (e). And it would be further agreed that admittance of its non-compliance may be seen in this McDonnell Douglas Service

Bulletin A52–35, which "recommends" the installation of such a viewing window.

On 19 June 1972, even though no DC–10s had been delivered to foreign airlines, there certainly had been a number of orders taken by McDonnell Douglas. What would Douglas's thoughts be about these instructions and urgent notices that they were now "recommending" be placed on the cargo doors, when a purchaser of a DC–10 did not have English as their mother tongue? Moreover, what about all these vital notices and the comprehension of same, when the DC–10 visited an airport in a country other than an English-speaking one? These questions were never considered at the time. With the sticking of these notices and instructions all over the place, both the F.A.A. and McDonnell Douglas evidently considered the door's operation automatically and magically becoming "idiot-proof".

During the afternoon of Monday, 19 June 1972, the F.A.A. also sent a telegram to all DC–10 operators, impressing upon them the importance of incorporating the "recommended" McDonnell Douglas modifications which had been circulated that very morning.

At the time, there were 36 DC–10s in service: American Airlines had 13; Continental Airlines 4; National Airlines 5; and United Air Lines were in possession of 14.

But the implications of this "strongly recommended" Service Bulletin and its modifications were never realized by either the Douglas Aircraft Company's designers or the Federal Aviation Administration's engineers. It is quite evident that those responsible at the F.A.A. had never considered the contortions that the door closing operator would have to go through in order to carry out the "recommended" instructions.

Imagine, if you will, being employed as a DC–10 door closer. Imagine also that your employer had actually made these "recommended" modifications. Now imagine yourself at the door at the start of the closing operations.

You are either on a ladder, or standing on the top of a baggage loading moving belt which has been repositioned to the area where the door operation controls are located. Looking at the plan of the door, you must be standing high up on something, over on the left-hand side.

The dispatcher is yelling at you to get the door closed. And so, with fingers on the operating switch you wait, while the electric motor drives it flush with the rest of the fuselage. It is closed – but it is still not locked. Now you have to keep the switch depressed for this magical five to seven seconds. You wait. You count to seven, and then hope and pray.

The next thing to be done is to descend the ladder or baggage loading unit, and reposition it to somewhere near the centre of the door. You reclimb the ladder and pull "down" the door-lock handle into a grooved rut leaving the handle flush with the outside of the door. You then partially descend the ladder in order to see into the peephole, whether the door is locked. If you do not move the ladder, it is IMPOSSIBLE TO SEE WHETHER IT IS LOCKED OR NOT. The operating switch is out on the left-hand side of the door – the door locking handle is about three feet to the right in the upper left-hand section of the door itself – and the "let's check and see if it's locked" viewing window is approximately another foot further to the right and about three feet below the level of the switch. So you MUST reposition whatever you were standing on from that originally needed to reach the operating switch, to a position from which you can see into the small viewing window.

Now imagine it is dark – night-time. Standing on the ladder, you now have to use an electric torch or – heaven forbid – a match, in order to see inside this one-inch diameter hole to check whether the door is locked. This in itself, is a tricky process for which (as was revealed by officials of the National Transportation Safety Board and confirmed by a DC–10 airline operator) *you would need special training*. There is no light source inside the hole, so after dark, light has to be shone in from the outside. The trick is: how do you do this with your head in the way?

If, for one reason or another, you eventually discovered that the door was *not* completely locked, the whole process would have to be carried out in reverse. Get down from the ladder – alter its position – climb the ladder – activate the door closing switch for a few more magical seconds – get down the ladder – move its position – get up the ladder – try and pull "down" the door-lock handle – shine light into the peephole – and hope to God it is locked this time.

These "compounding" problems are the ones so difficult to understand. Could not the brainwave of a designer who had, possibly, suddenly discovered that the door was *not* in compliance with the regulations because it did not have this viewing window in the first place, and who then set about designing such a thing, have realized the contortions and frustrations that the operator would have to go through in order to properly check that locking had taken place?

The door's design was faulty right from the start. Now they "compound" the problems by adding further to the difficulties of closing the thing.

Would this not have been the right time to say: "Let's start from scratch and design a door that moves *inwards* so that all these problems may be corrected in the one big hit"? But no. That wastes space and makes their aircraft less "economically" inviting to customers. So they did not do this, but added further to the door problems.

CHAPTER 26

THREE of the four airlines operating DC–10s reported to the F.A.A. that they had effected the modifications as "recommended" by McDonnell Douglas in their Service Bulletin A52–35 (the one that called for the viewing window and all the notices). American Airlines said it had completed all the work on 19 June 1972; National Airlines reported their work done by 21 June 1972; United Air Lines modified their final one on 23 June 1972; leaving Continental Airlines, which took until 25 June to have the work finished.

Straight after lunch on 23 June 1972, a letter was hand-delivered to F.A.A. administrator Shaffer's office. It was from John Reed, the chairman of the National Transportation Safety Board.

SAFETY RECOMMENDATIONS A-72-97 & 98.

Dear Mr. Shaffer,

The National Transportation Safety Board is investigating an accident involving an American Airlines McDonnell Douglas DC–10–10, N103AA, which occurred shortly after take-off from Detroit Metropolitan – Wayne County Airport, on 12 June 1972.

The aft left cargo door opened while the aircraft was at approximately 12,000 feet. The cabin floor over this cargo compartment then failed as a result of depressurization loading, and the floor dropped partially into the cargo compartment. This displacement of the floor caused serious disruption to the control cables which are routed through the floor beams to the empennage control systems and the engine controls. With the exception of the right rudder pedal cable, all of the cables on the left side of the fuselage broke. The cables running on the right side were also damaged – the cable guides tore from their attachments to the floor beams, and the cables were deflected downwards by the floor structure.

Preliminary investigation has revealed that the door latches were not driven fully closed, and the lock-pins which should have prevented the latches from opening, were not in place. The reason these door latches were not driven over centre to the fully closed position has not yet been determined. However, although the Safety Board believes this was relevant and a contributing factor, we are more concerned over the failure of the door safety features to preclude dispatch with an unlocked door. The design of the door should have precluded the dispatch of the aircraft with this door improperly closed. With the lock-pins not engaged, a small vent door on the cargo door should have remained open, preventing normal pressurization of the aircraft. Also, the vent door handle should not have stowed.

However, tests conducted at the Douglas Aircraft Company have demonstrated that the vent door can be closed and the handle stowed, without the lock-pins engaged. Apparently, some combination of deflection of the operating mechanisms and tolerances permitted such operation when a force of approximately 120 pounds was applied to the operating handle. During these tests our investigators observed bending of the sliding lock-tube which caused the cap end of the tube to make contact with the pilot indicator switch actuator arm. This arm moves through a very small arc to actuate the switch, and the Board believes that a similar contact on the accident aircraft door probably actuated the switch and gave the pilots a "door safe" indication on the annunciator light panel.

Finally, the Board believes that sudden loss of pressure in this cargo compartment for any reason should not jeopardize the safety of the flight. In this case, the loss of the door and the resultant cabin floor failure caused an unwanted rudder input, severely restricted the elevator control available to the crew and disrupted the No. 2 engine controls.

We are aware of the inspection procedures currently in effect to ensure safety of operations of the DC–10 as well as the existing safety features of the door design. Nevertheless, in order to preclude the recurrence of similar accidents, the Safety Board recommends that the Federal Aviation Administration:

1. Require a modification to the DC–10 cargo door locking system to make it physically impossible to position the external locking handle and vent door to their normal door locked positions unless the locking pins are fully engaged.
2. Require the installation of relief vents between the cabin and aft cargo compartment to minimize the pressure loading on the cabin flooring in the event of sudden depressurization of the cargo compartment.

These two recommendations are precise and explicit. Fix up the faulty design in the door – fix up the faulty design in the floor. Of course, at this stage the National Transportation Safety Board did not know that this was the *second* time a door had exploded open because the latching mechanism was not properly engaged, nor did they realize that this was the *second* occasion when the floor had either collapsed or warped.

The N.T.S.B.'s investigation into the Windsor accident continued while both the Board and the industry as a whole, waited to see what type of reaction would be forthcoming.

CHAPTER 27

ARVIN Basnight, the head of the F.A.A.'s Los Angeles office, was later to testify before the Senate sub-committee that it was unclear to him why the message that would have resulted in mandatory action being taken to see that DC–10 cargo doors were modified was destroyed save for one copy.

The opportunity arose, on 30 June 1972, for Basnight to ask why no mandatory order was issued when he happened to be talking to Shaffer. Basnight told Shaffer that the enforceable method – that is the issuance of an Airworthiness Directive – was the "preferred tool" for the F.A.A. to use in order to guarantee that the modifications were carried out. Shaffer listened to his opinion, and as Basnight recalled later: "he left it undecided". He was under the impression that Shaffer would give the matter due consideration.

CHAPTER 28

On 3 July 1972, still more paperwork, with still more "recommended" modifications, arrived on the desks of officials of the four airlines flying the DC–10s.

McDonnell Douglas Service Bulletin 52–37 – DC–10 SC 812 – was headed:

DOORS – CARGO – MODIFY AND ADJUST DOOR MECHANISM ASSEMBLY.

At the time, this "begging" Service Bulletin was issued, it was supposed to only affect 39 operational aircraft. Of the fuselage numbers 1 through to 45 that had already been delivered to airline customers, the whole lot needed modification by their owners. Of the remainder, McDonnell Douglas said:

. . . these will be modified prior to delivery or included in a subsequent revision to this Service Bulletin.

The reasons for Douglas's "pleading" with their customers to incorporate these modifications were also given.

In layman's language this modification was supposed to correct a situation where a baggage handler could, by pure brute force, jam the door handle into the "locked" position. If force was used on the handle prior to this modification the lockpin tube (A), a small pipe that connects the lock-pin bar with the door handle, would be bent and could possibly break. The handle (B) would then move into the stowed "locked" position and during the process of bending, the end of the horizontal vent door bar (C) would touch the microswitch (D) which operates the door warning light in the cockpit, and when it did, the light would go out giving a false indication to the pilots that the door was closed and locked. But before the lockpin tube broke, it would exert pressure on the bell-crank at the end of the horizontal bar (E), and this too would break. Naturally, all this would happen when the door's main locking-grips (F) were not properly in position, and therefore a highly dangerous situation existed.

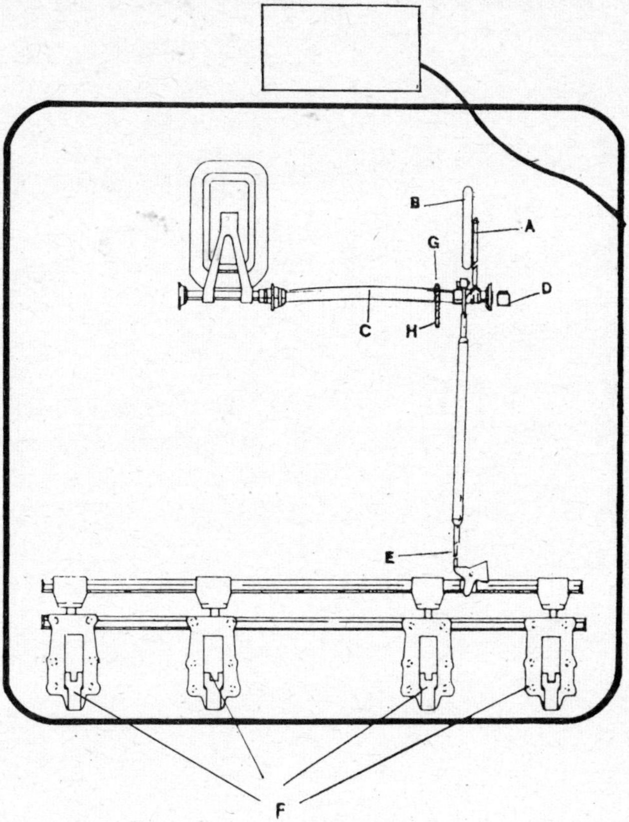

The modification kit contained 36 shims, 3 strikers, 1 support plate (G), 18 rivets, 4 nuts, 9 nutplates, 1 jumper, 1 clamp, 10 washers, 13 screws and a strengthening plate (H). McDonnell Douglas estimated that it would take 10.7 man-hours of the airlines' mechanics time to complete the work.

A "no charge" purchase order for kits under the provisions of this Bulletin had to be received not later than 90 days after the issue date of the "recommendation". Any purchase order received after that period would be charged for.

This Bulletin itself, contained all the information Douglas would wish to know. It listed which operators had the affected equipment. It detailed how many aircraft each company had that was involved. It also showed which part numbers were needed. The whole lot of the information was contained on this piece of paper.

Douglas sent the Bulletin to the airlines (already listed) and then expected *them* to fill out forms, re-list all the parts necessary, and return the information to the factory, after receipt of which the manufacturer would then advise them when they could obtain the kits. BUT, if you were a naughty airline that takes longer than 90 days to fill out all the forms and work out all the numbers, you would have to pay for the modification kit and its delivery.

If Douglas were genuinely interested in seeing their customers modify these aircraft – for safety's sake – why not simply send out the required number of complete kits, installation instructions and even engineers IN THE FIRST PLACE?

CHAPTER 29

On 7 July 1972, Shaffer replied to the National Transportation Safety Board's recommendations of 23rd June.

Dear John,

This is to acknowledge your Safety Recommendations A–72–97 and 98, issued on the 6 July 1972, which included recommendations resulting from an inadvertant cargo door opening reported on a Douglas DC–10–10 airplane on the 12 June 1972.

All operators of the DC–10–10 airplanes are currently performing 100 hour functional checks on the cargo door system and will incorporate necessary modifications in accordance with McDonnell Douglas Service Bulletins 52–27 and A52–35 within 300 hours. These modifications pertain to improvements in the inspection and operation of the locking and vent mechanisms.

Additional modifications to the cargo door locking and pressurization systems are being considered as part of a continued investigation effort. While a preliminary investigation indicates that it may not be feasible to provide complete venting between cabin and cargo compartments, your recommendations will be considered with respect to further action taken.

The wording of this letter suggests that actions would be taken which were in fact not taken but this was not realized because of Shaffer's stated intent that: "... additional modifications ... are being considered".

With reference to the middle paragraph:

All operators of DC–10–10 airplanes ... will incorporate necessary modifications ... (in) Service Bulletins 52–27 and A52–35 ...

Firstly, this sentence is worded in such a fashion that one would assume these two Service Bulletins are to be reissued as Airworthiness Directives, and thereby made mandatory. The use of the words: "... will incorporate necessary modifications. ..", is wholly misleading as without a mandatory order there is no compulsion on anybody to have the work done.

Shaffer, in his letter, is saying in so many words that the National Transportation Safety Board's recommendations for mandatory action are being, and will be, completely ignored, even though they call for steps to be taken in order to guarantee that modifications to both the door *and the floor* would be made before there was a disaster of enormous proportions.

Shaffer on that day had *still another* N.T.S.B. letter, with *still more* of their recommendations in front of him. This one was dated 25 February 1972, and it read:

Dear John,

The National Transportation Safety Board is currently investigating a fatal accident which occurred near Fairfield, Idaho, on the 20 February 1972, involving a Beech Model 65B–80 airplane.

Preliminary investigations of this accident have disclosed that a separation of the aircraft's left outer wing panel occurred in flight and that the separation was associated with an intense fire in the left engine. The fire had progressed from the engine's nacelle through the wheel well, and subsequently resulted in sufficient heat being gained by the wing spars to weaken them to the point of failure.

The Board has also been advised of a similar fatal accident which occurred near Alice Springs, Australia, on the 20 January 1972, involving a Beech Model 65–80 airplane. Information received by the Board from the Civil Aviation Authority of Australia indicates that there was an engine nacelle fire preceding the in-flight separation of the airplane's wing.

The Board is concerned that flammable fluids which can be ignited in the engine compartments, can result in uncontrolled fires causing extensive damage to the aircraft structure and loss of life.

In order to preclude the recurrence of similar accidents, the Board recommends that the Federal Aviation Administration:

1. Initiate a review of the certification and design criteria of Beech Model 65 and similar aircraft, as applicable to powerplant installation fire protection provisions.
2. If it is found that these criteria are inadequate, initiate corrective action to ensure adequate fire protection of the powerplant installation and the adjacent airplane structure.
3. Review the design and certification criteria of Part 23 of the Federal Aviation Regulations applicable to the fire protection of powerplant installations for adequacy and effect regulatory changes if deemed appropriate.
4. Advise all owners and operators of Beech Model 65 airplanes of the occurrence of two accidents wherein an engine fire resulted in separation of the airplanes' wings within 2 to 3 minutes. Operators should further be advised to take all possible measures to insure integrity and airworthiness of the powerplant installation and reduce the probability of an in-flight fire.

Shaffer's reply was to dismiss the first three recommendations. The fourth one was carried out – but only to a certain extent. The N.T.S.B. recommended advising all owners and operators of the *specific aircraft* that had been found prone to in-flight fires, namely the Beech 65, by letter about these two accidents.

Shaffer's interesting solution to this problem was to "carry out the Board's intent", by not only sending the letter to Beech 65 owners, *but* to owners of other aircraft. And so that no one could pinpoint more exacting information on (1) the type of aircraft involved in the accidents, and (2) the fact that both crashes happened to the same make and model of machine, the letter was appropriately worded:

> Dear Sir,
> Our records show that you are the owner and/or operator of a light twin-engined airplane which is powered by mechanically supercharged or turbo supercharged engines. It is extremely important that these kinds of airplanes, which are powered by a sophisticated type of reciprocating engine . . .

And so it waffled on. Finally the recipients came to the part where one would *expect* the "intent" to be included:

> Two recent fatal accidents occurred on light twin-engined airplanes with mechanically supercharged engines. Preliminary investigations show an engine failure initiated these fires. The fires progressed into the wheel wells . . .

Not a single word anywhere in this nearly two full pages of close typing that both the disasters happened to the one type of aircraft – the Beech 65. Because of this, it could easily be assumed by Beech 65 owners when they read this – IF they received it – that the two accidents occurred to makes and models, other than the one they owned.

And besides, by wording the letter in such a fashion, Shaffer could say in his of 7 July 1972, to John Reed of the National Transportation Safety Board, that he had actually done something the Board had recommended.

A closer inspection of the facts, however, indicates otherwise.

The fact is that many other accidents were being treated in a similar fashion by Shaffer.

CHAPTER 30

It was shortly after Shaffer's misleading DC–10 letter to the N.T.S.B., when he sent another reply to yet another set of the Board's recommendations on 13 July 1972. This time, it was the suggestion that all Beechcraft airplanes be modified mandatorily by the issuance of Airworthiness Directives, to make sure that *their* cargo doors *also* did not come open in flight, as had occurred, as we saw, on 19 May 1972, just 24 days before the DC–10 accident.

Shaffer wrote:

We wish to advise you that corrective action has been taken to assure a safe and proper use of the nose cargo door actuating system on Beechcraft 65–80 airplanes. As a result of the cited Albuquerque accident which investigation indicated was caused by a non-latched nose cargo door, the F.A.A. issued a Safety Alert to all owners and operators on the 31 May 1972. This Alert covered the need for positive door closure and rigging of door actuating mechanisms in accordance with the manufacturer's instructions. Subsequent examination of the door's three-pin latching design indicated that if properly maintained and fully secured by the operator, such provisions as specified under Federal Air Regulations 23.787 (b) would continue to be satisfied.

With respect to the Beechcraft 99 inadvertant door opening incident, proposed rulemaking action is being considered to require a third cargo door latch device which is in addition to the two existing latches now in use. This new double failure protection device is provided in accordance with Beech Service Instruction 051–113, sent to all owners and operators on the 20 June 1972. We feel that the added protection precludes need for internal tie-down provisions, as the door, when fully latched, is sufficient to prevent cargo from becoming a hazard.

Federal Air Regulation 135.117 presently requires that cargo be carried in an approved cargo rack, bin, or compartment, installed in the aircraft and that it be secured by means approved by the Administrator. With the proposed modification to the Beech 99, and with proper maintenance, the cargo compartments of both the Beech 65 and the Beech 99 should adequately restrain cargo, and fulfill the requirements of the regulation. Therefore, we do not concur that rulemaking under Part 135 be recommended.

So another set of N.T.S.B. safety recommendations are turned down by Shaffer. It will again be noted how the discrepancies either should not exist because the Federal Air Regulations say so, or should have been remedied by the hit-or-miss method of "requests" to the operators, under the issuance of such magnanimous pieces of paper called "Safety Alerts" or "Service Instructions", neither of which are legally enforcible.

On 9 August 1972, Benjamin O. Davis, the assistant secretary for Safety and Consumer Affairs, wrote a very pointed letter to Shaffer.

During staff review of your proposed reply to the N.T.S.B. regarding Safety Recommendations A-72-78 through 81, a few questions were raised which you may wish to explore further.

The N.T.S.B. recommendations, without mention of aircraft type, present the proposal that – when serious risk of safety to flight is threatened by inadvertant opening of cargo doors, redundant safety features should be provided to guard against this hazard.

Your reply does not address the reasonability of this proposal in general. It only discusses the two types of aircraft involved in the mishaps cited. Shouldn't this concept of redundant protection for catastrophic type hazards be specified for all aircraft?

In your specific reply regarding the Beechcraft 65–80, the proposal of the N.T.S.B. is rejected on the basis that, 'If properly maintained and fully secured', the door will prevent the shifting of cargo. That type of reasoning might apply to any operating system, but it does not answer the question of the consequences of either not maintaining or not operating the door properly.

Now Shaffer's reasoning was being questioned – and not just by the N.T.S.B.

On 27 July 1972, Continental Airlines was again in trouble with their DC–10 engines. At 1305 hours, during a climb-out from Los Angeles International Airport, slight damage was caused to the machine, with one person suffering minor injuries, when the number 3 engine (the one under the right wing), fell off in flight.

The McDonnell Douglas printing and mailing departments were at it again, and a Service Bulletin – 52–33 – was on its way to DC–10 operators. The number of aircraft affected this time was 54; manufacturing numbers 1 through 60, omitting machines 1, 2, 28, 36, 46 and 53. Aircraft number 1 had been delivered to American Airlines according to Service Bulletin 52–37 (dated 3 July 1972), but was now inexplicably omitted from the later list. Number 2 was McDonnell Douglas's factory machine. DC–10s numbered 28 and 36 as well as number 53, were yet to be delivered to Northwest Orient Airlines. And number 46 was a missing quantity, for later delivery to K.L.M. – presumably to be certificated by the civil air department that was well in the knowledge that there were design faults in the floor and control cable placements.

This Service Bulletin would have been the first one that would have been received by Turkish Airlines. It shows that when it was issued they had taken delivery of their two aircraft – fuselage numbers 29 and 33 – but this was not true – and there has been no confirmation that this "recommended" modification had been carried out.

The official reason for it being issued was:

A design investigation of the cargo door latching system at the Douglas Aircraft Company reveals that there is a possiblity of the door latch failing to complete the actuating cycle. This possibility is attributed to the loading personnel failing to hold the door power switch in the *closed* position for from four to seven seconds after the door contacts the lower jamb, as required. Installation of the nameplate and revising the 'unlatch' indicator light wiring to provide a visual indication that the door latch actuating cycle is complete, will minimize the possibility of improper cargo door latching.

Description: This modification installs nameplates on forward, centre and aft cargo door control panels and revises the cargo door control wiring.

Compliance: Recommended.

Approval: This Service Bulletin is F.A.A. Approved.

As can be seen in the reasons for such a modification being necessary, prior to this time the person who closed the cargo door had a hit-or-miss guessing game. The idea was that he held the operating switch in the "closed" position, and once the door was seated in its sill, the switch still had to be held in this position for the magical four to seven seconds. During this time, but unseen to the operator, the locking grips were being electrically driven around over the lock-bars to complete the process. But with distractions all around, plus possibly an anguished airline dispatcher yelling out for the doors to be quickly closed because the flight was running late, or for any of many other reasons, it is evident that the people shutting the doors were not holding the switch in the "close" position for a long enough period of time, and therefore from the outside, it appeared shut, but certainly was not fully locked. This modification was hopefully going to correct this.

The airlines affected by this Service Bulletin were: American Airlines, Continental Airlines, Delta Air Lines, K.L.M., Lufthansa, National Airlines, Swissair, Turkish Airlines and United Air Lines.

To go with the addition of the "door closed indicating light" was what McDonnell Douglas called a "nameplate". This was supplied in the kit, in the English language, to those who took the trouble of filling in all the order forms and applying for the modifications. It said:

HOLD SWITCH IN CLOSE POSITION UNTIL ABOVE COMES ON.

It was a small notice, designed to be placed on the operator's control panel between the door operating switch and the new light that illuminated when the door was fully locked after the magical four to seven seconds wait.

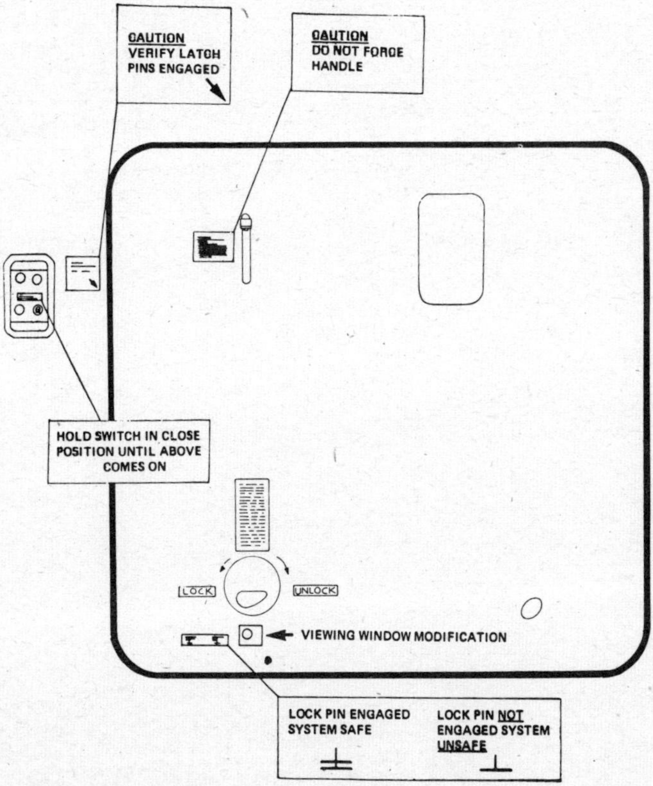

The one problem with this modification was that the intensity of the red plastic cover to the new "door locked" identification light, was such that it would be difficult to see were the panel, at the time of closing, bathed in sunshine, for example. However, one must presume that the F.A.A. were aware of this problem as the Service Bulletin said that it was F.A.A. Approved.

CHAPTER 31

STILL more paperwork arrived on the airlines' engineering directors' desks. This time it was McDonnell Douglas Service Bulletin 52–43. For some strange reason 52–44 was sent around on 22 September 1972, while this one, numerically out of order, was posted on 5 October 1972. This could indicate that Douglas had prepared it well in advance, but had retarded its issuance.

Again, aircraft manufacturing numbers 1 through to 60 were affected, but whereas Service Bulletin 52–44 stated that Turkish Airlines, two aircraft, as yet undelivered, needed modification, this one, 52–43, omitted them from the lists, and therefore this document *would not have been sent to that carrier*. Does this mean that the modification was done on these aircraft well in advance, or does it mean that these two machines were forgotten?

The reason for this "recommendation's" circulation was said to be:

> An investigation at the Douglas Aircraft Company of the cargo door warning system revealed that any of the three cargo doors could be closed, but not latched, without a warning light indication. This condition is possibly due to the CLOSED LIMIT and LOCK LIMIT warning switches being wired in parallel. If either switch fails to return to the 'reset' position, which is the common failure mode, the malfunction is hidden as long as the other switch functions. Wiring the door limit switches in series and the addition of slaved relays will provide a fail-safe warning circuit that will ensure a warning light indication when a cargo door is not properly latched. . . .

The date on this Bulletin was 5 October 1972. It is also trusted that the positive wording – no ifs or buts – is also remembered for later reference.

All of the Service Bulletins that had been issued so far had been F.A.A. Approved, had all been associated with the aircraft's cargo doors with not a single mention about the design faults in the floor; not a single mention about the design stupidity of running the control cables and wires through the floor beams. Had not anyone cared to notice these other two major problems? Had they not read the National Transportation Safety Board's recommendations?

CHAPTER 32

Of all the pieces of paper being spewed forth by the McDonnell Douglas engineering division, only two of the many Service Bulletins played any really major part in correcting the door locking problems in the DC–10. They were Service Bulletin 52–37, released on 3 July 1972 (which, by adding some support plates and other devices, made it necessary for the person closing the door to apply over 400 lbs. of pressure on the handle in order to stow it, *if* the lock mechanisms were not properly in place), and Service Bulletin 52–40, published on 27 October 1972, that introduced a mechanism whereby the door locking handle would *itself* indicate to the operator whether or not the locking procedures had been properly affected.

Service Bulletin 52–40 lists all the aircraft that had, up to that time, been delivered to the airline customers, but it also included, strangely enough, the – now – three Turkish Airlines machines, fuselage numbers 29, 33 and 78, though these were still in the possession of McDonnell Douglas.

Reason: One operator has reported that a cargo door came open and separated from the aircraft during flight. Subsequent investigation at the Douglas Aircraft Company revealed that the vent door can be closed, with the cargo door latches not properly latched, by forcing the handle closed and springing the mechanism. A cargo door closed in this manner appears to be properly closed and latched but it can open when cabin pressure is applied. Installation of a load limiting vent door handle link assembly will prevent the vent door from being forced into the closed position.

What this means, is that with the new assembly, if the cargo door operator attempted to force the door-closed handle into the stowed position with the door improperly locked, once he had applied about 50 lbs. pressure on the handle, the door mechanism would "de-clutch" itself, the handle would stow, but the small vent door would remain "open", giving the operator a guaranteed indication that the door was improperly locked.

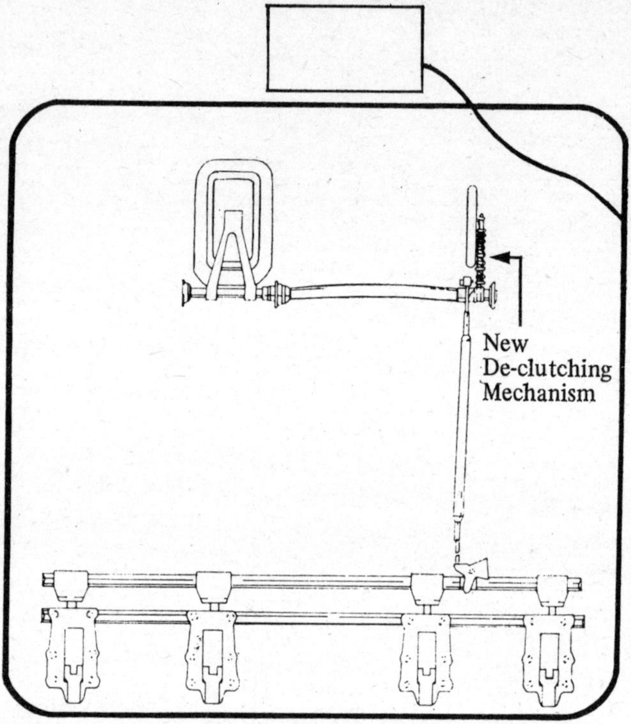

With this modification, the operator would have very quickly been able to ascertain whether the door was correctly shut, whereas all the other Service Bulletins (other than 52–37, increasing the handle pressure) in no way contributed to an overall positive correction of the door locking design problems. The others seem to have been "piecemeal" modifications.

CHAPTER 33

THE 8 November 1972 was the day after Richard Nixon had been re-elected to the presidency with a resounding majority vote. The celebration parties were over and many of his followers were back in their offices, including John Hixon Shaffer, when the telephone rang; it was one of Nixon's aides at the other end.

As is normal, following a presidential election, each of the senior government staff that were appointed by the resident, automatically submits his resignation. Normally, these are not accepted, being only a formality, and the staff member retains his post. Nixon's aide phoned Shaffer and told him to submit a "fluffy" letter, along the lines that it had been a pleasure to serve the administration – which in Shaffer's case, it must have been.

The formality was duly sent to Nixon, and much to Shaffer's shock, it was accepted on 10 December 1972, giving Shaffer the three months contractural period to work out, before finally leaving the F.A.A.

CHAPTER 34

THE following February, the National Transportation Safety Board was getting worried about the most dangerous aspects of the DC–10's design which, in their opinion, were *not* being tackled properly by the F.A.A.

They were fully aware that a whole "gaggle" of "recommended" modifications were vomiting forth in a seemingly never ending stream. They were also aware of the distribution of these Service Bulletins. Since the accident on 12 June 1972, a stack of self-assured, "this-is-the-final-answer-to-our-problems"-type Bulletins had been issued. A recounting of this list tends to indicate the utter confusion and evident inability of both the F.A.A. and McDonnell Douglas, to correct major design deficiencies by these hit-or-miss methods.

There was quite an impressive line-up of pieces of paper being sent around – in amongst the thousands of others – but they were all concentrated on only one small part of the problem *as it was evidenced* in the American Airlines, Windsor, Ontario, accident.

On 23 February 1973, John Reed, the chairman of the National Transportation Safety Board, wrote to Shaffer at the F.A.A.

This is in reply to your acknowledgement of the 7 July 1972, of the National Transportation Safety Board's recommendations A–72–97 and 98, in which you noted the actions which have been taken to prevent the recurrence of similar accidents.

With respect to Recommendation A–72–98 (regarding the venting of the floor to stop it collapsing) you observed that a preliminary investigation indicated that it may not be feasible to provide complete venting between cabin and cargo compartments.

When your investigation is complete, the Board would appreciate knowing if the installation of vents, similar to those on other DC–10 cargo holds, is feasible in the aft cargo compartment.

If complete venting is not possible, partial venting would be beneficial. Such venting could prevent the collapse of the aft cabin floor, it could reduce the amount of floor deflection, and attendant control cable damage in a DC–10.

One really could not have been more explicit about this. However Shaffer's period in office as the F.A.A.'s administrator was rapidly drawing to a close, and his interests in anything to do with airline passenger safety therefore would in no way improve from his past performance.

CHAPTER 35

THE 16 March 1973, according to the F.A.A., was the last day on which John Hixon Shaffer was the responsible administrator.

One strict ruling in American governmental circuits is that employees are *not* permitted to hold "second jobs". This means that they are not allowed to work in government employment during the daytime, and then have a second job that occupies them in the evenings. It also means that a government employee cannot be a director of a privately-owned company. And it would be a scandal if, for example, it was found that a government employee was also a director of a company over which he and his department were supposed to have some semblance of control.

On 13 March 1973, Shaffer became a director of the manufacturing company Beech Aircraft Corporation. This was confirmed by the Beechcraft office in Washington D.C. This would tend to indicate that at least for a brief period, Shaffer was employed in two positions: administrator of the Federal Aviation Administration, and a director of the Beech Aircraft Corporation.

But now with hindsight, it is interesting that some products made by the Beech Aircraft Corporation had been strongly criticized, at least for the past three years – the time during which Shaffer was the F.A.A. administrator – and that in the same period, were involved in numerous legal suits some of which ended with Beechcraft paying out millions of dollars to relatives of persons killed in accidents involving these machines.

It is also interesting that a Beechcraft 65–B 80 Queen Air, was involved in practically a similar occurrence to the American Airlines DC–10 – *just twenty-four days before* – when a cargo door opened in flight. And it is equally as interesting that Shaffer treated the Beechcraft door problem in exactly the same fashion as he treated the DC–10, saying that no mandatory modifications were necessary.

1. We now know that there was a "gentlemen's agreement" between Shaffer and Douglas regarding the DC–10.

2. There was a whole string of major problems with Beech aircraft and yet not one of them produced the necessary Airworthiness Directives.
3. Shaffer ordered the issuing of an inadequate letter to *all* small twin-engined supercharged aircraft owners, which completely omitted saying that there were major problems with the Beech 65 machines.
4. Shaffer signed at least one letter as administrator of the F.A.A. on the same date on which he became a director of the Beech Aircraft Corporation.
5. Negotiations MUST have been taking place while Shaffer was *still* in the position of administrator of the F.A.A., and while he was still responsible for the regulation and operation of safe civil air standards.

CHAPTER 36

ALEXANDER Porter Butterfield was sworn in as the new administrator of the Federal Aviation Administration on 14 March 1973. Butterfield, like Shaffer before him, was a political appointee made by the star of the Watergate Hearings, Richard Milhouse Nixon. Up to the time of his F.A.A. appointment, he had been on the President's personal staff and was one of the very few not to be indicted on criminal charges following the discovery of the Watergate conspiracy.

Butterfield was born in Pensacola, Florida, on 6 April 1926, to the now retired naval aviator Admiral Horace P. Butterfield, and his wife. He travelled extensively throughout his early years before making his home in Coronado, California. He attended the University of California where he gained a Bachelor of Science degree "cum laude" from the University of Maryland. Later he earned a Master of Science degree in the field of international affairs, from the George Washington University. He was also a graduate of the National War College.

In 1949, Butterfield joined the United States Air Force. As a pilot, he logged some 7,800 hours, just over 5,000 of which were in 34 different types and models of fighter aircraft. He also flew for several years with the Sky Blazers, a European-based USAF jet acrobatic team which (ironically) won the international competition at the 1953 Paris Air Show.

Butterfield's staff positions included a period of duty as an academic instructor at the U.S. Air Force Academy, and three years in Hawaii as the senior aide to the late General Emmett "Rosy" O'Donnell, the then commander-in-chief of the Pacific Air Forces. Later, in Washington, he served as a policy planner at the Pentagon; then as military assistant to the special assistant to the Secretary of Defence.

In addition, Butterfield was operations officer of an Air Defence fighter-interceptor squadron at the McGhee-Tyson Air Force Base, Knoxville, Tennessee; commander of the fighter squadron at the Kadena Air Base, Okinawa; and three times commander of the USAF's low and medium level air reconnaissance operations in south-east Asia. His last military assignment was in Australia, where he was the USAF's F–111 project officer and also senior U.S. Defence Department representative. Most people will, of course, remember the complete mess that reverberated around the world with regards to the "excellence" of these aircraft.

Alexander Butterfield is a rated command pilot and a qualified parachutist. He holds the Legion of Merit, the Distinguished Flying Cross, several air medals and miscellaneous commendations, campaign and service awards.

On 21 January 1969, Butterfield gained the position of deputy assistant to the President.

Practically no one had heard of Alexander Porter Butterfield until the afternoon of 16 July 1973, just over four months after Nixon had nominated him for administrator of the F.A.A., when, suddenly, his name entered the portals of history and the results of his actions, whether they were done conscientiously, morally, stupidly, or for spite, were to remain with us for many months to come.

However, before we look into this aspect of Butterfield's character, let us jump forward a year to the time he was appearing before the U.S. Senate's subcommittee hearings into the reasons for the ignoring of the fatal design faults in the DC–10 which resulted in the 346 deaths near Paris on 3 March 1974. At this hearing, Butterfield was asked these questions by Senator Howard Cannon, the subcommittee's chairman:

> (Q) When did you first become aware of the cargo door and the decompression problems of the DC–10? (A) The accident occurred on Sunday, 3 March, and I first became aware of the possibility of the decompression . . .
>
> (Q) No. I am talking about the problems generally of the DC–10 not just this latest accident. (A) In conjunction with this latest accident, Mr. Chairman, on 5 March, from Mr. Bakke.
>
> (Q) So you had no knowledge of this problem until this most recent accident? (A) Not that I can recollect. No, sir.

Butterfield was under oath when he was being questioned. We therefore must assume that his answers were honest.

On 23 February 1973, John Reed, the chairman of the Safety Board wrote to Shaffer. The letter was answered on 15 March

1973. It was signed Alexander P. Butterfield, administrator, and it read:

Dear John,
This is in reply to your letter dated the 23 February 1973, which clarified Safety Recommendation A–72–98 in which it was recommended that relief vents be installed in the DC–10 floor between the cabin and the aft cargo compartment.
We have requested that the manufacturer reassesses the DC–10 with regard to the effects on safety of probable large openings in the pressurized fuselage. The manufacturer is to consider rerouting of vital systems, reinforcement of the floor as well as incorporation of additional venting between compartments. These alternatives will include consideration of various degrees of venting as recommended in your letter.
Sincerely,

From this reply, it must be assumed that either Butterfield was well aware of the design problems in the DC–10 as the second paragraph above his signature so stated, or he signs letters, or at the very least had signed this letter without knowing the full details of the problem and the issues involved. It seems evident that he had signed and authorized something without checking out the information, as admitted by Butterfield below.

As the Aviation Act specifically places the responsibility for safety in the hands of the administrator, and as the administrator, having just entered or having been in office for some time, has signed a letter that deals with a known fault in an aircraft's design, thus the reasons for his statement: ". . . We have requested that the manufacturer reassess the DC–10 *with regard to the effects on safety* . . .", should we assume that this person is capable of carrying out his position, and especially to: ". . . exercis(ing) and perform(ing) his powers and duties under this Act (the Federal Aviation Act of 1958, Section 601 (b)), in such a manner as will best tend to reduce or eliminate the possibility of, or recurrence of, accidents in air transportation . . ."

Immediately after being questioned about Butterfield's knowledge of the DC–10 problems, Senator Cannon continued:

(Q) Now in March of 1973, one day after you were sworn in as administrator, you signed a letter addressed to chairman Reed of the National Transportation Safety Board concerning recommendations the Safety Board had made as a result of the DC–10 incident over Windsor, Ontario. Did you have any knowledge of the precise problem at that time? (A) No. I recalled the letter when I saw it again, but I did not, at the time that Mr. Bakke spoke to me, and I did not have intimate knowledge of the issue at the time I signed that letter.

(Q) Now in a letter which you addressed to me on 19 March 1974, you said on Page 2: "Upon appeal to the administrator by the manufacturer a decision was made to forego the issuance of Airworthiness Directives and instead to seek the modification through voluntary compliance by operators of McDonnell Douglas Service Bulletins". What is the basis of your knowledge regarding that particular incident? (A) Only hearsay, Mr. Chairman. I learned that from my staff on, I guess, 6 March, or possibly 7 March. And as I say, it is only an understanding.

(Q) You also say in your letter, and I quote: "On 6 March 1974, when I learned that certain of these modifications had not been carried out, and that there were some uncertainties in the records available to us with regard to the status of DC–10 aircraft, I directed that the door modification program be formalized through Airworthiness Directives and made the subject of an immediate F.A.A. Regulatory Surveillance Program." Now, what did you learn about modifications that had not been carried out, that you referred to in that statement? (A) Nothing specific. Only that there had been some exceptions and that the previous request was not a Directive but in fact was a request, and required only a voluntary compliance and there had been some exceptions to the compliance, and that some of the records were fuzzy with regards to compliance.

(Q) And what were the uncertainties you referred to in the records, and whose records are you referring to now? The aircraft records themselves or what? (A) I would have to defer to Mr. Bakke to give you the details of that.

So now we see that there has been an official occasion in a legally set-up inquiry at which Butterfield publicly admitted that he writes and signs correspondence about which, he does not know the details of the subject matter being discussed.

Returning to the earlier occurrence on 16 July 1973, we see Butterfield again in the witness chair:

SENATOR SAM J. ERVIN JR. – DEMOCRAT – NORTH CAROLINA – CHAIRMAN OF THE WATERGATE SENATE COMMITTEE INQUIRY.

(Q) Suppose for the record that you state your name and present occupation. (A) My name is Alexander Porter Butterfield. I am the administrator of the Federal Aviation Administration.

SAMUEL DASH – CHIEF COUNSEL AND STAFF DIRECTOR OF THE WATERGATE SENATE INQUIRY.

Mr. Chairman. At a staff interview with Mr. Butterfield on Friday (13 July 1973), some very significant information was elicited and was attended by the majority members of the staff and the minority members. The information was elicited by a minority staff member. Therefore, I would like to change the usual routine of the questioning and ask minority counsel to begin the questioning of Mr. Butterfield.

FRED D. THOMPSON – CHIEF MINORITY COUNSEL OF THE WATERGATE SENATE INQUIRY.

(Q) Mr. Butterfield. I understand you were previously an employee of the White House. Is that correct? (A) That is correct. I would like to preface my remarks, if I may, Mr. Thompson, with a statement. Although I do not have a statement as such, I would simply like to remind the committee membership that whereas I appear voluntarily this afternoon, I appear with only three hours notice and without time to arrange for permanent counsel or for assistance by a temporary counsel.

(Q) During what period of time were you employed at the White House, Mr. Butterfield? (A) I was at the White House as a deputy assistant to the President from the first day of the Nixon administration, the 21 January 1969, until noon on 14 March 1973.

(Q) And what were your duties at the White House? (A) My duties were many and varied. I was in charge of administration. That is to say that the staff secretary who is the day to day administrator at the White House, reported directly to me. And of course, I reported to Mr. Haldeman, as did everyone. In addition to administration, I was responsible for the management and ultimate supervision of the Office Of Presidential Papers and the Office Of Special Files. Both of these pertained to the collection of documents which will eventually go to the Nixon Library. Thirdly, I was in charge of security at the White House in so far as liaison with the Secret Service and the Executive Protection Service is concerned, and in so far as F.B.I. background investigations for prospective Presidential appointees is concerned.

A fourth duty was that I was secretary to the Cabinet. I was additionally, the liaison between the President and the Office of the President and all the various support units. By that I mean, the office of the military assistant to the President and the Office of White House visitors; again the Secret Service; the Executive Protection Service; the residence staff; Mrs. Nixon's staff; I served as sort of a conduit between all these elements and the Office of the President. Finally, I was in charge of the smooth running of the President's official day, both in Washington D.C. and at the Western White House at San Clemente.

(Q) Mr. Butterfield, are you aware of the installation of any listening devices in the Oval Office of the President? (A) I was aware of listening devices. Yes, sir.

(Q) When were these devices placed in the Oval Office? (A) Approximately the summer of 1970.

(Q) Are you aware of any devices that were installed in the Executive Office Building Office of the President? (A) Yes, sir. At that time.

(Q) Were they installed at the same time? (A) They were installed at the same time.

(Q) Would you tell us a little bit about how these devices worked. How they were activated, for example? (A) They were installed, of course, for historical purposes, to record the President's business, and they were installed in his two offices; the Oval Office and the E.O.B. (Executive Office Building) Office. Within the West Wing of the White House there are several, at least three, perhaps four – the three I know of – boxes called Presidential Locator Boxes. These are square boxes, approximately ten by ten inches, and on them are several locations which would tell where the President might have been at any time. Locations such as the residence – that is one of them – the South Grounds is another; the Oval Office is another; the E.O.B. Office is still another; the West Wing, meaning the West Wing of the White House, is another; and that, I think, is the last one. When the President moves – the East Wing is still another and I think that covers all the locations indicated on the box. When the President moves from his Oval Office, for instance, to his Executive Office Building Office, and he departs the West Wing and crosses the street, it is my understanding that the Secret Service agents, members of the Executive Protection Committee who cover him, it is my understanding that there are four, five or six of them; when he moves across the street, one of them covers the central location . . .

(Q) Now in the interview which you had with the staff of this committee on Friday, 13 July 1973, at 2.15 p.m., the interview, as reported by this staff, concludes with: "Butterfield stated this is all something I know the President did not want revealed, but you ask me and I feel it is something you ought to know in your investigations. I was told no one was to know about the information I have told you". Is that a correct quotation? (A) I could not say that it was correct, but there was some reluctance on my part to reveal information which I felt could have a number of serious repercussions with regard to foreign governments. It is very obvious that this could be embarrassing to our government.

This was the first indication that there were such things as "Watergate Tapes", and it was from this point onwards in the inquiry that things started to hot up for Nixon and his cohorts.

During his first year at the Federal Aviation Administration, nothing changed from the Shaffer days. No improvements were implemented in civil air safety standards. Butterfield retained Shaffer's illogical methods of correcting known faults in both aircraft design and operations by using, in the main, the unpublished hit-or-miss approaches. In his period as administrator, practically none of the myriad of faults that were permitted to persist in day-to-day civil air operations, had been mandatorily corrected.

It was seen from his evidence to the 1974 Senate subcommittee hearings that Butterfield had been writing letters and signing statements, containing information and details, of which he did not have full knowledge. He publicly admitted under oath, that he signed a letter dated 15 March 1973, the subject matter of which dealt with a design fault in an aircraft that had already been seen as playing a major part in two earlier accidents, and the serious and urgent attention of which was needed in order to avoid the third repetition. Yet it also seems that he did not have "intimate knowledge" of what he was signing. He then admitted to writing a letter dated 19 March 1974, in which he implied that when he learned how important the whole matter was to air passenger safety (*after* 346 people had died because nothing concrete had been done about this situation), he ordered that immediate action be taken – action that should have been instigated nearly two years earlier at the very latest. Yet when further questioned about the specifics of this information, he was unable to supply them.

The sad fact is that the measures which Butterfield's letter of 15 March 1973, suggested be taken, were not implemented. And one year later, there was no time left – 346 people were already dead.

CHAPTER 37

On 10 August 1973, a National Airlines DC–10 was involved in an accident while in-flight between New Orleans, Louisiana, and Miami, Florida. The preliminary investigation indicated that a hostess sustained serious injuries when she became lodged in an elevator shaft, between the top of an upward-moving food-service cart and the ceiling of the shaft. She reportedly entered the lift-well and lay atop of it in an attempt to release a malfunctioning locking system which had prevented removal of the cart from the elevator at the upper (passenger) deck level of the cabin.

As a result, the National Transportation Safety Board recommended that the Federal Aviation Administration:

Notify on a priority basis all operators of DC–10 type and other wide-bodied aircraft equipped with similar elevator/service cart equipment of the need of strict adherence to prescribed procedures, both for the operation of food service galley elevators and for the correction of malfunctions related thereto. Widest possible dissemination of the details of this mishap should be made to illustrate the consequences of unauthorized operating procedures.

This was to be the start of an exceptional flow of correspondence on this particular subject, and was quickly picked up by the Air Line Pilots' Association which had been in argument with the F.A.A., the airlines, and the aircraft manufacturers about this very matter, for many months.

The injured hostess, Connie Wargo, was standing on the passenger cabin floor level (A), at the doors to the lift (B), in which service carts are sent up from the galley situated on the floor below. One of these carts had become jammed in the lift and it could not be unstuck by the normal method of depressing the handles on either end, which in turn was supposed to release its braking system.

In order to gain access to the brake handle which was at the back of the lift, she depressed the safety switch that prevents lift operation when the doors are open, and she then pressed the

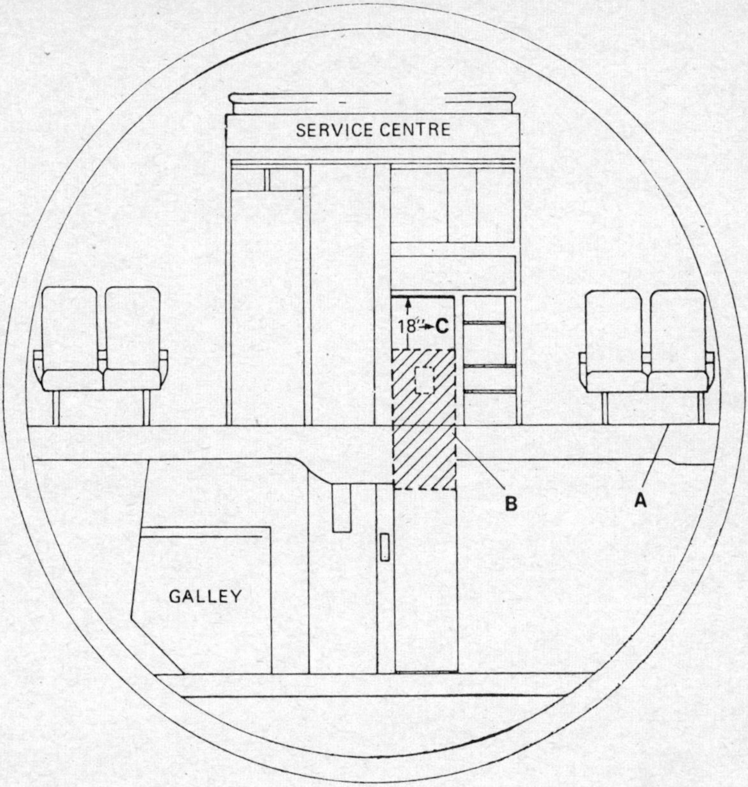

normal "down" switch to descend the elevator by approximately eighteen inches (C). Once the lift, with the cart inside it, had moved downwards, she then placed her head and the upper portion of her body into the lift-shaft in order to reach and depress the cart's brake handle at the rear of the lift, while another stewardess, Judy Mahoney, depressed the cart's brake handle at the front. They then attempted to dislodge it. But as they were doing this, the downstairs girl, Bonnie Campbell, noticed that the cart lift was not in its proper position, and so she pushed the "up" button in the galley. The lift then moved upwards, pinning Mrs. Wargo between the top of the service cart and the top of the lift shaft.

When Miss Campbell heard the screams, she pushed the "emergency stop", and the "down" switch, but the lift did not move. At the same time, the yelling attracted the attention of two other hostesses. One of them, Vivian Ford, also attempted to operate the lift by the normal "down" button, but when this failed to move it, and on hearing Mrs. Wargo's shouts to: "push the button", she did just that, but the lift did not budge.

Flight attendant Cynnie Sharp also attempted to operate the lift by pushing the "down" switch on the passenger cabin floor level, but to no avail. Neither of them activated the "interlock" at the same time as pushing the "down". As the lift could not be moved, Miss Sharp went to the cockpit to get help from the flight engineer. When he arrived, he also tried to operate the lift by using the normal "down" button. With still no luck, he opened the overhead access panel and pulled the food-lift circuit-breaker, with a view to lowering the lift by cutting the main drive chains. But after he had obtained tools from the cockpit, and while in the process of cutting the chains, he remembered the function of the door and lift interlock safety switches, and once he had reset the electric power and pressed both switches at the same time, the lift moved downwards and Mrs. Wargo was removed from her position lying prostrate across the top of the food cart, and then given first-aid.

According to National Airlines flight attendant training manual, dated 1 April 1973 (sic), Page 22, the following procedures are to be followed.

If cart is stuck in lift, send lift to lower galley – open dutch doors – reach to back of cart and depress both brake handles. If this fails then the mushrooms on the floor may be loose. Ask flight engineer to come back and tighten the mushrooms after he has removed cart.

National Airlines included a notice to all flight attendants on 10 August 1973, which read as follows:

In the event of any malfunction with the personnel lift or cart lift on the DC–10, do not try to correct the situation yourself by placing any part of your body in the lift. Contact the engineer immediately.

Mrs. Connie Wargo proved to be most uncooperative to the investigators. The only thing she would do to assist them was to give them a statement, about which she refused to be questioned. It was even written for her by a colleague.

Wargo's recollections, in a number of places did not agree with the investigator's reports, and indicated that she was well aware of "another method" of releasing stuck carts in lifts – possibly the one in the National Airlines Flight Attendant Training Manual, cited earlier.

Complaints about these lifts came in thick and fast. A flight supervisor was caught inside the personnel lift on 21 December

1971, and had to make her escape out of the emergency hatch. On 9 May 1972, a stewardess was in the same lift when the lowering mechanism broke and it crashed to the lower level. There were hundreds of incidents of the lifts stopping at other than floor levels, meaning that the food service carts had to be physically hauled out of them. Then on 18 and 19 January 1973, some passengers who had wanted to have a look at the downstairs galleys, were trapped inside the personnel lifts.

Though a good deal of correspondence was exchanged concerning the lift defects, it was yet another series of design problems that were permitted to exist in the DC–10, until finally someone was very seriously injured. It is not as though National Airlines, United Air Lines, or American Airlines did not know of the dangers according to evidence supplied. It is simply that they, like the Douglas Aircraft Company and the F.A.A., did absolutely nothing about it.

CHAPTER 38

McDonnell Douglas were at it once again, with the issuance of a further door modification contained in an enormous 60-page Service Bulletin – number 52–49.

One would have thought that the door troubles should have been straightened out by this. But evidently not.

Bulletin 52–49 listed 135 DC–10s affected by this "recommendation" (omitting fuselage number 2 – Douglas's own machine; number 119 which was simply missing off the list; and numbers 83 and 125 – evidently finished in manufacture, but not delivered). Twenty-three carriers (including Turkish Airlines) were affected.

One operator (said the Bulletin, rather belatedly) reported that a cargo door came open and separated from the aircraft during flight. Subsequent investigation at the Douglas Aircraft Company revealed that the vent door could be closed, with the cargo door latches not properly locked, by forcing the handle closed and preloading portions of the mechanism. A cargo door closed in this manner appears to be properly latched and locked but could open when cabin pressure is applied. DC–10 Service Bulletin A52–35 installed a lock mechanism viewing window to permit visual inspection to determine if cargo doors are latched and locked prior to flight. DC–10 Service Bulletin 52–37 readjusted the lockpin tube assembly to prevent forcing of the vent door handle to the closed position when the lockpins are not properly engaged. DC–10 Service Bulletin 52–40 also installed a load-limiting vent door handle link assembly to prevent forcing of the vent door handle to the closed position when the lockpins are not properly engaged. Installation of a revised 'closed loop' cargo door mechanism will eliminate the human factor error by ensuring prevention of vent door closures unless the cargo door latches are latched and locked.

What this meant was that prior to this "request" for a modification, the small vent port in the main cargo door was closed directly by the action of the operator pulling down the locking handle. The change meant that the vent door (A), should now be closed by the action of the main door's locking mechanism moving into place (B). Before this "suggestion", the operator could, by putting force behind the handle (C), bend the main locking-bar that connects it (D), which would mean that the lockpins (E) were not in place but the handle was stowed. And because the smaller vent door was closed by the action of the locking handle, the door would appear to be locked, but in fact, was not.

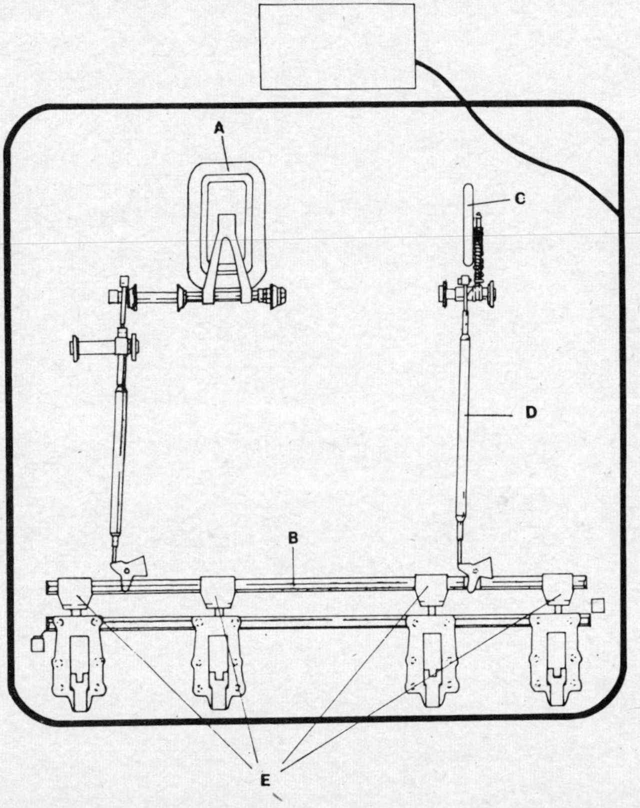

Douglas's main "safety feature" in this part of the door system was in this vent door operation. Naturally, if the vent door was not closed tight, the aircraft could not be pressurized and therefore the pilots, so they thought, would realize something was wrong.

With this modification to the system, the small vent door would supposedly not close until such a time as the lockpins were in place, and locked.

The relocation of the "door locked" switch that operates the light on the pilots' control panels, would mean that it was controlled by the lockpins themselves. If these were not in place locking the door's lock-grips, the light in the cockpit would remain "on". And the third part of this modification meant the installation of yet *another notice* – again, in English – for the operators to read and digest, and it said:

CAUTION: VERIFY LATCH PINS ENGAGED

with an arrow on it pointing to the base of the door.

It is worthwhile at this particular point, referring back to a statement contained in McDonnell Douglas Service Bulletin 52–37, which was sent around to DC–10 operators on 3 July 1972 – FIFTEEN MONTHS PREVIOUSLY.

Readjusting the lockpin tube assembly (the earlier Bulletin said), to prevent the vent door handle from being forced to the closed position when the lockpins are not engaged properly *WILL* provide a more positive and reliable cargo door closed, latched and locked system.

Why should it take fifteen months to discover that their earlier "requested" modification was not good enough and required re-modification? Does this mean that their July 1972 recommendation was rushed out with insufficient thought being given to it? Does this mean that someone had been working on the door design for those fifteen months, and that it had taken all that time for a solution to be found? Or would it possibly mean that *still further* complaints were being received at the Douglas Aircraft Company from airline customers, and they were *still* having trouble with the locking of cargo doors?

Any one of these above possibilities could be correct. But the most telling things of all were the methods used by the McDonnell Douglas Corporation in trying to effect design fault changes in their products. It is quite evident, and there was a glaring example, that when these had been uncovered, Douglas had no idea at all about how to correct the situation positively and quickly. And besides, the F.A.A. – being none the wiser – had evidently also condoned McDonnell Douglas's efforts.

Detroit International Airport, 12 June 1972. *(Right)*. The remains of the aft cargo compartment door of the American Airlines DC-10, half ripped off. *(Below)*. The devastation inside and the remains of the collapsed passenger cabin floor. The broken flight and engine control cables that ran through the floor beams can be seen.

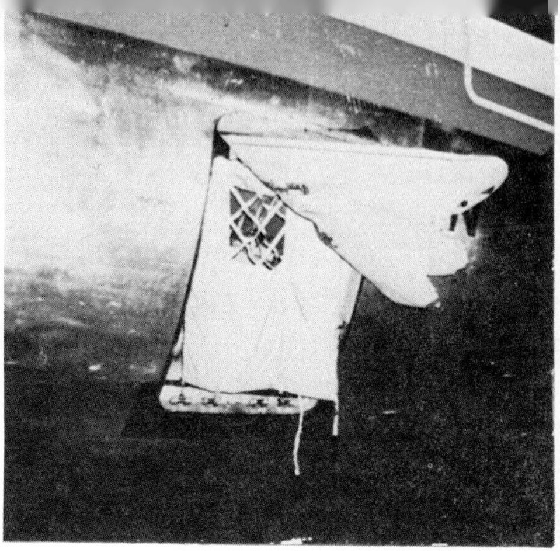

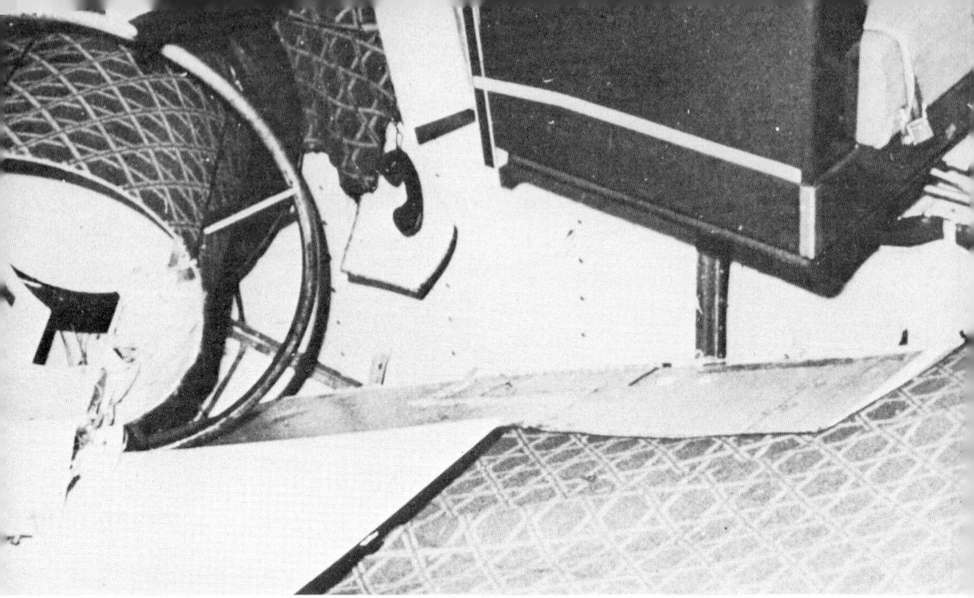

THY TÜRK HAVA YOLLARI – TURKISH AIRL

(Above left). Interior of the American Airlines DC-10 where the floor collapsed, and *(above)* the rear section of the passenger cabin in the vicinity of the floor hole.

(Left). This photograph is one of the very few now in existence of the actual Turkish Airlines DC-10 registered number TC-JAV that crashed, killing 346 people just outside Paris on the 3 March 1974. It was taken only days before the disaster, on one of its regular visits to London Heathrow Airport. Ironically, this machine had been held up on numerous past occasions, with difficulties in closing and locking the aft cargo door - the one that caused the accident - which can be clearly seen to the left of, and just below, the rear passenger door.

(Overleaf). Same aircraft - different situation.

NATI

(Far left). The National Airlines DC-10 an engine of which disintegrated in mid-air near Albuquerque. Six months later, this same aircraft was again involved in an almost identical accident. *(Left)*. In the top left-hand corner is the shattered window through which a male passenger was sucked to his death.

(Below). The Iberia Air Lines DC-10 that crashed, landing at Boston Logan Airport. Note that the words Iberia Air Lines of Spain, on the side fuselage, and the aircraft's name near the nose have been painted out. *(Inset below)*. The broken fuselage is in the same position as the floor that collapsed in both the Windsor, Ontario, and Paris accidents.

(Top). The rear interior passenger cabin in the Iberia Air Lines DC-10 accident at Boston. *(Below).* Five people were trapped in the rear section of the passenger cabin when the floor collapsed and was thrown upwards.

CHAPTER 39

SOON after 1600 hours on 3 November 1973, there was trouble in the air, about a hundred miles from Albuquerque Airport, New Mexico.

(For those who wish to follow the drama piece by piece, here is a list of the participants:

NA27 – National Airlines, Flight 27, a DC–10 bound from Houston, Texas, to Las Vegas, Nevada, and was overflying the Albuquerque area.

NA22 – National Airlines, Flight 22, a DC–8 also overpassing Albuquerque.

AA54 – American Airlines, Flight 54, another DC–10 with a scheduled stop-over at Albuquerque.

FTWC – Fort Worth Air Route Traffic Control Centre.

EPHI – Al Paso High Altitude Air Traffic Control Sector.

ROHI – Rosewall High Altitude Air Traffic Control Sector.

ALHI – Albuquerque High Altitude Air Traffic Control Sector.

ZUHI – Zuni High Altitude Air Traffic Control Sector.

ALWA – Albuquerque West Area Air Traffic Control Centre.

ALEA – Albuquerque East Area Air Traffic Control Centre.

ACIC – Assistant Chief of Air Traffic Control, Albuquerque.

ALAR – Albuquerque Arrival Air Traffic Controller.

ALAP – Albuquerque Approach Controller.)

The drama developed over two different radio frequencies.

The first, indicated by this typestyle is Albuquerque Air Route Traffic Control radio channel.

The second, indicated by this typestyle is the Albuquerque Airport tower.

Conversations solely between air traffic controllers are indicated in this typestyle.

1618:35	**NA27**	**Albuquerque Centre. This is National Twenty-seven. Three nine zero** (39,000 feet altitude).
	ALHI	**National Twenty-seven. Flight level three nine zero. Ident** (ification, by pushing the button

		on your transponder so that we know who and where you are).
1639:50	ALHI	National Twenty-seven. I show you squawking (transmitting on your identification transponder) 'May-Day' (the international emergency signal). Is this intentional? National Twenty-seven. Albuquerque
	ALHI	American Five Four. How do you read this transmitter?
	AA54	Five by five (very loud and clear).
	ALHI	Thank you sir. National Twenty-seven, Albuquerque.
1640:35	ZUHI	National Twenty-two. Albuquerque.
	NA22	Two Two.
	ZUHI	National Twenty-two. Er, could you, er, . . . Any chance of your getting a hold of National Twenty-seven? He's er, (on radio frequency) one thirty-five oh five.
	NA22	OK. And if I get him, what then?
	ZUHI	Well, he's supposed to be on (radio frequency) one thirty-five oh five. He's squawking (transmitting) emergency. And er, the sector above us . . . er . . . he's at, er, thirty-seven, er, thirty-five, and he can't talk to him. And right now, he's (NA27 is) at your twelve o'clock position (directly in front of you) forty miles, and he's northwestbound.
	NA22	Thirty-five oh five, and you want us to tell him to go to what frequency?
	ZUHI	Well, if you can find out, er, if he's having any difficulty or not. He's squawking emergency.
	NA22	I'll do it. Thirty-five oh five.
1641:00	ALHI	Desk – er, seventy. (Two controllers calling one another.)
	ALWA	Yeah.
	ALHI	Do you see the "May-Day", forty-five miles southwest of Albuquerque?
	ALWA	Yes.
	ALHI	OK. That's a National aircraft, a heavy DC–10. We can't talk to him. He's squawking "emergency". He's supposed to be going to Winslow. It doesn't look like he's going that way. He might . . .
	ALWA	. . . looks like he's in a right turn.
	ALHI	Yeah. You'd better keep everything out of the way. I don't know what his altitude is or where he's going.
	ALWA	Alright.

1641:50	**ALHI**	**Approach Control. Albuquerque Centre on line two.**
	ALAP	**Go ahead.**
	ALHI	**Roger. Do you have contact, or have you talked to a National Twenty-seven?**
	ALAP	**National Twenty-seven?**
	ALHI	**Yeah.**
	ALAP	**Negative.**
	ALHI	**Alright. He's a DC–10. He's supposedly** (heading for) **Socorro direct** (via) **Winslow. At the present time, he's about forty-five to fifty** (miles) **southwest of Albuquerque VORTAC** (a radio beacon used for navigation). **He changed . . . er, he's "emergency" squawk and we can't talk to him. He seems to be in a right turn.**
	ALAP	**OK. He's a DC–10?**
	ALHI	**Yeah. A DC–10.**
	ALAP	**OK. We'll run our** (radar) **scope out there** (to find where he is and to keep track of him).
	ALHI	**OK. Find out what his altitude is. We don't know but, er, all we do know is he's left our** (radio) **frequency or something, and he's squawking "emergency".**
	ALAP	**OK.** (His flight plan was) **Socorro direct to Winslow.**
	ALHI	**That was his original route. Right. He's changed to "emergency" and left our frequency.**
	ALAP	**Where is he now though?**
	ALHI	**Oh, vector one ninety. About fifty miles . . . forty-five miles southwest of Albuquerque.**
	ALAP	**We don't see a thing out there from the VORTAC** (in the vicinity of the radio beacon on our radar). **I can't go out about forty-five miles from the VORTAC southwest.**
1642:50	**ALHI**	**Well, he's fifty miles southwest of the VOR** (radio beacon).
	ALAP	**OK. Fifty.**
1642:55	*NA27*	*Centre –* (this is) *National Twenty-seven. How do you read?*
	NA27	*Centre – National Twenty-seven. How do you read 'emergency' frequency?*
	NA22	*National Twenty-seven – National Twenty-two.* (You are on radio frequency) *one twenty-one five.*

	NA27	*OK. National Twenty-two. I'm executing an emergency descent. We've lost number 3 engine* (the one under the right wing) *and, er, we're badly rolling to the right. We have a moderate, er, shaking.*
	NA27	*We're in moderate turbulence right now. Would you inform the southwest* (radio centre) *please?*
	NA22	*Yes, sir. We're talking to the centre ourselves.*
	ALHI	**We'll call you back Albuquerque Approach, when he comes back, and if we see what he's going to do.**
	ALAP	**Alright.**
	ALHI	**Right now he seems to be going, er, straightening out, and heading straight north.**
	ALAP	**Straight north?**
	ALHI	**Yeah. Looks like it now.**
	ALHI	**We'll give you a call.**
	ALAP	**OK.**
	ALHI	**National Twenty-seven. Squawk . . . er . . . Stand-by on the transponder** (the identification radar unit) **if the squawk is intentional for "May-Day".**
	NA22	*Twenty-seven. Do you read Twenty-two?*
	NA27	*Abandon the airways* (keep well away). *We're down at, er, one thirteen point five right now* (13,500 feet altitude). *We're going to take her on down to about ten thousand* (10,000 feet).
	NA27	*We're shaking pretty badly. Unable to shut the number 3 engine down* (the one under the right wing). *We've got the handle off* (we've retarded the throttle control for that engine) *but we can't get the, er . . .*
	NA27	*. . . can't get the fuel lever down.* (They cannot shut off the fuel supply being pumped to that engine.)
1644:10	**AA54**	**Er, Centre – American Fifty-four. Where was National Twenty-seven's last position?**
	ALHI	**Roger. Right now, he is, er, sixty** (miles) **southwest of Albuquerque. Look's like he's turned inbound towards Albuquerque.**
	AA54	**OK.**
	ALHI	**National Twenty-seven. We're monitoring** (radio frequency) **one twenty-one five at the Centre.**
	NA22	**Centre – National Twenty-two. Thirty-five eight** (35,800 feet altitude).

	ZUHI	National Two Two. Roger. Er, were you able to raise your company (on your special National Airlines radio frequency)?
	NA22	We got him. He's lost number 3 engine and he's, er, . . . er, get him on (radio frequency) one twenty-one five. He's transmitting on "emergency", er, er, frequency now.
	ZUHI	National Twenty-two. Roger. Thank you very much.
	NA22	Gettin' his aircraft down to ten thousand feet. He's making an emergency descent.
	NA22	His aircraft is, er, shaking pretty badly. He's unable to shut the engine down. He's making an emergency descent to ten thousand (feet). Did you get that Centre?
	ZUHI	Yes, sir. Got that. Thank you.
	NA22	He's still on (radio frequency) one twenty-one five.
	ZUHI	OK. Thank you.
1644:20	ALWA	Seventeen?
	ALAP	Yeah. Why don't you get another aircraft to talk to him? Albuquerque looks his closest bet (at somewhere to land). We got good rescue equipment you might tell him.
	ALWA	Yeah. Well we don't know, er . . . they're . . . They're not talking (to) high altitude . . . working him. They just pointed him out to me.
	ALAP	We're getting a seventy-seven hundred squawk now (an emergency signal on their identification transponders), er, it's fifty miles west of the VORTAC (the radio beacon) at thirteen five (13,500 feet altitude).
	ALWA	At thirteen five? He's down that low?
	ALAP	Er, thirteen four, now.
	ALWA	Thirteen four?
	ALAP	OK. He's about five miles south of vector twelve (another radio beacon) and we're gonna have to get him . . .
1644:40	*NA22*	*OK. Have you got an intended point of landing?*
	NA27	*I prefer Albuquerque right now. That's all I can do. We've got her slowed down, er, pretty good and, er, I'll give you an estimate in a few minutes.*
	NA22	*OK. I've already contacted the Centre for ya.*

		I advised them you're making an emergency descent to ten thousand feet. I advised them of your condition. We'll keep advised with the Centre, eh? In touch with the Centre and in touch with you. Standby (on frequency) *one twenty-five one five. Don't shift frequencies.*
	NA27	*OK. Give us a vector* (a heading towards the airport), *also we need it immediately. We're getting no navigational facilities* (on our radio receivers), *right now.*
	NA22	*OK. A vector to where?*
	NA27	*Albuquerque Airport.*
	NA22	*OK.*
1644:40	**ZUHI**	**Seventeen?**
	ALWA	**Alright then. You got that** (earlier information) **alright?**
	ALAP	**OK. If you can talk to him, why not send him over** (have him change frequencies) **to** (the) **Approach** (Controller) **on, er, either** (radio frequencies) **one, er, two three point niner, or one two three point one – if you can ever get a hold of him, or get another aircraft to talk to him.**
	ALWA	**Alright.**
	ZUHI	**Hey, seventeen** (ALWA), **while you get there, I've got that, er, National Twenty-seven. He was broadcasting on** (radio frequency) **one twenty-one five. He's going down to ten thousand feet now.**
	ALWA	**One twenty-one five?**
	ZUHI	**Yeah. He's lost number 3 engine. He's coming back to Albuquerque.**
	ALWA	**Alright, then.**
1645:10	*ALAP*	*National Twenty-seven, this is Albuquerque Approach Control. I am reading you loud and clear. How do you hear me?*
	NA27	*OK. Loud and clear Albuquerque. Go ahead.*
	ALAP	*National Twenty-seven. Fly* (on a) *heading of zero eight zero* (degrees). *Descend at your discretion for vectors* (directions for flight, turning) *left. Runway two six* (is the one we suggest you land on). *Wind, two nine zero degrees at two zero.* (This indicates that there was a strong wind blowing from the west north west.) *Altimeter* (setting) *three zero zero one* (inches of mercury on the barometric altimeter).

	NA27	*OK. Thirty oh one* (is a check of the altimeter setting), *and two eight zero* (degrees) *heading.*
	ALAP	*Zero eight zero.*
	NA27	*OK. What heading? Two eight zero?*
	ALAP	*Heading zero eight zero.*
1645:20	**ALWA**	**Do you have VHF** (very high frequency radio) **guard, er, capability?**
	ALAP	**Yeah. We're talking to him.**
	ALWA	**OK. Thank you.**
1645:30	**NA22**	**Centre. This is National Twenty-two. Do you read?**
	ZUHI	**National Twenty-two. Read you loud and clear.**
	NA22	**Albuquerque Centre has got him** (is speaking to National Twenty-Seven) **on** (frequency) **one twenty-one five guard frequency, and, er, they seem to be OK. He has asked him to give them a vector.**
	ZUHI	**OK. Very good. Thank you very much sir.**
1645:55	*ALAP*	*National Twenty-seven. You're forty-eight miles southwest of Albuquerque Airport.*
	NA27	*Say again?*
	ALAP	*Your position is four eight miles west of the Albuquerque Airport.*
	NA27	*Got it. Fine.*
1646:05	**ALWA**	**Seventeen?**
	ALHI	**Er, seventeen. Have you tried to contact that guy on** (frequency) **one twenty-one five?**
	ALWA	**No. Approach Control is working him.**
	ALHI	**Oh. Approach Control is working him.**
	ALWA	**That's what they told me. Er, just a while ago they said they were talking to the National.**
	ALHI	**Oh. Oh, great. OK.**
	ALWA	**And, er, they said altitude readout said he was at thirteen three, thirteen four** (13,300 or 13,400 feet altitude).
	ALHI	**OK. And that, er, I heard the part about the number 3** (engine) **out** (of action) **and he was landing at Albuquerque.**
	ALWA	**OK.**
	ALHI	**OK. Thank you.**
1646:25	*ALAP*	*National Twenty-seven. Turn left to a heading of zero seven zero. I show you tracking a heading of about a hundred degrees.*
	NA27	*OK. We're just doing the best we can. If we get it within twenty or thirty degrees that will be fine for us right now.*

	ALAP	*Fine.*
	NA27	*And we're trying to get up here* (for a landing at Albuquerque).
	ALAP	*OK. When you have a chance, give me the, er, number of people, er, on board. And the amount of fuel.*
	NA27	*OK. I'll get it worked out for you here in a minute.*
1646:35	**ALWA**	**Approach. Line two. Just want to verify once again, you are talking to the National.**
	ALAP	**OK. What do you want to verify?**
	ALWA	**That you are talking to him.**
	ALAP	**Yes, we are.**
	ALWA	**Alright.**
	ALAP	**We're vectoring** (leading) **him to the airport.**
	ALWA	**Alright. I'll leave you alone.**
1647:00	**ALHI**	**Approach?**
	ALAP	**Go ahead.**
	ALHI	**Are you working that National Twenty-seven?**
	ALAP	**Yeah. I've got him. He's eleven six** (11,600 feet altitude), **and I'll work him in.**
	ALHI	**OK. Fine. I wasn't sure.**
1647:55	*ALAP*	*National Twenty-seven. You've got traffic* (another aircraft) *at eleven o'clock* (ahead of you and slightly to your left), *five miles* (distance), *westbound. Indicates VFR below ten thousand.* (This means that there is clear weather in which visual flight may be conducted below 10,000 feet.)
	NA27	*OK. We're at one one thousand* (11,000 feet) *and let us know if he* (the other aircraft) *get's real close, then we'll make a turn.*
	ALAP	*OK. He'll pass to your left, and no problems.*
	NA27	*OK.*
1648:15	*NA22*	*Er. Twenty-seven from Twenty-two.*
	NA27	*I don't have time for you right now buddy. Er, Centre, we'd like to go to another frequency and get off guard.*
	ALAP	*National Twenty-seven.* (Change frequency to) *One twenty-one point one. Go ahead.*
	ALAP	*One twenty-one point one. National Twenty-two. You can come on this frequency also.*
	NA27	*National Twenty-seven.* (The new radio frequency is) *one twenty-one point one. Go ahead?*

	ALAP	*National Twenty-seven is loud and clear. National Twenty-two are you on this frequency?*
	NA22	*Roger. We're on. Twenty-two. Just want to tell Twenty-seven we've notified the company* (National Airlines of the troubles) *so don't bother with it.*
	NA27	*Well don't worry. We won't bother about that.*
	NA22	*OK. Well, I took care of it.*
1648:50	*NA27*	*Well thank you buddy.*
1649:00	*ALAP*	*National Twenty-seven. You are clear of traffic* (you are clear of the other aircraft). *You can descend at your discretion to eight thousand* (8,000 feet altitude).
	NA27	*OK. Understand pilot's discretion down to eight* (thousand feet). *Would you at least advise or verify the altimeter* (setting) *again?*
	ALAP	*Altimeter* (setting is) *three zero two* (inches of mercury on the barometric altimeter).
	NA27	*OK. And understand we're going to be cleared into runway, er, two six. Is that correct?*
	ALAP	*That is correct. The wind is two eight zero degrees at one five* (it is a wind of fifteen knots from the west northwest); *and it's* (the runway is) *thirteen thousand feet long.*
1649:20	*NA27*	*OK.*
1649:45	*ALAP*	*National Twenty-seven. The field elevation* (the runway height above sea level) *is fifty-three fifty-two* (5,352 feet).
	NA27	*Twenty-seven. Roger.*
1650:55	*ALAP*	*National Twenty-seven. Turn left to a heading of zero five zero* (degrees).
	NA27	*Zero five zero* (degrees) *heading.*
1651:10	*NA27*	*And, er, National Twenty-seven. We've definitely declared an emergency. We want all your gear* (all the emergency equipment) *out there. Our cabin is filled with smoke. We're losing pressure with the number 1 hydraulic system, and, er, we want everything there* (all the emergency services). *We want a straight in, as quick as we can.*
	ALAP	*National Twenty-seven. Roger. Runway three five is available, however it's ten thousand feet long and I'd recommend runway two six, with a left base leg entry* (in order to get to it you would fly down past the airport and turn back to land).
	NA27	*Twenty-six.*
	NA27	*And we'd like to know the surface wind.*

	ALAP	*The wind is two seven zero degrees at one five knots.*
	NA27	*OK.*
	ALAP	*You know how many people are on board yet?*
1651:55	*NA27*	*Stand by.*
1652:10	*NA27*	*A hundred and fifteen on board.*
	ALAP	*National Twenty-seven. Say again?*
	NA27	*We've got a hundred and fifteen on board.*
1652:15	*ALAP*	*Thank you.*
1652:20	*NA27*	*And say the range of the field for us please* (tell us how far we are from the airport.)
	ALAP	*OK. You're sixteen* (miles) *southwest of the airport now.*
	NA27	*OK. And we're out of ninety-two hundred* (9,200 feet altitude descending) *and, just coasting on down.*
1652:30	*ALAP*	*OK.*
1652:45	*ALAP*	*National Twenty-seven. You should be able to see the airport now. We've got the runway two six lights on. Let me know when you have them in sight, at twelve o'clock, fourteen miles* (they are straight in front of you at fourteen miles distance).
1653:15	*ALAP*	*National Twenty-seven. Descend and maintain seven thousand five hundred* (7,500 feet altitude). *You're ten* (miles) *southwest of the airport.*
1653:20	*NA27*	*Down to seventy-five hundred. OK.*
1653:30	*ALAP*	*National Twenty-seven. There's mountains ten* (miles) *east of Albuquerque. You'll be making a five mile base leg* (a turn five miles from the runway, back on to the runway heading) *to two six. Nor more, er? No further east than that?*
1653:40	*NA27*	*OK. Twenty-seven.*
1653:50	*NA27*	*And be advised, er, we'll probably be evacuating personnel when we get on the deck.* (There'll probably be an emergency evacuation once we've landed and stopped).
	ALAP	*How much fuel* (do you have) *on board?*
	NA27	*Say again?*
	ALAP	*How much fuel on board?*
	NA27	*We've got about thirty-five to forty thousand* (pounds weight of fuel).
1654:05	*ALAP*	*OK.*
1654:10	*NA27*	*OK. Twenty-seven. Give us the field position right now.*

	ALAP	*OK. You're now six* (miles) *southwest of the airport on a downwind.*
	NA27	*Six southwest. OK.*
	ALAP	*I'll go ahead and keep you* (on this frequency) *'till final approach.*
1654:25	NA27	*OK. We'd appreciate that.*
1654:55	ALAP	*National Twenty-seven. If you lose, er, radio* (contact) *with me, plan to turn left, heading three six zero* (degrees) *in the next two miles, for your left base* (turn) *to* (runway) *two six.*
1655:00	NA27	*OK.*
1655:15	ALAP	*National Twenty-seven. Turn left now on to heading three five zero* (degrees).
1655:20	NA27	*OK. Three five zero* (degrees).
1655:55	ALAP	*National Twenty-seven. Turn further left heading three two zero* (degrees).
	NA27	*Testing . . . one . . . two . . .*
	ALAP	*Go ahead, National Twenty-seven. Descend to* (landing) *pattern altitude now. Er, seven thousand now* (descend to 7,000 feet now).
	ALAP	*Heading* (now) *three two zero* (degrees).
1656:25	ALAP	*National Twenty-seven. Turn left to a heading of two six zero* (degrees). *You're turning in on a five mile final* (approach) *now to runway two six.*
1656:50	ALAP	*National Twenty-seven. You're cleared to land on runway two six. The wind is now two seven zero degrees, at one eight* (18 knots).
1657:15	NA27	*And be advised, sir, we with, er, stewardess on board – broken leg, and we don't know what else is damaged right now if you can hear me.*
1657:25	ALAP	*OK.*
1658:00	ALAP	*Wind* (is) *from two seven zero degrees, at one seven* (17 knots).

What happened was, when they were approximately 35 miles west of the Socorro radio beacon, the number 3 jet fan assembly disintegrated, causing an immediate and total loss of power on that engine. Pieces of the interior of its workings ripped their way into the fuselage of the DC–10 tearing a number of large holes in its skin and the rapid loss of pressurization inside. One piece of engine blade flew into and smashed a cabin window and the passenger who was seated next to it was sucked out through the hole even though he was asleep with his seat inclined back and the seat-belt loosely done up around his waist. Before he finally left the aircraft, another passenger had tried to save him when he was half-way out the window by grabbing his ankles, but the suction was too great.

However, other parts of the exploding engine's fan blades and discs, also cut through some of the electric wiring inside the aircraft, and incredibly, smashed their way into the casing of the number 1 engine, located under the *left* wing, right over on the other side of the fuselage.

Many things went wrong and were found to be wrong with this machine. For example, the investigators, on checking back through the pilots' log-book entries with reference to the engine that had exploded, found that between 8 August 1973, and 12 September 1973, there had been 15 problems reported about it. On 13 September 1973, it was taken off the aircraft at National Airlines Miami Base for some major repairs. On 23 September 1973, it was placed on the accident jet in position number 3 (under the right wing), and between that time and the near disaster, a further 26 faults were logged by the pilots.

Also, the investigators noted, in the remains of the wrecked engine, that all of the bolts that had held the front covering in place, were outside the tolerances laid down by the manufacturer. On 5 November 1973, 2 days after the accident, McDonnell Douglas sent around an Engineering Dispatch "recommending" an inspection of these engine cowls. National Airlines complied with the "request" and 6 more discrepancies were found in the engines on their ten DC–10s. On 7 November 1973, the Federal Aviation Administration issued an Airworthiness Directive making the earlier Douglas "recommendations" mandatory by law. This time, National found *still further* minor problems in most of their aircraft.

When the disintegration occurred near Albuquerque, the loss of pressurization in the passenger cabin manifested itself as a loud explosion. It filled with a blue-grey smoke which irritated the throats and caused a burning sensation in people's eyes. The lift, that operated between the passenger cabin and the galley situated underneath, dropped down to the lower position, and its doors burst open inwards. The lights in the cabin suddenly went very bright, and the "No Smoking" signs came on. At about this time several passengers lost consciousness for a short period.

The oxygen masks were not immediately available. Some attendants and passengers forced open the oxygen compartment doors in the seatbacks in front of them, in order to gain access to the gas, while the head stewardess telephoned the cockpit and asked the flight engineer to manually eject the masks. Most of them then fell down, except in one large section of the cabin where their operation was affected by the dislocation of the electricity supplies.

Two stewardesses were in the lower galley when the decompression occurred. This is separated from the forward cargo compartment by two sliding doors. Two major penetrations of the aircraft's hull occurred in this cargo compartment by flying pieces of engine blades. These holes caused an extensive rush of air which sucked the lift down from the upstairs position and also caused the doors of the compartments in which the meal carts were stored, to fly open, with the trolleys falling out and bouncing around the area of the galley.

As the oxygen masks did not immediately drop, the two stewardesses in the lower galley tried to reach portable oxygen bottles which were kept nearby, but they lost consciousness before they could get to them.

Upstairs in the passenger cabin, four large ceiling panels in the centre of the rear compartment fell down on top of people's heads. But an enormous amount of confusion and panic occurred because the oxygen mask doors would not rapidly open.

This accident and the oxygen problems, made public an alarming situation that was allowed to be incorporated in all DC–10s and Lockheed 1011 TriStars. Most other planes have an emergency oxygen system (most British Airways Tridents and BAC 1–11s incidentally, have *no* oxygen installations) that is supplied by pipes from large and centralized cylinders. In the DC–10 and TriStar, the F.A.A. has permitted the manufacturers to incorporate a "chemical reaction" method, with a small metal flask in each mask compartment. With this system, the doors to these compartments are supposed to open immediately and automatically on loss of pressurization, so long as the aircraft is above 10,000 feet altitude. When they are opened and the masks are presented to the passengers, the first person to remove his from each compartment and pull the "supply" cord, triggers off an explosion inside this flask containing the chemicals, and within about ten seconds, the heat from this explosion starts a fire, and a reaction gives off small amounts of oxygen.

(For the technically minded, this is what happens: Sodium Chlorate, when heated to 487° *Centigrade*, decomposes and yields approximately 45% of its weight as gaseous oxygen. The source of heat to start this reaction is firstly an explosive substance – Cordite, for instance – which then causes the following: $2FE+O_2$ – gives – $2FeO+$Heat. $2NaClO_3+$Heat – gives – $2NaCl+3O_2$.)

It all sounds so simple (it is also so economical for both the aircraft manufacturer and airline operator), except for a few minor points which quite evidently had been conveniently overlooked by the certification authorities, including the F.A.A.

Firstly, there is no effective heat-shield around the cylinder in which the explosion and flame is causing the chemicals to react. Between one and two minutes after the activation of the firing mechanism, the cylinder reaches about 260° *Centigrade* (500°F), and this heat penetrates the head-rest of the person seated in front, in which the oxygen system is mounted. If the cylinder is too close to the head-rest, it could scorch it, though at the time of writing one does not know whether there have actually been any fires caused by this.

Secondly, if the passenger pulls the activation lanyard too sharply, not only will the explosion charge firing mechanism work, but the whole chemical cylinder could be easily pulled out of its flimsy mounts, and therefore we have a metal bottle rapidly reaching anything up to 260° *Centigrade*, rolling around the floor, or landing on someone's lap. In the Albuquerque accident, there were a number of cases of passengers' hands being burned because they innocently tried to either pick up or move these cylinders after they had fallen out of the containers and when they were at those temperatures.

Thirdly, once the chemical reaction has been started, there is no way of stopping it or turning the oxygen supply off. The unit continues developing the gas until it has run its full course and the flame extinguishes itself – about 15 minutes after being fired.

This is a highly dangerous situation that the F.A.A. and the other civil air departments which "automatically certified" these aircraft (namely the British, Spanish, Dutch, German, Zairois, Gabonese, Belgians, Swiss, New Zealanders and Turks) permitted McDonnell Douglas and the Lockheed Aircraft Corporations to use these oxygen systems. The fact is they were much cheaper to install, easier to maintain, and lighter in overall weight.

Using such a system as is in the DC–10 and Lockheed TriStar, is inviting fires in these aircraft; the possibility of a fault occurring in the explosive substances which would cause panic; the possibility of an explosion with an unwanted highly oxygenated atmosphere that would last for minutes after the emergency was over; not to mention the scalds and skin burns and the smell of smouldering upholstry that would be evidenced either if the cylinder was pulled loose accidentally by the unknowledgeable passenger, whose consciousness at that time, depended on the speed and agility of activating the unit and getting the mask on his face. But what would happen if the aircraft flew through turbulence and was bumped around causing these cylinders to break loose from their flimsy holds and fly through the air as dangerous, boiling hot projectiles?

The oxygen system in the DC–10 and Lockheed TriStar are as dangerous as the DC–10's floor design, the cargo hold door design, and many other McDonnell Douglas money-saving features.

Butterfield was in charge at the time of this Albuquerque accident, but in a similar fashion to Shaffer, he did not adequately look after passenger interests. Butterfield failed to order prompt correction of a large number of slowly-being-revealed design faults in not only the DC–10, but in other F.A.A. certificated aircraft.

So now that we know that the Albuquerque accident and near major catastrophe also brought glaringly to light either badly designed or badly maintained engine cowlings (referring back to the American Airlines lecture on how the DC–10 came into being as a design partnership between American, McDonnell Douglas and others (unnamed)), what would happen if a repetition was to occur? Surely passengers could immediately point the finger at Butterfield's incompetence, or inability?

On Monday, 8 July 1974, practically exactly the same thing recurred. The engine cowling from the number 1 engine (the one under the left wing) came off a National Airlines DC–10 in flight, smashed into the leading edge of the left wing, bounced over the top of it, collided with the rear part of the passenger fuselage before being sucked into and disintegrating the rear number 2 engine.

MOREOVER, the aircraft involved in the near catastrophic disaster not far from Tampa, Florida, on 8 July 1974, *was the same one* as was involved in the Albuquerque accident on 3 November 1973, N60 NA.

Now surely, one is entitled to ask, in those eight months, National Airlines maintenance department should have improved? Surely, Butterfield should have taken action to make sure that the Albuquerque incident could not be repeated? Surely the F.A.A. should have seen that this money-saving feature, dreamed up by the consortia involved in the DC–10 design, was certainly saving money, but even if it had not yet caused an accident evidently sooner or later might do so, at the possible cost of hundreds of passengers' lives?

There is only one answer to correcting these problems – decertificate the DC–10 and correct all the design problems. If this is not done, passengers should be made well aware that their lives are in danger the moment they step inside these machines.

CHAPTER 40

At 1542 hours on 17 December 1973, Flight 933, an Iberia Air Lines DC–10, was making its landing approach at Boston, Logan Airport. The field was covered in a blanket of low cloud with its base at an altitude of 300 feet above the runway. As the machine descended under the cloud, Captain Jesus Caleron Gaztelu felt that they were not properly aligned with the strip and so he discontinued with the automatic approach and made the adjustments himself. As he was doing this, the aircraft descended below the minimum glide path and both Co-pilot Alfredo Perez Vega, and Flight Engineer Celodonio Martin Santes, drew his attention to their low altitude.

Gaztelu tried to pull the nose of the DC–10 up, and he also moved the throttles forward, but it was too late to save the aircraft colliding with the lead-up approach lights quite some distance before the start of the runway. The machine rebounded and then came down on to the runway and skidded for about 3,000 feet on its belly, partially on the concrete surface, but most of the time off to one side. Before it came to rest with all its landing gears either ripped off or collapsed, the fuselage snapped apart just forward of the rear passenger doors. A fire broke out under the left wing and rapidly spread to the tail section. The fire was fed from a ruptured left-wing fuel tank that was spewing kerosene all over the surrounding area. However, using over 1,000 gallons of a high density foam, the fire-fighters were able to fairly quickly bring it under control.

As soon as the aircraft came to a stop, the 167 people on board started an emergency evacuation, but of the eight exits built into the DC–10 only four of them could be used. The number 1 Right exit could not be operated because of a fault in the mechanism. The number 3 Left exit was not opened because the wing fire was directly outside it. And both of the number 4 exits at the back were made useless by the utter devastation that existed inside.

The DC–10's floor collapsed in exactly the same area that had caused the near-disastrous catastrophe over Windsor, Ontario, Canada on 12 June 1972. This time it was displaced upwards for a distance of about three feet, causing multiple failures of seat-tracks and seat restraint mechanisms. However, no seats were completely detached from the floor.

In this rear section of the cabin, five occupants were trapped because the passenger aisles were completely blocked by wreckage. Four of these people, two cabin staff and two passengers, had to evacuate the aircraft through a break in the ceiling and then, after walking towards one of the extremities on top of the machine, had to slide or jump off it to the ground. One of these passengers received a fracture of the left ankle and a compression fracture of the second lumbar vertebra. The other suffered a fracture of the right ankle. One of the crewmembers who had to jump, fractured her left pubic ramus as well as receiving a fracture of the acetabulum. The fifth person in the rear of the DC–10 had to be rescued later by the crew.

Of the people on board, three received serious injuries; thirteen suffered minor injuries; and the remainder got away with their lives.

The interior of the cabin was a complete shambles. All along the aircraft, ceiling panels had dropped down, some on to passengers' heads; the floor failed under seat 27D and this allowed dirt and dust to be sprayed up into people's faces; and there was this massive floor failure under the last three rows of seats, with the entire ceiling over the top being completely destroyed.

A few days after the accident, the investigators received a most interesting letter from a Richard T. Campbell of Reston, Virginia, and to give air passengers an idea as to what happens after disasters such as this, some quotes from it may be appropriate.

> Those of us who were sitting in the back of the airplane were in the area where the fuselage broke open and therefore we were covered in mud, rocks and debris. The debris being the whole interior trim of the fuselage. The way the overhead panels came down, the luggage racks fell off the walls and all of the other bric-a-brac inside the airplane came loose, it must have been installed with used chewing gum. It was extremely fortunate that we all got out without many more being physically injured.
>
> We had been through a crash where the bits and pieces and racks and panels had flown through the aircraft, at 1542 hours in the afternoon, on Runway 33 at Boston. And it wasn't until nearly 1745, that us survivors arrived at a warm and secure environment at a hotel *on the Boston airport facility*. That's over two hours later.
>
> We flew in on Iberia Air Lines who are represented in Boston by the

Irish Airlines; to the Boston Airport, run by the Boston Airport Authority, and therefore it was only out of the smart thinking and good hearts of Eastern Air Lines that some passengers were given blankets to keep themselves warm.

The airport reaction to get the fire engines to put out and protect the multi-million-dollar airplane and any people in it, was excellent, and they certainly should be commended on how quick they accomplished it. They should also be commended that they had ambulances on the scene of the crash equally as fast as the fire engines, to handle the dead and seriously injured.

But they had no program, and I repeat, *no program*, for the physically uninjured survivors. There were 167 people on this airplane. Sixteen injured subtracted from 167 leaves 151 people in shirt sleeves, in some cases without shoes, wandering around in a freezing rain with absolutely no program as to what to do with them.

Many minutes after we had been standing in the freezing rain, some airport delivery vans began to arrive. It took the fire engines about three minutes to get there. It took these delivery vans about 20 minutes.

We loaded ourselves into a van, sitting on the floor and on top of each other, and any other way we could get inside out of the cold and wet. We then sat there because the driver had no idea where to take us.

We were finally driven to the fire station and literally herded in there like cattle. When we arrived at the station, which was between 1600 and 1615 hours, there were absolutely no provisions there for survivors. There was no program as to how to get the survivors off the runway. There was no provision as to where to take them to get them off the runway. When the improvization used got us off the runway, it was to the fire station where they had no provisions for survivors.

And lo and behold, who showed up first – not the airport's emergency people – but the Boston television reporters with their one million watt lamps, cameras and interviewers who immediately flooded us with their lights and demanded answers to their questions.

As you might imagine, we, and the other hundred plus people, were in no condition to be interrogated by overbearing television people. Those of us who weren't injured, were scared to death about those of our families who were injured, and we were all cold and wet and needed some peace and quiet.

In the long run, the television newsmen with their bright lights and the newspapermen with their demanding questions got absolutely useless information. They kept asking us what happened, what caused it? We were just riding on an airplane that was about to land. How the hell do we know what happened. We crashed – that's all we know.

Eventually, they showed up with buses to take us from the fire station to the air terminal, where we had the "privelege" of standing in a line. With my daughter (who was 16-years old) holding up fantastically under the situation, my wife who was in pain and my six-years old son who was completely bewildered and in shock, and I who was forcing myself to "keep going", we got into line for Passport Control. When we got to the front of the line, the officer asked us for our Passports. I told him they were somewhere on the aircraft. I told him all we did was to get ourselves off. He then told me that we would have to get into another line. Before leaving him we were handed a form with a bunch of very small boxes on it which under normal condition I could have filled out, but because of the condition I was in, I kept writing larger than the size of the boxes. When it was finally filled out, I was told to go and see a man in an office. On arriving there he looked at the form, stamped it and told us to go through a doorway. So my family and I went through this door and we found ourselves out in the general terminal.

We were covered with mud, and all we could see was nothing but the normal airline check-in counters. We were again accosted by newspaper and radio reporters with recording machines, wanting us to tell them what happened.

We wandered around the terminal, along with the 150-odd other people, some of whom were fortunate enough to have Eastern Air Lines blankets around them, for quite a few minutes. Then an announcement was made on the public address system that all passengers on Iberia Flight 933, should go to a certain boarding Gate. When we arrived there, they couldn't understand what all these people, covered with mud, were doing at that Gate number, so we went back to the lobby and discovered that they had meant the door opposite the Gate.

Now we have some 150 people – many of them soaking wet from standing out on the runway – going outside the terminal on to a roadway in front of the building, with the temperature at about 20°F, and piling into buses. Up to this time, nobody – absolutely nobody – had given us any information as to what might be happening to us, where we could go, what we should do or any other guidance or comfort. And I would like to point out, that I and many other of the passengers don't need comfort under normal travel conditions, but it sure as hell wouldn't hurt to have a little under these kinds of circumstances.

We were then taken to the Hilton Hotel which is on the property of Boston Airport. When we arrived there, we were herded into a large conference room, with a big table with soft drinks and coffee on it. This might sound great to tea-totalling adults, but I tell you, those of us who now and then have a drink, sure as hell wouldn't have minded having one right then.

There were many children and quite a few babies on this flight. It was approaching Christmas-time and many American families stationed in Spain, were coming home for Christmas. There was no milk for the babies. Finally, after almost half-an-hour, the hotel found some milk, but nobody could come up with a baby bottle. And it's very difficult to get a few months' old baby to drink milk out of a glass. Some diapers were also badly needed.

The people who were in shock, such as my wife and many others, were stretched out on the floor of this conference room provided by the hotel, and we stayed there for another hour or more, continually asking questions like; What's going to happen to us? Where are we going to be put? Where are we supposed to go? We got no answers. Finally, at about 2000 hours, we were given a room in the hotel.

To conclude this side-tracking, one would just like to add that Irish Airlines (or Aer Lingus as they are more correctly known), and Iberia are members of the international fare price-fixing cartel I.A.T.A. (the International Air Transport Association), whose motto is: "Quality In Service". It should also be pointed out that besides forcing passengers to pay extortionate fares, I.A.T.A. members are the ones involved in the biggest – and the most – crashes, yet they carry about half of the world's air passengers.

The interesting point about this Boston Airport, Iberia Air Lines, DC–10 crash, is that the aircraft broke apart approximately between fuselage stations 1530 and 1850. It was between these points that the floor disintegrated upwards. In the Windsor, Ontario accident, the floor collapsed downwards between fuselage stations 1801 and 1921. This means that the exact same floor section collapsed in both crashes. At the time of writing, there is no solid explanation *why* the Boston DC–10's floor disintegrated in an upward direction. Neither is there an explanation as to *why* it was three feet higher than normal alignment. This crash was, of course, still under investigation.

CHAPTER 41

By 26 December 1973, the stewardesses versus officialdom fight was starting to "hot up", in relation to these new-fangled devices known as downstairs galleys, that had first appeared with the DC–10s and Lockheed TriStars, and were being incorporated into some Boeing 747s, presumably to create even more passenger space.

Prior to their introduction, the Federal Aviation Administration had, quite evidently, very little knowledge about the problems that would arise. They had permitted them to be installed in the hope that not too much would go wrong, and that they could then start to legislate as experience continued. But of course, as soon as they were introduced, the airlines realized that they were on to an extra ten to fifteen passenger-seat bonanza, and were therefore effecting "quiet modifications" *en masse*.

The F.A.A. finally circulated an Air Carrier Operations Bulletin, which detailed the difficulties as were known:

SUBJECT: Problems associated with the introduction of lower deck galleys in the wide-bodied jets.

With the development of the jumbo jets came the lower deck galleys. Although a lot of work went into the design and concept of having a galley in the lower section of the aircraft, very little thought was given to policies and procedures required for the new environment. Consequently, isolated problems appeared, and not being corrected, soon became common. Some of these problems eventually resulted in injuries to cabin attendants.

The following problems or discrepancies in procedures are listed, discussed and recommendations made to help the Principal Operations Inspectors to identify problem areas in his assigned air carriers' procedures, and to help him take remedial actions.

1. *Problem:* Passengers having access to lower galleys.

Discussion: We have received numerous complaints that passengers have been going down into the lower deck galleys. They have either let themselves down in the lifts or have actually been taken down by flight attendants. There is no justification for passengers to be in the galleys where they would interfere with the flight attendant duties In addition, there is no provision for oxygen masks and safety belts for extra persons.

Recommendation: That the air carriers incorporate into their manuals, prohibition against passengers being allowed on the lower deck and so placard each lift.

2. *Problem:* Excessive numbers of cabin attendants on the lower deck.

Discussion: Some air carriers have conducted training and/or instructions on the lower deck during a flight with 5 or 6 flight attendant trainees. They have also allowed "deadheading" crewmembers to occupy or visit the lower deck in flight.

Recommendation: Only two flight attendants should be allowed on the lower deck at any time during the flight due to the number of oxygen masks and seat-belts. One additional person may be allowed for instruction, evaluation, or inspection duties.

3. *Problem:* Insufficient communications systems.

Discussion: It is very difficult to hear the public address system announcements in the lower galleys. The reason for this is aerodynamic noise and noise emitted by the air compressors which are located nearby. The flight attendant working in the lower galley cannot hear a captain's warning of clear air turbulence, or a 10-minute warning of beginning of descent. In addition, there have been reports of numerous failures of the interconnecting telephone systems. Flight attendants in the galleys have to rely on others to pass on the warnings.

Recommendation: Increase the volume of the public address system in the lower galleys. Incorporate flight attendant procedures regardless if the communication is operable or not, to ensure that all warnings are passed to and acknowledged by the persons in the lower galleys.

4. *Problem:* Inadequate emergency equipment, procedures and training for combatting fires in the lower deck galleys.

Discussion: Recent *en route* inspections by headquarters personnel revealed a non-conformity throughout the aviation industry in equipment and training for flight attendants who have to work in the lower deck galleys. In one instance, there were adequate procedures and the attendant knew them well. However, the oxygen equipment that the flight attendant would have to wear, did not include a full-face oxygen mask. Needless to say, if there was a very smoky condition, the flight attendant could not see in order to complete her emergency procedures. In another case, there was a full-face oxygen mask in the lower compartment, but the flight attendant had no idea how to combat the fire. When asked how she would handle the emergency, she responded: "I would call the flight engineer". In no case were there any procedures for removing injured attendants from the lower deck.

Recommendation: Principal Operations Inspectors should review their assigned air carrier emergency procedures pertaining to the lower deck, to ensure that procedures and equipment are adequate and that training includes removal of injured flight attendants in the lower

section. They should take into consideration the advisability of having the flight engineer away from his duty station during an emergency.
5. *Problem:* Inoperable lifts' procedures.

Discussion: The lift operations between the different aircraft (DC–10, L–1011, and B–747) are not compatible. In one instance, if the personnel lift is inoperable, no food service will be performed during the flight. In another instance, the flight attendants may go down but the service is limited to the number of carts that can be delivered and stowed in the passenger cabin. Additional problems arise when lifts become inoperable during the flight.

Recommendation: Inspectors shall review assigned air carrier flight attendants' procedures to assure that there are adequate instructions throughout their system on dispatching aircraft with personnel or cart lifts inoperable. Additionally, they should have a procedure in the event these lifts become inoperable during the flight. Further assurance should be sought to determine that each carrier is keeping flight attendants informed on conditions and procedures which are set forth in the instructions that affect them.
6. *Problem:* Inadequate stowage of carts.

Discussion: Some airlines do not have a sufficient number of mushrooms (tie-down devices) in the cabin in order that each serving cart may be "tied down" in the event of turbulence. These carts can weigh up to 250 lb, and should be anchored when not being transferred to or from the cart lifts.

Recommendation: Inspectors should assure that the air carrier procedures and training includes instructions to the flight attendants that all carts must be properly stowed for take-offs, landings and whenever they are not being moved from one location to another.
7. *Problem:* Retractable mushrooms (tie-downs) inoperable.

Discussion: The retractable mushrooms in the lower galley have been found inoperable. They are either jammed in the "down" position or when lifted to the "up" position will fall back down when the cart is placed over it. The automatic brakes on the cart are insufficient to keep it from moving about during take-offs and landings.

Recommendation: Inspection should be conducted periodically in the lower galley to see that the mushrooms are operable and that the procedures requiring each cart to be tied down or stowed are adhered to.

The above problems require immediate action. Inspectors should make a concentrated effort to review their assigned carriers' procedures. To ensure that the above recommendations are implemented, this area should be emphasized during *en route* inspections.

But still, yet again, no enforcible rules regarding safety or rather the lack of it, have been introduced because of these instances cited. In reality, the arguments regarding safety and the lower deck galley started in February 1973. In a letter dated 28 March 1973, the F.A.A. acknowledged that there were major safety problems in this area in convertible freighter DC–10s and Pacific Southwest Airlines Lockheed TriStars. These aircraft are not equipped with an emergency egress from the lower galley through a hatchway in the forward passenger compartment area, as most other machines are. But hatch, or no hatch, emergency escape from the downstairs level is a major problem, practically unthought of prior to this correspondence.

In normal passenger carrying aircraft, the way into and out of the lower galley as we have seen, is via a "personnel lift" located directly alongside the cart lift, the latter of which is there to transport meal trolleys from the galley to the passenger cabin, and vice versa. If the personnel lift becomes inoperative, the hostesses in the lower galley would have to squeeze themselves into the cart lift. If this lift is inoperable, food and carts would have to be transported in the personnel lift. If both are inoperable, and the lifts are at the lower level, there is an emergency ladder inside the personnel lift shaft up which the hostesses could climb. But if the lift is stopped at the upper level, the floor of it does not permit this ladder to be used, and so the only means of escape is up another emergency ladder and through a hatchway in the passenger cabin aisle floor, towards the centre of the first-class compartment.

As may be recalled, in the American Airlines DC–10 accident near Windsor, Ontario, on 12 June 1972, this hatch cover flew up and hit a woman passenger in the face.

However, a letter from the Air Line Pilots' Association, dated 15 March 1974, reveals some other mishaps:

> We are aware that there are two known incidents of persons falling into the escape hatches and if the normal procedures are continued, more serious injuries are likely to result. A flight attendant fell into the lower compartment in November 1973, during the investigation of National Airlines DC–10 accident near Albuquerque; and a passenger stepped into an open hatch on a Boeing 747 on 30 April 1972, on American Airlines, during a meal service when the lifts were inoperative. The person was injured even though the flight service director did catch the passenger's foot as he began to fall through the hatch. We feel that notification to carriers and flight attendants is necessary to restrict the use of the escape hatch or the cart lift for anything other than that for which they were designed.

The ridiculousness of this lower galley/lifts situation is, surely, something that no one had thought about. They were evidently installed as means of increasing passenger fare revenue, without any of the fault contingencies being taken into account. The airlines, through their own fare price-fixing cartel I.A.T.A. have already reduced most passenger services to a level equal to second-class rail travel, yet it must be pointed out that the latter is a far cheaper and safer proposition, and this may be readily seen in a study of I.A.T.A. members major disasters and accidents.

The cabin staffs point out that in the convertible freighters, however – that is, in aircraft which are carrying half passengers and half freight, or in which all the seats may be rapidly removed to change the machine from one carrying passengers to one that can then carry freight – the cargo itself prohibits the use of the forward floor hatch emergency exit from the downstairs galley, and therefore if the lifts jam at the upper passenger cabin level, the hostesses working down below have no means of escape.

As you know, all wide-bodied aircraft have a manual back-up means of escape for occupants of the lower deck galleys in addition to the normal means of entrance and exit through some of the galley lifts. In addition, the DC–10–10 and –30 provides a means of escape through the cart lift shaft. An important point to remember, however, is that this lift must be in the "down" position if this particular escape method is to be used, (A.L.P.A. reiterated in further correspondence).

As you also know, all areas and compartments within the cabin of the aircraft where occupants are carried, and all cockpits of all aircraft, have a manual means of escape. This alternative way is required by the Federal Air Regulations to provide the level of safety needed to assure escape if necessary.

You are also aware that no Federal Regulation governs the escape provisions of occupants of the lower galley area.

Based on the installation of the escape system designed for all other wide-bodied aircraft, and the lack of regulations governing lower deck galleys, a manual means of escape must be installed in the DC–10 Convertible Freighter comparable to that of other wide-bodied aircraft.

At the time of writing, the argument continues.

CHAPTER 42

THE worst air crash in Turkish civil aviation history (up to that time) occurred on 27 January 1974.

A Fokker Fellowship F–28, a short-haul, small-capacity passenger twin-jet aircraft, crashed moments after taking off from Izmir Airport, on a scheduled flight to Istanbul. Of the 73 persons on board, 63 died instantly, and all the others suffered critical injuries.

Reports at the time said that the aircraft was less than 50 feet above the runway after lifting off, when it crashed back on to the ground and burst into flames. The machine was owned and operated by Turkish Airlines.

Police, who were quickly on the scene, reported that most of the passengers were Turks, with a minor sprinkling of one or two other nationalities.

The accident was the second in January 1974, involving a Fokker Fellowship F–28. The other was on the first of the month in Northern Italy. On that occasion, the Itavia-owned aircraft was coming in for a landing at Turin Airport when it crashed to the ground short of the runway, killing all 43 persons on board. The Fellowship is made by the Fokker Aircraft Company of Amsterdam, and the authority that originally certificated this machine was the same government department for which a person called Schipper worked – the same Schipper, as you may remember, that pointed out to the F.A.A. the dangers in the floor and control cables placement in the DC–10.

Meanwhile, the "letter war" was continuing between the Federal Aviation Administration and the Douglas Aircraft Company. The F.A.A. had originally written to Douglas on 2 February 1973, following some "pestering" letters from the National Transportation Safety Board asking what they were planning to do about the A–72–98 recommendation calling for a modification to be made to the rear passenger cabin floor in the DC–10. The Board, in their investigation into the June 1972,

American Airlines accident, realized that because there was no significant venting to permit rapid internal pressure equalization, the floor could collapse and the flight control cables running through it would then be affected.

As a result of the N.T.S.B.'s recommendations sent to them on 23 June 1972, the F.A.A. via Richard S. Sliff, the chief of the aircraft engineering division in Los Angeles, wrote to the senior engineer at the Douglas Aircraft Company on the subject of: "Explosive Decompression Effects In Jumbo Jet Airplanes". One part of the letter was as follows:

> While we recognize that it may not be feasible to completely design for all aspects of such massive failures, we believe that loss of an airplane because the floor collapsed under such conditions and disrupted vital systems, would be unacceptable.

The reply to this correspondence was signed by a William C. Starloff, the manager of F.A.A. liaison at Douglas, and was dated 15 March 1973:

> 1. Your referenced letter requests that we reassess our DC–10 Jumbo jet design with regard to the sudden pressure loss in any compartment and its effects on the cabin floor and vital aircraft systems. This request is based on the probability of pressure losses due to the loss of non-plug (outward opening) cargo doors or small bomb explosions in passenger or cargo compartments. We feel that the (outward opening) cargo doors and small bomb explosions require different criteria considerations and therefore should be discussed separately.
>
> 2. Douglas believes that the design of the DC–10 (outward opening) cargo door and its latching systems has been properly evaluated and meets the requirements of F.A.R. 25 which requires that pressure cabin floors and partitions be designed for openings not shown to be extremely remote. Since the DC–10 incident (of 12 June 1972), the cargo door latching system and the vent door system has been improved through redesign. The redesigned systems were accepted by the F.A.A. and foreign certification agencies based on the redundancy and safety of these modifications. A review of in-flight service aircraft, including wide-bodied jets, and future aircraft show a considerable increase in floor and fuselage weight if the airframe is designed for a large cargo door opening in flight. In our opinion this is unwarranted due to the excellent safety record of structure in service aircraft. By virtue of our investigation it should be only necessary to show that (outward opening) cargo doors are designed so that an inadvertent failure of the doors is extremely remote because they meet multiple fail-safe requirements.
>
> 3. Douglas believes that it is beyond the scope of the designer to design for all types of small bomb explosions. The request for the aircraft structure to be designed to any criteria regarding small bomb explosions is impractical because it will add weight to the aircraft and

not increase safety. An investigation was conducted for the DC–10–30 aircraft and DC–8–63 aircraft at an altitude of 30,000 feet. It was determined that the DC–10 compartment can withstand a larger hole in the shell than the DC–8 structure for a differential pressure of 6.5 lb per square inch, even though the strength of the DC–8 floor structure is greater (710 lb/ft^2) than the DC–10 (450 lb/ft^2). The larger hole can be sustained in the shell of the DC–10 because the ratio of the upper compartment volume to the lower compartment volume is smaller than the DC–8 and consequently a lower Delta pressure. For the same venting area, the aircraft with the smaller volume in the lower compartment is most critical, since an equal sized hole in the skin will lower the pressure more rapidly in the smaller compartment. For a given hole size in the fuselage skin the DC–8 develops a higher Delta pressure across the floor due to the small lower compartment and slightly larger ratio of upper to lower compartment volumes.

4. Control systems have their own set of requirements such as multiple failure analyses, fault analyses and separation of systems due to engine disc failures. In our judgment, those requirements are adequate to protect control systems under all known conditions.

5. A reassessment of the DC–10 design with regards to the effects on safety for (outward opening) doors and small bomb explosions shows that the present standards and levels of substantiation are adequate.

This letter of course, did not answer too many of the original questions. Moreover it appears to be designed as one that would "flatter you with science".

The air pressure inside an aircraft's fuselage whether it be a DC–10, a DC–8 or roughly anything else, must be maintained within certain restrictive limitations in order that we earthbound humans be not too greatly affected when climbing to and flying at altitude. The higher the machine goes, the less air density outside, and therefore the greater the pressure needed inside the cabin in order to affect us the least.

Aircraft are built to "height limitations"; that is to say, the strength of the fuselage can only withstand so much pressure being built-up inside it in order to keep us in this comfortable environment, without the whole design having to be so strong that it would be uneconomical to get it up into the air. Therefore, the only thing that materially changes in the two aircraft examples (the DC–10 and the DC–8), is the actual area inside the fuselage that has to be pressurized.

Both these machines fly at approximately the same speed. Both fly approximately within the same height (strength) limitations. Both maintain approximately the same "pressurization comfort" for the passengers during the flight. The only major difference is the size. So imagine using a DC–8–63 as a "basic design" and then scaling it up in size to that of the DC–10. The pressure loading inside both aircraft against the outer fuselage is approximately the same; the only difference is that the larger of the two has a greater pressurized area. Therefore as far as the internal pressure is concerned, the two machines are roughly the same.

However, with the larger fuselage of the DC–10 and therefore its greater capacity for carrying payload, the structure of this aircraft, logic tells us, should be stronger than that of the smaller, lighter DC–8. And what do we find in reality? Douglas have admitted in that letter that the floor strength in the DC–10 is just over half that of the DC–8 (450 lb per square foot, as against 710 lb per square foot). But using rough figures, the cabin floor area of the DC–10 is, in relation, about one third greater than that of the smaller aircraft.

Now looking at the aft cargo compartments in both these machines, the DC–8's, with a capacity of approximately 320 cubic feet, has a door 1,584 square inches in size. The DC–10 on the other hand, has a capacity of 870 cubic feet, with a door size of 2,112 square inches. Reducing these to comparable figures, we see that the door in the DC–10's compartment is, in relation, approximately half the size of the door in the DC–8. Theoretically therefore, the air inside the DC–10's hold (imagining that the doors came off both machines at the same altitude) would create a slower rush escaping out of the smaller opening, but there would be a greater area of pressure now on the rear passenger cabin floor which must be added to the original greater weight already being taken by the floor because of its greater size. Yet, as we have seen, the floor strength is about half that of the other aircraft.

It is not only the size of the hole through which the pressure is escaping that has to be taken into account, but also the greater floor area and therefore the heavier weight on it, added to which must be the greater area of pressure being exerted as the air in the compartment underneath escapes.

Douglas's attempted explanation simply does not seem to stand up to scrutiny. What is more, their reply says nothing about the other important question asked by the F.A.A. with regard to the vital control cables that run through the floor beams holding this weaker floor in its position. The F.A.A. must have been fully aware of the fact that of the three jumbo aircraft in passenger carrying services (the Boeing 747, the Lockheed TriStar 1011, and the DC–10), the DC–10 and the TriStar have those controls running *through* the floor. In the Boeing 747, the cables and controls from the cockpit to the tail are routed above the ceiling of the passenger cabin, well out of the way, though Douglas have pointed out that the 747 has hydraulic lines in the floor.

However the F.A.A. appeared to have accepted the Douglas reply *to a certain extent*, as the next piece of correspondence on the subject was written by a A. F. Madayag and R. S. Sliff of their Western Region, just one year later, on 13 February 1974.

Gentlemen,

We believe that a problem may exist on current jumbo jet airplanes and for future designs which warrants assessment of the capability of those transports to withstand relatively large openings in different portions of the pressurized fuselage. In order to explore this area further, there is a need to review the technical basis for your views expressed in your reference letter. Consequently, please advise us of the technical basis of your views on explosive decompression effects, including the following:

(1) Information on the detail design and operation of any (outward opening) doors.
(2) The existing compartment volumes, venting areas (including details of vents) and floor strength.
(3) The variation of floor weight and design pressure differential.
(4) The variation of pressure differential (at cruising altitude) with vent area and unrestricted blow-out area in various compartments.
(5) The variation of weight penalty with additional vent area.
(6) The design details of areas that would have to be modified to provide additional venting.
(7) The location of vital systems in the floor.

We would appreciate your assessment of these data with regard to integrity of jumbo jet configurations as compared to other jet transports and the need for and the cost benefits involved in increasing the level of safety in current and future transports with regard to explosive decompression effects. This information is needed for evaluation for possible changes in standards.

In addition we would appreciate receiving a probability analysis, as would be required of new aircraft by F.A.R. 25.1309 Amendment 25–23, evaluating whether your various cargo door designs, including the "closed loop" Phase IV system, assure that door opening is extremely improbable. The analysis should be realistic and include consideration of the possibility of production, maintenance and operational errors.

This letter was asking the same questions as the other, written one year earlier, only this time they pertained to "future aircraft designs" more than the more explicit references to the DC–10. Besides, even though the F.A.A. was now talking in terms of "future aircraft", the whole list of problem analyses related directly back to the DC–10 near catastrophic accident over Windsor, Ontario, on 12 June 1972. Moreover, it appears adequately clear in Question 5 of their letter, that were they to insist upon Douglas incorporating a workable pressure equalization venting system between the aft cargo compartment and the passenger cabin in the DC–10 design, this would mean that the whole of the floor would have to be strengthened and as a result, the machine would become less economically inviting to potential airline customers.

William C. Starloff, Douglas manager for F.A.A. liaison, replied on 25 February 1974:

1. We have reviewed the request in your letter . . . dated 13 February 1974, and have determined that the questions asked would require much clarification before an attempt could be made to answer them. Even with clarification, we do not have the manpower available at this time to undertake the study, nor are we in a position to accept this burden alone.

2. We believe the noted questions involve the whole transport aircraft industry; therefore, it is suggested that the F.A.A. consider the possibility of funding this study because of the magnitude of the effort required.

3. We will not plan any future action on your 13 February 1974 letter until we hear from you further.

From their reply, it now appears that the Douglas Aircraft Company, a division of the McDonnell Douglas Corporation, is saying that they neither have the manpower, the interest, nor the money to conduct such a study which relates directly back to the design they incorporated in the DC–10. Moreover, they emphatically state that they consider the American government should pay for the costs involved in finding out more about why these faults exist in their aircraft. Mind you, the F.A.A. are not wholly blame-free in this matter. After all it did take them just on one year to follow up their original unanswered questions and because of this evident laxity, they decided to, this time, relate the specific problems in the DC–10 to be all-embracing on the industry and its future developments.

As we now know, within three weeks of the exchange of these letters, the exact same problems were to resurface, only this time with disastrous consequences. Following the Paris catastrophe, John Brizendine, the president of the Douglas Aircraft Company claimed they had been working on studies of these problems:

> . . . at a modest or low level of intensity activity. We have increased the level of activity now on it, at the bequest of the F.A.A., the press, Congress, and as one who has already read this letter, I think far too much emphasis is being placed on the words suggesting that a funded study might be appropriate, and we are working on the issue seriously.
>
> One of our people wrote the letter, and perhaps certainly a more delicate choice of words could have been used which would have caused us much less embarrassment and harassment from the press and other places on the subject because of those words. They are unfortunate in their application.

So now is Brizendine suggesting that we should not consider everything that is written on McDonnell Douglas paper, as being official?

However, early in November 1974, McDonnell Douglas announced triumphantly to the press that they had decided to make major improvements in new DC–10 aircraft.

Douglas announced that they were *actually going to strengthen the floors* of the DC–10s – BUT – only in aircraft coming off the production lines, starting with fuselage 162.

In giving an excuse as to WHY they had suddenly decided the floors needed strengthening, a spokesman said:

> There are other, more possible causes of rapid decompression in any pressurized aircraft, such as explosions, collision and machinery disintegration. It is against these potential occurrences, however remotely possible, that our changes are designed to increase protection.

Well then, what about Starloff's excuses to the F.A.A.? And what about the 160 DC–10s with their original weak floors? Are these also going to be remedied by magic?

CHAPTER 43

In Europe, 3 March 1974, was an ordinary type of late-winter's day, one conducive to the British staying for a few extra hours in bed reading the Sunday newspapers; or collecting the family into the car for a drive out into the countryside, with the possible added attraction of a picnic lunch, complete with ample supplies of "*vin ordinaire*", as is a Parisian's pleasure.

Paris has a number of forests within a not too great driving distance from the city centre, and possibly because of the atmospheric pollution or simply the crowds, even in that relatively small city, a day out in the bushlands was one of the pleasures of life.

The weather that morning was not too bad. A slight drizzle in London; a high level, high density cloud layer over most of the Continent. Winter's cold was still with us with a goodly portion of the French countryside blanketed by a light carpeting of snow.

Back in London, with the newspapers read, and the table set for the traditional Sunday roast lunch, the next part of the day's activities was commenced for many of the home population, in tuning their radios in to the 1300 hours, BBC–4 radio broadcasts of overseas and home news roundup.

The one o'clock time signals pipped their way past.

BBC Radio 4. *The World This Weekend*. First the news headlines, this Sunday lunchtime:

Reports are coming in that an airliner with 180 passengers on board has crashed in a forest near Paris.

As Mr. Thorpe, the Liberal Party leader, ponders his next move – advice from a leading Liberal – he says the party would reject any pact with the Conservatives.

In Spain, the Bishop of Bilbao has been ordered to leave the country.

We're just getting news of a serious air crash in France. An airliner has come down in the *Ermenonville* Forest about 15 miles north of Paris's *Le Bourget* Airport. First reports suggest that there may have been as many as 180 people on board, but this has not been confirmed. It's thought the airliner might be Turkish.

This was the first news to break in Europe simultaneously on the BBC and the French government-controlled radio networks of O.R.T.F. The accident had occurred just 22 minutes earlier – at 1238 hours.

As the BBC's programme continued, mainly with local political news as it was then only days after a general election in which the ruling Conservative Party had lost its mandate, there was, up to that time, a lot of talk of the Liberal Party (with a minute number of members in Parliament, but with a huge section of the population's vote behind them) joining a coalition with either of the two larger political groups – the Conservatives or the Labour Party.

However, right at the very end of the programme, telephone lines had been arranged from the London studios to Paris, and reporter Robert Williams asked Ian McDougall on the Continent, what was known at that time about the crash.

Well, all we know so far is that a Turkish airliner has crashed about twenty miles northeast of Paris in a forest there called *Ermenonville*. The casualty list is described in one report from Turkish Airlines as being 334 people on board including a crew of 11. An earlier report said that 180 people were on board. If, of course, this figure of 334 is confirmed, this will be the worst ever air crash involving a single aircraft.

Search parties have already found a large number of bodies strewn around the wreck and it is believed that there are no survivors.

It appears that nobody living on the ground was hurt by the plane which apparently crashed in a reasonably isolated spot, and immediately the news was put out over the radio, the French police asked everybody to stay right away from the area so as not to hinder the rescue operations at all.

Robert Williams: How far out is it? How far out from *Le Bourget* was it?

Ian McDougall: Well this is . . . this is not near. The plane took off from *Orly* at 1230 hours, and that was half an hour late already. It crashed very shortly afterwards, north – *Orly* is in the south of Paris – it crashed north of Paris; north of *Le Bourget* Airport, and north of the brand-new huge airport which is going to be opened this coming week by President Pompidou.

Because of British Sunday radio programming, the news bulletins and flashes were few and far between.

This is BBC Radio 2. The time is half past two.

Here is a news flash. More about the air disaster in France. Turkish Airlines now say there were 345 people on board the DC–10, Flight 981, which came down near Paris on its way from Istanbul to London. The airline says the majority were Turkish, French, Japanese and British. It appears there were no survivors.

This is BBC Radio 3. The time is half past two.

Here is a news flash. There's been an air disaster in France – it looks like the worst in history. Turkish Airlines say there were 334 passengers and 11 crew on board a DC–10, Flight 981, which came down near Paris on its way from Istanbul to London. The airline says that people of a number of nationalities were on board, the majority were Turkish, French, Japanese and British. It appears there were no survivors.

The main intial details were repeated on Radio 4 at 1430 hours, and on BBC Radio 1 at 1500 hours:

Here is a news flash. We've got more details about the air crash near Paris in which 345 people are thought to have died. It's now reported that 200 of them were British Airways' passengers who had to switch flights in *Orly* because of industrial action at Heathrow. The plane which crashed was a DC–10, Flight 981, of Turkish Airlines.

Ironically, British Airways European flights had been cancelled on this very Sunday because of a strike called by maintenance engineers at their main London, Heathrow, engineering division. Also, ironically, British Airways had had two major disasters between July 1971, and June 1972, in which 181 people died, and in both cases, maintenance deficiencies had been found in the crashed aircraft. Now these people were on strike for more money.

During the afternoon, the Turkish Minister for Transport, Ferda Guley, made a statement to the Turkish press, which was picked up by the other world news outlets and given airing far in excess of its known news value. This person said that in his opinion, the cause of the disaster was possibly a bomb. Someone of his standing and position should have known better. The news media, likewise, should have realized that a transport minister would, naturally, wish to divert attention away from the airline – the Turkish government-owned national carrier – having any involvement in the world's worst air holocaust.

This person was wrong – or wrongly advised – to make such a bland statement within a couple of hours of the disaster, and practically before the crash investigators had arrived on the scene.

The first major item of importance to clues as to what might have caused the catastrophe, was not contained in a news bulletin as such, but was in an interview on the telephone broadcast over O.R.T.F. the usually moribund French government censored and controlled radio network, between a reporter in the studio and a farmer's wife, who, to her great surprise, had discovered three aircraft seats and six bodies in one of her husband's newly ploughed fields. There was nothing else to be found – just the three sets of seats (two banks of three passenger seats and one flight cabin crew's double jumpseat), and nearby, six bodies; four males and two females. The seats were smashed to pieces, but were still recognizable as such. The bodies were still recognizable – but only just – and the crew's seat was fairly well intact except for massive ground impact damage. The peculiarity was that there was nothing else. Surely something like a piece of fuselage skin, for instance, would also have to be nearby, in order to make a hole, large enough for such objects as seats to fall from the aircraft? But no. There was nothing else to be seen.

The next development came at 1600 hours, on BBC Radio 4:

It has now been confirmed that there are no survivors from the DC–10 airliner which crashed near Paris this afternoon with 345 people on board.

The plane was Flight 981 of Turkish Airlines, on its way from Istanbul to London, with a stop at Paris, *Orly*. Our Paris correspondent says it's thought that those on board include a large number of British subjects who boarded it at *Orly* – more than 200 people who were due to travel to London by British Airways, changed flights because of industrial action at Heathrow Airport, London.

And another item of news just coming in. We're getting reports from Beirut that an airliner of British Airways has been hijacked while on its way from Beirut to London. One report says that the pilot was forced to fly the plane to Athens. Another report says that it was over Athens that the hijacking happened.

No one, not even the reporters on the French radio network, had realized the significance of the three airway's seats and the six bodies being found nine miles back along the flight path before the main wreckage. They used this fact to support the Turkish minister who claimed that there had more than likely been a bomb on board, yet none of them asked themselves *why* no other wreckage was in the immediate vicinity, other than that which had been found?

As the day progressed, the hundreds of different news agency reports were being compiled into the main evening radio bulletin:

BBC Radio news at 6 o'clock.

The worst air disaster in history. All 345 people on board a DC–10 airliner were killed when it crashed near Paris. It's thought that many of those who died were British.

And there's been a hijacking of an airliner belonging to British Airways on its way from Beirut to London. It's thought that there are 90 passengers on board, and that the plane has landed at Amsterdam.

At Home, Mr. Thorpe the leader of the Liberal Party, has been consulting leading Liberals about whether they should do a deal with the Conservatives. Mr. Heath has also been seeing senior colleagues in Downing Street. Mr. Wilson returned to London tonight.

The Israeli Prime Minister, Mrs. Meir, has said she's giving up the idea of trying to form a new government.

Three hundred and forty-five people, more than half of them believed to be British, have been killed in the worst air crash in history. Their plane, a giant DC–10 of Turkish Airlines, plunged into a forest soon after taking off from *Orly* Airport in Paris. It was on the last leg of a flight from Istanbul to London and there were no survivors. From Paris, Ian McDougall reports:

Turkish Airlines Flight number 981, a DC–10 bound from Turkey to London via Paris, took off from *Orly* Airport, just south of Paris, at 1230 hours European time, and some 15 minutes later, crashed into a wooded area known as the *Forêt de Ermenonville*, 25 miles northeast of the capital. Officials at *Orly* say no message of any kind was received from the pilot. The plane simply disappeared from the radar screen.

Eyewitnesses have described hearing an enormous explosion but it's not clear exactly where the plane was when it happened. One eyewitness says the plane's nose suddenly dipped.

Seven bodies and a piece of aircraft have been found in fields nearly ten miles from the scene of the crash. The 216 passengers who boarded the plane in Paris, most of them British, but some Japanese and French as well, were searched before embarkation, while those who had already boarded in Turkey, remained in the plane during the stop-over. At the scene of the crash itself, bodies, mostly beyond individual identification, were spread over a wide area amidst mud and slush in between the gashed and uprooted trees.

The French Minister of Transport has gone to the scene. The Turkish Transport Minister has been quoted by the French radio as saying that the crash could have been due to sabotage, but there's no proof.

John Starr went to the scene of the crash. A few minutes ago he sent us this report:

A mile or more of the Forest of *Ermenonville* is hived down. The pine trees are shattered stumps, sometimes 10 or 15 feet high. And even in the late afternoon, the little forest paths were full of stretcher bearers, toiling to bring back more of the 340-odd victims. There's snow on the ground in the forest, and it adds to the impression of some terrible battlefield because it's stained with blood as are the wrecked seat, and the jackets and the jumpers and the other bits of clothing, much of it 50 feet high in the trees on either side of the flattened paths.

So violent was the explosion that some of the bodies have been found six miles away from here.

As I was leaving the scene a few minutes ago, the *Gendarmerie* captain was giving a team of specialists instructions to bring in every piece of paper, jewellery, letters, and other clues to the identities of the victims, because identification is obviously going to be difficult.

A thousand people are working in the rescue operation. Hundreds of them are women and girls from the local volunteer organizations.

At London's Heathrow Airport, friends and relatives who have turned up to meet the plane have been waiting for news. The complete passenger list is expected to be issued tonight. Here's Robert Bufton:

It's not clear exactly how many people on the plane were British. Thomson Holidays have said they had forty holidaymakers who boarded the plane at Istanbul, and British Airways who have set up a special accident control room at Heathrow Airport, told me a few minutes ago that they had now received a passenger list but it is so far incomplete with only 250 names on it.

It's believed however, that a large proportion of these names are in fact British, though they're not being released until relatives have first been informed. The accident room expects to receive a complete list later tonight. But seven children are feared to be among the dead. British Airways said some of the victims would normally have been their own passengers, but for the strike at Heathrow which grounded flights and meant that foreign airlines were only available to fly people from Paris to London. Since before midday, about forty people have waited at Heathrow for more news to filter through – they're either friends or relatives of some of the victims.

The narrow escape for members of the England Rugby Union team who drew with France yesterday. They'd planned to come home on the Turkish Airlines plane. They were turned away at the last moment because the plane was full. They left later on an Air France flight.

Although the airliner was carrying so many people, it was not overloaded. And as our Air Correspondent, Reginald Turnill, reports, the big jets have an excellent safety record:

So far, we don't know whether the pilot was able to tell *Orly* Airport what sort of trouble he was in. He was airborne for just nine minutes between take-off and disaster. And though 333 of the 345 seats on this all-tourist plane were occupied, there was no question of overloading. He would have been carrying little fuel for the short haul from Paris to London.

The all-American DC–10, first of the airbuses, has been involved in several incidents, but no major accidents since starting passenger services two and a half years ago. Mass air travel began five years ago with the Boeing 747 Jumbo jets. And airlines and insurers have felt that one day a major disaster like this was inevitable. But the Jumbo jets and airbuses have in fact established far higher safety standards than previous planes.

Later news contained a round-up of the earlier telexed items that had been confirmed by the French investigating officials, including the following guess at what had actually happened:

It's still not known what caused the crash, but first reports suggest that it was a mid-air explosion. Some bodies were found several miles from the crash site.

However, one new piece of information was that French television had reported that the plane's undercarriage had been lowered and there was the suggestion that the pilot was trying to make an emergency landing at the not too far distant *Roissy* Airport, due to be opened five days hence. They also said that part of the aircraft's flight data recorder had been recovered and it was hoped that this would give some clue as to what had happened. BBC Radio news once again:

Our correspondent Jack Starr was at the scene soon after the crash. Here's what he saw:

I'm standing just beside the cabin of the wrecked red, white and black pieces of metal shafts everywhere. Up in the trees on either side of the wide swathe cut by the plane, there are tragic little bits of clothing. Coats, pullovers, jackets, boots and alas, even at this late hour in the afternoon, there are still bodies there.

The little lanes through this wood are filling with stretcher parties – an endless procession of stretcher bearers – and still there's need for more. And strangely enough, apart from the sound of helicopters overhead, there is a remarkable silence about the whole rescue operation, possibly because the enormity of this crash has shocked people into silence.

At Heathrow Airport, London, a crowd of friends and relatives of the people on the flight have been waiting throughout the day for news, but it's thought the complete passenger list may not be out for some time because of the problems of identification. Here's Robert Bufton:

It's still far from clear exactly how many people who died in the disaster were British. Tonight, British Airways, who've set up a special accident room at Heathrow Airport, indicated it could be several days before they complete their work of informing relatives. So far, they've an incomplete list to go on, with about 250 to 300 names. But a high proportion of these do appear to be British. The list contains surnames and initials only, and in many cases there is no way of telling whether they're male or female, and this is making it all the more difficult for officials to make contact with relatives.

The list so far, I understand, has people who boarded the plane at the three stages; at its original departure point at Ankara, the capital of Turkey; at Istanbul; and in Paris; and airline officials are hoping to complete the list by late tonight.

Thomson Holidays say they had forty holidaymakers on the plane from many parts of Britain. It's feared too that seven children are amongst the dead. It had been thought at one time, that the England International rugby team, returning from yesterday's game in Paris, may have been on the plane, but tonight it was confirmed that they had returned on another flight.

Investigation teams are being flown out from the American makers of the DC–10 to help with the on-the-spot investigations. They'll be paying particular attention to the actual evidence, and eyewitness reports of a mid-air explosion before the crash. Here's our Air Correspondent, Reginald Turnill:

It is almost certain tonight that a mid-air explosion destroyed the DC–10, nine minutes after it had taken off from *Orly*. French air traffic controllers had watched it climb to 13,000 feet, saw the height read-out vanish from the radar screen – a catastrophe so sudden, the passengers never knew what happened, and the pilot apparently had no chance to send a distress call. Most likely it was sabotage. A bomb in the tail. But another possibility being urgently examined by the accident investigators is that one of its big jet engines blew up. There've been several previous engine failures on DC–10s, and in one case last November, a wing engine exploded rupturing the cabin, and one passenger was sucked out and killed. But although this DC–10 was filled up at Paris with British Airways passengers, stranded following the Heathrow strike, there's no question of the plane being overloaded, although 333 of its 345 seats were occupied – it was only lightly loaded with fuel for the short hop from Paris to London.

With mass travel rapidly increasing, and hundreds of these huge Jumbo jets and airbuses now in service, it did seem inevitable that one day, the airlines and insurers would have to face a disaster like this one. The appalling loss of life is likely to cost the aviation insurers over £30,000,000 ($72,000,000).

There seemed to be no doubt in that air correspondent's mind that anything other than sabotage was a possibility. His use of the words: "It was almost certain tonight that a mid-air explosion destroyed the DC–10 . . . Most likely it was sabotage. A bomb in the tail". (For the record, he did change his mind *five days later* when the investigators confirmed his presumptions were completely wrong.)

Meanwhile, the final radio bulletin of the evening was transmitted from BBC Radio 3, at 2330 hours. Again, a recapping of the main stories of the day, with the lead still the DC–10 air disaster. *Orly* Airport by this time, had said that there was no radio contact between themselves and the pilot; the French television confirmed that there were eyewitnesses who saw the undercarriage lowered and then raised again, causing them to think that an attempt was being made to try and land at the brand-new Paris airport nearby. It also included a final report from the scene of the devastation by Jack Starr:

The Forest of *Ermenonville*, where French families go for weekend walks, is a nightmare of search-lights and stretcher bearers and twisted pieces of metal. More than 340 people were aboard the DC–10 that took off only a few minutes before the crash. The agony of death is now written in the pathetic debris of clothing and books and papers scattered among the stubble of trees, cut off close to the ground by the final terrible descent of the plane, and in the high branches on either side of its final flight path.

There are many British labels, hand books, and papers to point at the tragedy of British people who couldn't fly back to London the way they'd planned, and took another way. All of them are being stored and examined tonight in make-shift identity centres, and while the ambulances and stretcher bearers and the first-aid teams are still working, the French crime brigade have begun an investigation into the possibility of sabotage. Paris is known to have housed extremists of many nations, or many nationalities, and the likelihood of sabotage has increased by reports of what appear to have been an explosion and a trail of smoke from the tail of the Turkish aircraft.

French government ministers have been on the spot in time to see the smoking wreckage. It's certain that France will enforce a most thorough explanation of this most terrible air crash.

CHAPTER 44

THE long hours of the night gave one a lot of opportunity for considering the information gleaned on the different broadcasts heard since the lunchtime crash of the day before. The intriguing thing was, if what the farmer's wife had said was correct; if she had indeed discovered the mangled remains of the three airline seats and six bodies; how did all this paraphernalia get out of the machine without it completely disintegrating on the spot?

I had already put forward proposals to a publisher for a book which would tell the stories of sixteen air crashes, all of which had been caused in one way or another, by design faults that had not been uncovered before the airplanes were given certificates of airworthiness, even though in a number of instances, the faults had made themselves glaringly evident in saved incidents long before the major disasters had occurred.

One of these planned stories dealt with the DC–10 and the door/floor/control cables' problems, and therefore one had a head-start, so to speak, having already partially researched this particular incident. Up to this time, even though the British and French news media were giving background details of DC–10 occurrences – the three engine disintegrations that had occurred were the only ones cited – none of them included the most potentially dangerous design fault of the lot: the details of the American Airlines accident on 12 June 1972, over Windsor, Ontario. To the European press, this near disaster simply had not occurred.

On digging out the earlier American Airlines accident report that was buried in amongst hundreds of similar documents, at about three o'clock in the morning, re-reading the information – applying the thesis to the Paris disaster as was then known – it became adequately evident that the 346 deaths had more than likely been caused by a repetition of circumstances seen in the earlier American Airlines accident, though this time, possibly the control lines to the tail had been completely snapped or jammed in such a fashion by the floor collapse, that it caused the aircraft to go immediately out of control. If this had occurred, we would then have the results we knew about – the three seats, together with possibly a few smaller bits and pieces, some of which would be the remains of the battered aft cargo compartment door. There would then be a large area over which the aircraft would still be flying, though completely out of control, and then the "main wreckage area".

The only drawback to this thesis at that time was that there had been no reports of any wreckage, other than the six bodies and the three sets of seats (such as the door), being found away from the main crash site.

The following morning just before nine o'clock, the BBC radio current affairs department were on the telephone asking for some comment about DC–10s and Paris crashes. And so we arranged for a recording session at 1030 hours at the studios in central London.

Naturally, if one wanted to speak with authority, one must be up to date with matters at hand. The radio was tuned to BBC–4, for the nine o'clock news.

First, the air disaster near Paris. A searching inquiry has begun into what caused the DC–10 of Turkish Airlines to crash, killing every single person aboard. A six-man team of investigators leaves Turkey today to join the French experts, in efforts to find out whether the worst disaster in aviation history was an accident or sabotage.

The Turkish authorities now say that there were 347 passengers and crew – three more than were reported earlier. More than 200 of the victims were British.

Turkish Airlines have categorically denied reports that they suspected five of the passengers were Palestinian guerrillas. According to these reports, the five had planned to sabotage a British airliner on the Paris to London route. By chance, they were transferred to a Turkish DC–10, and the bombs they were carrying exploded. A Turkish Airlines official has described all of this as just another wild theory. Police checks at Istanbul Airport are very stringent, he said.

Here's our reporter Stephen Jessell:

Contacting the next-of-kin is proving difficult. In some cases the passenger list only gives surnames and initials, or just the surnames. Sometimes the address given is that of a victim and neighbours have to be called in to help trace relatives.

Where the ticket was booked through a travel agent, the authorities have had to wait until records can be consulted later today. But they've been helped by people who have contacted them.

Some details are known though. The dead include a Rugby Union team from Bury St. Edmunds, who went to Paris to see Saturday's

International. Eighteen of the party of twenty-one were on the plane. There were two school-boys in the original party, one of them a last-minute replacement for the team's captain.

At the scene of the crash in a forest where French families often go for weekend walks, arc-lights were set up as the search for clues continued throughout the night. The experts wanted to find a vital piece of evidence as John Starr reports:

They're still searching in the forest filled with the stubble of bare tree stumps, festooned for miles around with clothing, shoes, coats and dresses, and more pitiful remains, for the biggest part of the 'black box' (the flight data recorder) which could explain why the DC–10, with more than 350 people on board, could have crashed without warning five minutes or so after taking off from *Orly* Airport for London.

Sabotage has obviously been mentioned because there are reports that there was an explosion on board; that smoke was seen coming from the tail; and the *Orly* Airport controllers say that they were in touch with the pilot, they lost contact, but they didn't hear any distress call. And yet the French police say that all 216 passengers who came aboard at Paris – the majority apparently British and because of a strike in London changed their plans for going home, – had been searched. Others coming from Turkey stayed aboard the plane in transit.

With this most terrible of all air disasters, the French organized one of their greatest operations of attempted rescue. There must have been a thousand men and women and girl volunteers in the forest, seeking in vain for the living and in despair for the dead, and the little paths were filled with teams of stretcher bearers.

To back up the BBC's radio reportage, a copy of the French daily newspaper *Le Figaro* was obtained for comparison. In translating French prose to English, one has attempted to reproduce the subtleties of the French language into colloquial English, and therefore some slight differences will appear from an exact word-for-word translation.

The *Le Figaro* reporters who wrote the following were: Michel Bole-Richard, Guy Le Bolzer, Lucien Miard, and Jaques Dubessy:

The DC–10, Flight 981, took off at 1230 hours, 30 minutes late because of all the added passengers that had to board.

The start of the flight was quite normal and the pilot, after having been authorized to take off by the air traffic controllers, had a two-way radio conversation with the regional air traffic controller who was the person in charge of it at this stage. The pilot was told to carry out the normal procedures, with the aircraft flying away from the airport at 130 degrees magnetic, and heading out in the direction of the

Coulommiers Visual Omni Range beacon. But it did not fly over the village of *Coulommiers*, but instead headed towards the beacon at *Montdidier* in order to reach a height of 700 metres. The air traffic controller then changed his attention for a few seconds from the DC–10 in order to contact another aircraft.

Then the code, TUR 981, was suddenly heard on the radio. This was the voice identification sign for the DC–10. The tape recording back at the airport, which is not very clear, has been handed to the investigation commission. The air traffic controller reports having heard a discussion in a foreign language (possibly Turkish), intermingled with cries of panic and alarm signals. One of the signals was of 1,000 cycles tone which is one of the main automatic alarm sounds.

At the same time, the radar operator who was watching the aircraft, said that he saw something at about 1240 or 1241 hours, with the machine at 13,000 feet altitude, at the edge of the *Départments* of *Seine-et-Marne* and *Oise*.

Some of the technicians were of the opinion that an explosion, accidental or criminal, had occurred. The incident could have involved an explosion because of the fact that the bodies of six victims were recovered at *Saint-Pathus* in the *Départment* of *Seine-et-Marne*, but the aircraft continued for a further 17 kilometres.

However, the hypothesis of an explosion having occurred in the air seems to be refuted by the testimonies of several witnesses to the crash.

M. Lhote, of *Epinay-sur-Seine*, who works at the air traffic control at *Le Bourget* Airport, said:

> My wife, my children and myself were in our car driving to *Roissy*, when we saw an aircraft in a steep nose-down dive, which I realised was not normal. It was at about 800 metres altitude when I first saw it. I thought at the time that it was heading towards *Le Bourget*, but it was not in alignment with the airport. Also, it was coming down too fast. Possibly the pilot had the intention of landing at *Roissy?* The machine was not on fire; there were no flames or smoke; and I got the impression that it was in very serious difficulties. It later disappeared, and we saw an enormous cloud of smoke, evidently from an explosion. There wasn't an explosion on board the DC-10 before it disappeared from our sight and crashed to the ground.

Another witness, M. Carrel, who was working at a golf club not far from where the aircraft hit the ground, gave this version of what he saw:

> I saw the aircraft literally come down into the forest. It collided with the tops of the trees as it crashed. Within two or three seconds, I heard an explosion.
>
> There was a terrifying sound. All the windows at the golf club shook. Then, with some friends, we made our way towards where the accident had happened. When we arrived, we saw an enormous amount of small pieces of debris. But we all agreed that there was absolutely nothing we could do.

At *Saint-Pathus*, six bodies were found close by, fifteen kilometres before the main wreckage area. Mme. Dubois, a farmer's wife, said:

I found the bodies on my property. One female was lying in a bed of endives. She wasn't one of the hostesses but was a passenger who was wearing a white dresscoat. A hundred metres away in amongst the pear trees, there was a body of a man. A further hundred metres revealed another man and nearby a woman, and the remains of an airliner passenger seat. The other bodies were found in a field, at the edge of the *Route Nationale 330*. One of the dead was Turkish. Two others seemed to be Japanese.

At the scene in the late afternoon were, M. Jaques Chirac, the Minister of the Interior; M. Aymar Achille-Fould, the Secretary of State at the Ministry of Transport; M. Maurice Grimmaud, the Secretary General for Civil Aviation; and M. Lalanne, the *Prefect* of *Oise*. M. Achille-Fould said:

The finding of the bodies at *Saint-Pathus*, tends to prove that an explosion took place in the aircraft while it was flying in the air.

Asked about the circumstances of the accident, M. Maurice Grimaud said:

After it left *Orly*, the flight proceeded normally. There were no radio messages that it was in any difficulties. Therefore, the Secretary General of Civil Aviation has indicated that once they have recovered the Black Box flight recorder, they will then be able to answer all the questions that are posed.

It is 1345 hours, and the aircraft is smashed into small pieces in the forest at *La Baraque de Chaalis*, near *La Mer de Sable*. *Senlis* is eight kilometres away; Paris, 45 kilometres. The accident happened near the border of the *Départments* of *L'Oise* and *Seine-et-Marne*. The rescue teams arrived from all parts of Paris. They were very rapidly at the site of the tragedy. It is a spectacle of broken pieces, shattered trees, in amongst which there is not much visibly remaining that once was the DC–10. For a distance of one kilometre, the trees have been broken, bent and burned at the tops. In the branches high overhead, passengers' belongings may be seen. On the ground, the bodies, the pieces of engine and all the other parts of the aircraft, cover the soil.

In the snow, there are numerous pieces of clothing; bags; personal objects. Here, a pullover of blue and white wool. There, an overcoat. The pieces of the bodies are being placed into plastic bags. And then, one of the rescue chiefs calls from amongst the trees in an urgent voice: "*Attention. Il y a un sac à main sur un branche au-dessus de ta tête.*" (Watch out, there's a hand-bag hanging on a branch above your head.) Intact in amongst the wreckage there is a doll, books, childrens' games, a watch, and the equipment used by Rugby players.

Safety officials, rescuers, police and the CRS (the brutal French riot squad-cum-secret police), are searching. But it is not for survivors. There simply could not be any.

In a number of places, the wreckage is still smouldering, giving off puffs of smoke. Seats, and the interior fittings of the aircraft, have been practically destroyed by the small fires. The remains of the cockpit is facing back along the crash path. A small piece of a wing can be seen a few metres away from it. At the head of it, there is the coat of the pilot-in-command. Three hundred metres further on, a large piece of the fuselage with ten windows still intact. The eleventh is broken. All around in the trees, pieces of clothing, toiletries, a small pair of children's pyjamas, the torn out pages of a book. And on the ground, an intact copy of *The Wolf* lies amongst the wreckage.

The aircraft literally is in millions of pieces. Another three hundred metres separated the nose from the remainder of the wrecked DC–10. In the line of trajectory, the earth has been ploughed up. The grove of pine trees is in thousands of pieces. The soil all around is impregnated with kerosene. The dumbfounded rescuers are overcome with emotion. Security posts have been set up and manned by both the Red Cross and police back at the cross roads. In the air you can sense there has been a catastrophe. All rescue operations have been futile.

At 1600 hours, the chief doctor from the Paris Airport Authority, who is in charge of the medical aspects, said: "There is nothing we can do here."

It is night in the *Forêt d' Ermenonville*, and the tragic search is continuing with the area illuminated by arc-lamps. Police, CRS and the military have mapped out into squares, the area of the disaster, and their major preoccupation has been to make sure that no passing weekend tourists have been victims of the air crash. The ambulances were continuously transporting the bodies to the chapel, *Saint-Pierre-de-Senlis*, an ancient church built in the 13th Century, but today secularized. On the stretchers hang the white drapes and in the centre of the nave, the corpses are laid down. The curious – in large numbers unfortunately – are turned away.

Despite the requests by the authorities, the picnickers hindered the movements of the rescuers.

All of the dead were taken to the mortuary chapel of *Senlis*, and then 120 bodies were transferred to *Meaux*.

A POSSIBLE CAUSE OF THE ACCIDENT IS ADVANCED BY THE TURKISH MINISTER OF TRANSPORT.

Soon after the catastrophe, the Turkish Minister of Transport, said in Ankara that the accident was possibly caused by an act of sabotage.

M. Ferda Guley said that he has not ruled out this possibility and he speculated that it could have been the work of a revolutionary group of Turks.

The repression of terrorism started after the military *coup* on the 12 March 1971, and was conducted with a vengeance. Over 4,000 'leftists' are in prison in Turkey; 1,700 people have been convicted by military tribunals, and at the moment, twelve are awaiting the carrying out of the death penalty. Three students were hanged in 1972.

Hunted down, the various 'leftist' groups started clandestine activities. Many relentless operations were carried out, the most famous being the kidnapping of three British technicians working on the Black Sea coast. The gun battle that ensued, resulted in fourteen guerrillas and police being killed. Several airplanes have also been hijacked, but all the time these events have happened within the Turkish borders.

On 20 December 1973, it was discovered for the first time that Turkish terrorists were forming an organization in a foreign country, in France at *Villiers-sur-Marne*, in the Paris suburbs. This was a clandestine Turkish/Palestinian organization, that planned to continue their 'home' operations.

'The Turkish Liberation Army' is the name of the 'leftist' group that is the most active. It's members have said that: "Action serves to perfect one's ideas", and so they formed a group which they called *Les Montagnards*. They are inspired in their activities by the methods used by the Palestinian terrorists. They are trained to carry out such activities in camps in both Lebanon and Syria, in company with members of the known Palestinian organization *SAIKA*. It is for this reason that one of their chiefs, Deniz Gezmis, left at the end of 1972, to participate in the kidnapping of the British technicians near the Black Sea.

NATIONAL MOURNING IN TURKEY.

The Prime Minister of Turkey, M. Bulent Ecevit, last night called in the members of his Cabinet to proclaim a day of national mourning because of the catastrophe in the *Forêt d' Ermenonville*.

In a voice filled with emotion, M. Ecevit expressed the profound sorrow and the sympathy of the Turkish Government to all the people concerned in the tragedy.

The nightclubs and the numerous restaurants were closed last evening. The radio and television stations interupted their programmes to play funeral music, and to give the latest information as it came in from France.

CONDOLENCES FROM M. POMPIDOU TO PRESIDENT KORUTURK AND QUEEN ELIZABETH.

M. George Pompidou last night sent telegrams of condolences to M. Sari Koruturk, the Turkish President, and to Queen Elizabeth of Britain.

To the Turkish President, M. Pompidou said:

> It is with deep sorrow that I learned of the aviation catastrophe which took so many human lives in the course of the accident to an aircraft of Turkish Airlines.
>
> I express, Your Excellency, my sentiments of deep sympathy and my sincerest condolences to the families of the victims of this tragic event: the worst in aviation history.

In his message to Queen Elizabeth, President Pompidou wrote . . .

> . . . of the large number of British subjects who perished in the Turkish Airlines' accident.
>
> In these tragic circumstances, I assure Your Majesty of my deepest sentiments and sympathy, and please convey to the families of the victims my sincerest condolences.

At 1030 hours, we started recording the BBC radio piece outlining the "door coming off/floor collapsing" theory, and soon afterwards, following a slight editing and being checked by the BBC lawyers, it was broadcast. This was the first occasion on which anyone, anywhere in the world, had said that the crash had more than likely been caused by a badly designed cargo door coming open.

Every single press outlet in Europe and America, except BBC Radio 4, maintained their more sensational presumptions. But if a bomb had caused a sizeable hole in the machine's fuselage, why did not the entire plane disintegrate there and then, on the spot, remembering that a large hole would have been needed in order that the seats and passengers had got out in the first place?

The engine explosion theory, also being cited as a possible cause in the press, if studied, was equally as spurious. If such an almighty engine disintegration had occurred as would create a hole big enough for these three sets of airline seats to be expelled, it would have also caused the engine and some other parts of the fuselage, such as the tail, to also detatch. Where were they?

These points had evidently slipped through the press people's minds as they were considering their front-page headlines. Besides, bombs or engine explosions are of greater news value than such a matter as a door coming off.

As the morning progressed, contact was made with BBC television news, Independent commercial Television News, the London Broadcasting Company. But they did not understand that the more likely cause of the Paris crash was the design fault in the door. They were not interested in the story. Each of them had their "experts" who were adamant that either a bomb or an engine disintegration had caused the disaster.

By midday, just twenty-four hours after the crash, the story had been relegated down the headlines. BBC Radio 4, once again:

Now yesterday's air disaster in France.

Experts say that it could be eight days before all the wreckage of the plane, belonging to Turkish Airlines, can be recovered from the *Ermenonville* Forest near Paris. All 347 people on board died – more than 200 of them British.

The French evening newspaper *Le Soir*, has had an anonymous telephone call from a man claiming to represent a group called 'The Liberal Front'. He said that they were responsible for the disaster, but so far, there's no clear evidence of sabotage.

For a report from the scene where the wreckage of the plane is being carefully sifted by investigators, here's John Starr:

There's bright sunshine in the forest now. The stretcher bearers have gone. The devastation of smashed trees and metal remains.

Much of the plane's 'black box' flight data recorder is understood to have been recovered, and the contents are under examination for clues to the disaster. Sabotage is not ruled out, especially as there is evidence pointing to an explosion in the rear of the plane which would have explained why some bodies were found six miles away from the main wreckage.

A man telephoned the newspaper *France Soir*, saying that the crash was the work of the commando *André*, and was organized by the 'Liberal Front Foreign Political Movement'. No one is sure what this is. He also warned of another operation in three or four days time against Air France. This could well be a hoax of course, but the warning has not been discarded.

French aviation experts are now going back over the plane's route in its last minutes of flight, to establish just where it began to disintegrate. Both Turkish and British officials have been here, and American experts are expected later in the day.

One report which is going around here is that an air traffic controller picked up an alarm sound from the plane just before it crashed. This contradicts earlier reports that the *Orly*-north control lost touch with the plane.

So far the British Airports Authority at Heathrow, London, haven't published a list of the dead. They say there has been some difficulty in contacting relatives. But it is known that a leading British trade union chief was one of the victims. He was Mr. Jim Conway, aged 58, the general secretary of the Amalgamated Union of Engineering Workers. A cultural attaché at the American Embassy in London, also died in the crash, along with his wife and two children.

The dead also included the former British athlete, John Cooper, who won two silver medals in the 1964 Olympic Games in Tokyo. And four members of the Davenport Rugby Club in Stockport, are also reported to have been on the plane.

The two Arabs who hijacked the VC–10 of British Airways and set it alight at Amsterdam Airport, have so far refused to disclose their motive. They also haven't said why they gave themselves up, after originally asking for the plane to be refuelled. The hijackers claim they are members of a group called 'The Arab National Youth for the Liberation of Palestine'. But the main umbrella terrorist group, 'The Palestine Liberation Organization', have disassociated themselves from the hijacking.

In the early afternoon, I received a telephone call from the British commercial television *Today* programme. Would I happen to have any further evidence to back up my theory of the crash being caused by a door coming open? Naturally, I did. There was the copy of the American Airlines accident report which told, in nearly every technical detail, how that near-disaster had developed and occurred. Would I care to discuss the matter further on television that evening? Of course.

Also appearing, was a member of the British Parliament, Norman Tebbir, who had for the past couple of years, interested himself in the subject of civil aviation safety and the associated problems, by joining the independent British committee that called themselves "The Air Safety Group". It was this Group which, back in 1961, had dared the airlines to take part in an experiment whereby one of their executives would stand in a pool of the more dangerous fuel, JP–4, some of them were using, and then light a match, while a member of the Group did the same thing standing in a pool of kerosene. Needless to say, the airlines found excuses that prohibited them from attending, let alone participating in the demonstration.

Norman Tebbir was completely open-minded about the causes of the Paris disaster, though he tended to lean towards the bomb or engine explosion theories as being the more plausible.

The appearance on the *Today* programme was transmitted "live", and a few minutes after it had ended, the telephone rang and it was a person claiming to be a VC–10 pilot (which would have meant that he more than likely worked for British Airways or British Caledonian Airlines). He was adamant at the "door coming off" theory, adding that he knew all about aircraft and pressurization, and at 12,700 feet altitude, there simply would not have been sufficient pressure built up in order to cause an explosive decompression. It was explained to him that the American Airlines, Windsor, Ontario near-disaster on 12 June 1972, had occurred at 11,750 feet – 1,000 feet *below* the Paris altitude – and that the accident report stated that there would have been a pressure of 6.2 tons weight against the partially locked cargo door. The accident report was also wrong, he claimed in reply.

This same person, or another hoaxer, must also have telephoned the air correspondent of the London *Times* (the subject of a large advertising campaign telling the public how interesting it was to read this newspaper), and the correspondent duly wrote:

> Among suggestions of possible causes of the disaster has been one that the catches to the cargo door failed, allowing the door to fly off as the airbus had reached 13,000 ft.
>
> This, it has been suggested might have resulted in a depressurization explosion, the collapse of the passenger cabin floor and the rupture of the control lines from the flight deck to the tail.

Air safety experts doubt whether the air outside the aircraft at 13,000 ft. is thin enough for a depressurization explosion of such force to occur. Their opinion still inclines more to a bomb or a runaway tail engine as the cause of the crash.

One wonders who these 'experts' were this correspondent consulted. Possibly gardening experts?

Meanwhile, later that evening European time, but late afternoon American west-coast time, McDonnell Douglas had evidently, for one reason or another best known to themselves, woken up to the fact that the world's worst air disaster could have been caused by this badly designed door coming open *yet again.* They sent a telegram to all airlines operating DC–10s and begged that the carriers enforce procedures that had been previously 'recommended' on the operations of this door. John Brizendine, the president of the Douglas Aircraft Company, later stated that at that time they were aware that a piece of the aft cargo compartment door had been found. However, the French investigators said they did not find it until the Tuesday morning – quite a number of hours after this telegram was sent. The cable continued:

This information is being reported to you at this time without implying any conclusions, to make you aware of the events as they evolve. The Douglas Aircraft Company strongly recommends that all operators enforce procedures to ensure that all cargo doors are fully latched prior to take off. Alert Service Bulletin A52–35, dated the 19 June 1972, is cited for your ready reference.

Meanwhile, none of the airlines which were operating DC–10s did anything outwardly evident that either could be regarded as steps to ensure that still another repetition would not occur, or to scare any of their intended passengers away from their DC-10 flights. These machines continued to be operated as though the Paris disaster simply had not occurred.

CHAPTER 45

EVEN though forty-eight hours had passed since the crash, the accident investigators could only match up about 50 names with 50 'bodies'. All the rest, and mainly the British passengers on board the DC–10, did not coincide. Passports were found in the rubble belonging to people alive and well in London; other passengers had swapped places with people who had already bought airline tickets; and hundreds of telephone calls were received from relatives of 'others' who were 'missing', whose names did not appear on the passenger manifests but who were travelling from Paris to London on that day. Some of the relatives of those killed in the disaster whose documentation was correct, on the other hand, were not advised for anything up to two to three WEEKS later. It took the French seventeen days to tell one American that her husband had been killed in the catastrophe even though his name, address and other details were in their possession for that length of time.

But the main problem concerned the British. British Airways had evidently carried quite a number of people on the outward journey to France who were travelling under 'assumed' names, or 'borrowed' passports. On top of this, the airline was not able to supply any other details of the passengers other than the names given. They were evidently therefore, hoping that once the news of the disaster had been broadcast, relatives of the victims would ring up wanting to know where they were, and in this fashion British Airways hoped to complete its lists. However, the air services between London and Paris were, on Sunday 3 March 1974, in a complete shambles because of the maintenance engineers' strike. This meant that most persons travelling on that day, were being chopped and changed to other carriers by British Airways' office in Paris, and just one of these happened to be the ill-fated Turkish Airlines flight. Any one of thousands of people holding British Airways tickets could have been on board the Turkish Airlines DC–10, and British Airways had not the faintest idea of who was and who was not on board.

But even though quite a number of air passengers were travelling under other people's names, it seems quite evident that the French immigration officials were not too particular in assuring themselves that the travellers were in possession of correct individual documentation. Surely, one asks, they should have realized *on the outward journey*, that people were travelling on other people's passports, *if* the proper checks had been made? They quite evidently were not, as was later discovered *after* the disaster.

Following a major air accident, there are certain logical ways for the human factors' investigators to go about identifying the victims. The French consider human factors' investigations as useless, as is evidenced in the point that they have no specialists in this field on their investigatory staff. The British, likewise, consider human factors to be an 'extravagance' and they, like the French, also have no specialists in this field. Usually, on occasions like this after a major disaster, the civil air accident investigation branches obtain human factors' 'assistance' from the country's air force personnel, but their lack of experience, lack of judgement, and general lack of overall knowledge of the subject results in the mess that was to be seen after the Paris catastrophe. There is also another possible reason for not having human factors' specialists, and that is that they may just happen to recommend improvements which, in the long run, could turn out to be derogatory to the airline that had the accident, remembering that in both cases of Britain and France (and Turkey for that matter), a large percentage of the industries are nationalized.

The desperate situation of body identification could be seen in three later newspaper reports. Firstly, the British *Sunday Times* of the 24 March 1974:

Three weeks after the crash of the Turkish Airlines DC–10 near Paris, the French police are still not certain how many people died. The death toll, which has been put at 346, including the crew of eleven, could rise to 351.

Because the aircraft, weighing 200 tons, crashed into the forest of *Ermenonville* at 420 miles an hour, many of the victims' bodies were smashed beyond recognition. The police believe that the Turkish Airlines' passenger list may have been incomplete and that up to five more children were on board and died with their parents. So far only 22 death certificates have been issued by the French authorities and it is unlikely that more than 120 of the passengers will ever be positively identified.

In an attempt to establish the names of the passengers more certainly, the police have set up an 'identity centre' in the headquarters of the *Gendarmerie* at *Senlis*, a few miles from the crash site. There, laid out in two rows are the personal belongings that were found scattered around the site and every day last week groups of relatives arrived to try to pick out a watch or a key or a photograph that might help to identify one of the victims.

The system is far from foolproof. Four days after the crash, a man who had been presumed dead because his passport was found at the scene and because his name was on the passenger list, walked into the identity centre. He explained that his fiancée had taken his seat at the last moment and was carrying the passport in her handbag.

Relatives who visited the centre and who are able to face the ordeal, are asked to visit a nearby chapel where the victims' clothes lie in bundles on 3 long trestle tables.

In the vestry are displayed the hundreds of letters, diaries and address books that were found at the scene. At first, police thought the letters might prove important clues, until they discovered that the doomed jet was carrying air mail from Turkey to London.

The second telling piece appeared in the London *Evening Standard* of the 16 April 1974 – 44 days after the disaster:

The widow of one of the victims of the Turkish Airlines DC–10 crash near Paris in March is to petition the French Ambassador in London, M. Jacques de Beaumarchais, in protest over 'the continuing delays by the French authorities to announce a burial date for the crash victims'.

Mrs. Shirley Backhouse, whose husband Steve was one of the 346 people who died, intends to hand in a petition to the French Embassy Chancery in Knightsbridge on Friday.

"I hope anyone who would like to be associated with this protest will join me then, either relatives of those who died, or representatives of rugby clubs or other organizations who lost members in the crash," Mrs. Backhouse, of Disley, Cheshire, said.

"Even now, more than six weeks after the crash, the French authorities have not fixed a date for the mass burial and memorial service which they have said will take place."

She pointed out what severe practical problems were being encountered by widows with children who could not obtain insurance or other benefits because no death certificates had been issued.

Mrs. Backhouse also intends to ask the French authorities to reconsider the citing of the proposed mass-burial. She considers the cemetry too near to Orly Airport.

What no one had told her was that one of the many strange French laws state that no unidentified person may be buried with those who have been identified. As there was not much left after the Paris crash, the people involved in the French investigation had got themselves in a 'twist', not knowing quite what to do. This, however does not excuse them from creating enormous amounts of unnecessary hardship.

On the 21 April 1974 – 49 days after the disaster – the British *Observer*, in an article dealing mainly with the victims' relatives compensation claims against the McDonnell Douglas Corporation (which was part of a subtle advertising 'service' they were offering, as will later be seen), also contained a couple of paragraphs that delved further into this matter:

As relatives' anger at the French handling of the burial arrangements for the DC–10 victims increased this weekend, it was revealed yesterday that a British identification team had been ordered from the crash site by French authorities. Authoritative sources in London maintain that if the team had been allowed to co-operate, this would have avoided the need for a mass burial, which is upsetting so many relatives.

The team, made up of medical personnel and experts in identifying clothing and jewellery, had gone to the scene in the *Ermenonville* Forest soon after the crash. While they were carrying out preliminary investigations, the unit was called before the head of the French forensic pathology institute.

He told them that their services were 'not required' and made it quite clear that they were to stay away. He said the French would be grateful for any information they could supply.

It is widely felt that the French action has unnecessarily prolonged the job of identification, causing relatives much extra suffering.

A number of relatives have told *The Observer* that they are prepared to form an association of the next-of-kin to deal with humanitarian problems which may arise.

As can readily be seen from those newspaper reports, the French accident investigators were using the ordinary police in their attempts to make the identifications. Unfortunately, air accidents are completely different catastrophes to incidents involving a couple of cars colliding, or a person falling out of an upstairs window, which are matters the police force treat as 'everyday affairs' and for which they are adequately trained to cope. But air disaster victims present completely different sets of problems – some of which are so subtle that the civilian police would not easily recognize them, and so they might be overlooked. The fact that the police were asking victims' relatives to try and identify ordinary air mail being carried on board the machine is a case at point. Any human factors' specialist would have, very soon after arriving at the scene, determined whether the letters were being carried by the victims or were, as was the case, freighted post.

Even three months after the disaster had occurred, the French were still perplexed over the identity of about a third of the passengers on board. People had telephoned the authorities saying that one of their relatives was missing, a person who was supposed to have been travelling between Paris and London on that day, yet whose name did not appear on the passenger manifest, nor had there been any identification document found in the wreckage tying in with the name reported.

What would be expected from a group of amateur air accident human factors' investigators when they were confronted with a scene of the bodies, aptly described by a British Consulate official at the rescue site to arriving relatives, as, 'mincemeat' (his word)?

One London man even went to Paris to try and identify some of his son's belongings after he had reported him missing and presumably attending the Rugby international the day prior to the disaster. His trip was in vain. He is now of the opinion that possibly his son could still be alive – but where?

The facts are that only six bodies, out of the 346 persons on board, were found. Those were at *St. Pathus*, seventeen kilometres back along the flight path. The remainder who finished up in the *Forêt de Ermenonville*, ended up as was described – 'mincemeat'.

However one slight point *The Observer* omitted to mention in its story (some stories are better when you leave out certain of the facts – they sound more dramatic), was that the group of 'experts' that went to Paris to assist with passenger identification, was headed by a Squadron Leader A. Cullen of the Royal Air Force's pathology branch. Of course, one country does not appreciate members of another country's armed forces arriving on its territory uninvited – and especially when it would be agreed that a military official would be just as good (or bad) at civil aviation human factors' accident investigations, as the ordinary local police.

The French were still supposedly involved in the investigation of a mid-air collision between two Spanish registered aircraft which occurred just 363 days before the DC–10 disaster, over their territory, with amateur civil air traffic controllers (in fact, the military) operating the national system. The collision occurred when two machines, allotted the same altitude, were approaching the same point at the same time. Immediately after the accident, the French Minister of Transport, quite evidently without much forethought or investigation, called a press conference and emphatically stated that the military air traffic controllers were in no way responsibile for the 68 deaths when one of the planes exploded on contact. You will notice that practically the same thing occurred immediately after the DC–10 disaster.

Yet sixteen months after this crash, the French had still not released the accident report. It was as if they were treating the whole matter in true 'Watergate' style – keeping quiet until the statute of limitations expires, before possibly revealing that their stand-by military controllers did play more than just a passing rôle in the deaths and destruction. The normal length of time taken to investigate such accidents by, for example, the National Transportation Safety Board, is nine months. Here it is, double that length of time and the French still remained resolutely quiet on the subject.

By not disseminating the fullest information in the quickest possible time, other accidents with similar causes could quite easily occur in the meantime, to other aircraft in other parts of the world.

The main cause of the passenger identification confusion must be referred back to British Airways. It was, in the main, the added passengers that were given seats on board the Turkish aircraft by British Airways around which most of the confusion lies. British Airways has in the past, made a lot of propaganda about their 'fantastic' world-wide computer system. Yet it is not so 'fantastic' as to be able to give passengers' addresses in the case of an accident. British Airways were also instrumental in the stopping of the amenity of passenger seat selection at the time of checking-in. British Airways were also closely tied in with the pressures, which were bowed to, placed upon the Civil Aviation Authority to have oxygen systems removed from inter-European jet aircraft; and a few years back they were also very much against the mandatory installation of cockpit voice recorders until, of course, one of their aircraft crashed and no one to this day knows exactly what happened.

CHAPTER 46

AT last the F.A.A. had to act. With 346 deaths, they had to do something. Airworthiness Directives should have been properly thought out and issued 21 months previously, straight after the Windsor, Ontario accident, to ensure that a repetition could not occur, though tracing back the history of the 'recommended' Service Bulletins and their modifications – even if these had been issued as Airworthiness Directives in the first place – we could still have seen the Paris disaster. They in no way corrected the overall design deficiences.

Now the F.A.A. had to be *seen* to act. They had to issue an Airworthiness Directive – this time. It was the 6 March 1974 – three days after the holocaust.

And so out it flowed. It was achieved so rapidly that it was sent by telegram to the carriers involved, as well as to McDonnell Douglas, and all the air attachés in Washington of the foreign countries with airlines using DC–10s, but it was not printed in the Federal Register until the following day.

The F.A.A. stated that they were issuing it: "... because of an apparent in-flight depressurization which may have been associated with a cargo door."

In order to prevent possible in-flight depressurization of the airplane that might result from the opening of an improperly secured cargo door, accomplish the following:

1. Prior to further flight, unless already accomplished, perform the modifications and functional checks related below as follows:

A. Install viewing window in all cargo doors. Install placards and perform functional checks of door system and operation with McDonnell Douglas Service Bulletin A52–35, dated 19 June 1972.

B. Modify the cargo door latch actuator wiring in accordance with McDonnell Douglas Service Bulletin 52–27, dated 30 May 1972, or later F.A.A. Approved revisions.

C. Except for those aircraft which may have been modified in accordance with McDonnell Douglas Service Bulletin 52–49 dated 25 October 1973, or later F.A.A. Approved revisions, replace strike plate and adjust the switches on all forward, centre and aft cargo doors and install a support and plate on the aft cargo door in accordance with McDonnell Douglas Service Bulletin 52–37, dated 3 July 1972 or later F.A.A. Approved revision.

2. Prior to each flight, a crewmember must check each cargo door as follows:

A. Ensure that the cargo restraint curtain is in its proper position.

B. Ensure that no foreign matter is in the exposed door locking mechanism.

C. Ensure that the locking mechanisms are in their proper positions.

D. Accomplish a visual inspection through the inspection port and for locking pin placement.

E. Ensure all cargo door warning lights are extinguished prior to taxiing.

3. If any abnormalities in pressurization are observed during flight, immediately depressurize the aircraft and land at the nearest suitable airport.

4. The checks required by this Airworthiness Directive are to be accomplished by flight crew, notwithstanding any requirement of Federal Aviation Regulations.

Simultaneous with the issuance of this Directive, inspectors assigned to U.S. DC–10 operators will institute action to amend approved air carrier training programmes to ensure flight crew members are thoroughly familiar with and indoctrinated in the operation of locked cargo doors.

At last the F.A.A. had done something decisive. But was it good enough? With all those grandiose words of 'modifications' and 'viewing windows' and all the rest, once this order had been complied with, THE ONLY DIFFERENCE between a DC–10 being operated under these conditions and the Turkish Airlines DC–10, smashed into a million-plus pieces just north of Paris, was Section 2 – A, B, C, and D. The Paris aircraft had been modified – had it not? The pilots, before the disaster, had not experienced any troubles with the pressurization system – had they? But what if the Turkish DC–10 door had been checked by a crewmember – there would be no difference then – would there? So therefore, if someone actually did skimp in carrying out Part 2 – A, B, C, and D, we could expect another DC–10 loaded with people to end up in exactly the same fashion as those near Paris. In other words, the F.A.A. Airworthiness Directive was *useless*, unless they had inspectors standing next to every DC–10, seeing that the crewmember whose job it was to check the doors for proper locking, did his job conscientiously. After the Airworthiness Directive *the door itself was still not idiot-proof.*

But the Airworthiness Directive solely deals with the door.

Nothing was said about the possibility of the floor collapsing – and even less was said about what would happen to the control cables that pass through the floor beams, if the former did eventuate.

Soon after receiving the Airworthiness Directive, National Airlines which had the DC–10 that was involved in the engine disintegration at high altitude on the 3 November 1973, *and a practically exact repetition just on six months later*, were visited by F.A.A. inspectors. They jumped on that particular aircraft – N60NA – which just happened to be nearby, and carried out a thorough inspection to see whether or not it had been modified. They discovered that this machine had *not* been modified, in that the support plate 'recommended' by McDonnell Douglas as part of their Service Bulletin 52–37 issued on the 3 July 1972 – 21 months previously – and which Shaffer had said under oath was the answer to all the design faults in the DC–10 aircraft, the modification that could be: ". . . done overnight . . .", *had not been done*. This particular aircraft however, had affected 438 flights from November 1973 until that date, and during the same period it had been incessantly held up in departures because of difficulties in closing the aft cargo compartment door. National Airlines claimed they 'accidentally' missed out making the modifications because the Service Bulletin *looked like any other normal piece of paper* received through the post, and the urgency of its instructions was not obvious. But then, National Airlines were unable to explain why they did not notice the deficiency when, surely, they must have seen that they had this modification part 'left over'. Each of the five kits ordered and received by National Airlines for their five DC–10s, contained 97 parts – four of which were the strike plate and the rivets needed to place it in position. The remaining 93 parts of the kit had been fitted onto this particular machine, but the most important part *was not*. Where were these parts that were left over? Why was this discrepancy not noticed?

But worse was yet to come. A telegram was received from Laker Airways in London, saying that on issuance of the Airworthiness Directive, they too made a complete inspection of their two aircraft – both of which were supposed to have been modified *by McDonnell Douglas prior to delivery* – indeed, the aircrafts' log books were signed off stating that *the modifications had been done*. But during their inspection, they too had discovered that fuselage number 47 had not been modified in an exact similar fashion to the National Airlines DC–10.

Then disaster struck. The F.A.A. next received notification from the National Transportation Safety Board's investigators at the scene of the Paris disaster. They advised that they had inspected the remaining twisted pieces of metal, originally the aft cargo door that had fallen off the Turkish Airlines DC–10 in flight, and it *likewise* had not been modified in that it too did not contain the support plate mentioned in Service Bulletin 52–37. Again, it was the Douglas Aircraft Company that was *supposed* to have made this modification prior to delivery to Turkish Airlines – and again, the aircraft's log book showed the modification as being completed and signed off by a Douglas engineer.

It was not as though the N.T.S.B. investigators had any doubts. In order that the modification be affected, three holes had to be drilled in the door and rivets had to be used. Even if the plate had come off during the fall to, or impact with the ground, the rivet holes or the rivets should have been in evidence. But they were not. Just pure, smooth, twisted metal.

There are reasonable grounds, in view of this questionable quality control procedure, to be concerned about its effects on other supposed modifications on other McDonnell Douglas made aircraft.

As the F.A.A. inspections of all American operated DC–10s continued, they found yet another aircraft belonging to yet another carrier that had not been fully modified, and on top of this, sixteen further minor discrepancies were located in the remainder of the fleet.

The support plate that was subsequently discovered 'missing' from these three machines, was supposed to have raised the pressure that would have been needed on the lock handle to be exerted by a door closing operator in order to *force* it into a flush position with the door while the lock-grips were not in place, from about 120 lbs. to something like 420 lbs. With this increase, it was felt by all concerned, that if someone did force the thing shut, they must have done it with full knowledge that the door was improperly closed.

The pressure previously required on the door close handle if everything was correctly done, was about 30 lbs. It has been estimated that at both Detroit Airport on the 12 June 1972, and at Paris *Orly* on the 3 March 1974, the persons that closed these doors must have placed about 120 lbs. pressure on the handles – four times that normally required – and in actual fact, the person involved at Detroit admitted getting his knee to it in order to force it into the closed position. The chap at Paris reportedly had no trouble in stowing the door's handle. But still, without the 52–37 modification and with a little more pressure than normal, you could get this handle down into the flush position with the outside of the door.

With the addition of the support plate – the one omitted on these three aircraft – it was contended that the pressure then required would be in the vicinity of 420 lbs., but presumably it would also be possible to dislodge the plate under such circumstances, and *then*, how would you know?

As we have already discussed, of the whole list of 'recommended' changes, there were only two modifications that had any real major effect on guaranteeing that these doors were properly closed and locked. They were: Service Bulletin 52–37, the one discovered to be 'missing' from the three DC–10s, and Service Bulletin 52–40 (which no one has talked about since it was issued, and was even ignored in the F.A.A.'s 'panic' Airworthiness Directive), issued on the 27 October 1972.

After the Paris crash, John Brizendine, the president of the Douglas Aircraft Company, made the following statement:

> . . . according to our manufacturing records, all Service Bulletins that had been issued to improve the cargo door latching mechanism had been incorporated in the Turkish Airlines aircraft prior to its delivery.

From the point of view of a part of Service Bulletin 52–37 having been omitted from incorporation, that statement is quite correct. Their records did show – falsely – that this modification had been affected. And as we have discussed, the only two modifications of any value in the saving of lives were those in Service Bulletins 52–37 and 52–40.

At the American Congress hearings into the affairs of both the F.A.A. and the McDonnell Douglas Corporation's involvement in the Paris disaster, the committee's astute counsel, Daniel J. Manelli, ensured that the above statement made by Brizendine was accepted by the committee under oath. This means that Brizendine had sworn that it was a correct statement. Mr. Manelli continued his questions:

> You are referring to the fact that the records indicate at least that the work (of incorporating the Service Bulletins) had been done (prior to the aircraft being delivered to Turkish Airlines).

This is a computer printout which pertains to fuselage number 29, which is the one that crashed. We obtained this from Douglas. I don't know what page this is, but down the page, about $1\frac{1}{2}$ inches, you see, the Service Bulletins are listed on the left. Service Bulletin 52–29 is at the top of the page, and you go down to Service Bulletin 52–40, which is the one we have not mentioned yet today. That Service Bulletin calls for the modification for the forward, centre and aft cargo door assemblies, and then it gives the man-hours and elapsed-hours, and then there is an 'X' in the last column. That indicates 'open', which I take it means that the work is not finished. (A) (Brizendine) Not accomplished.

(Q) This is Service Bulletin accountability effective on 20 March 1974. As of that date, it would indicate that 52–40 had never been accomplished on that airplane; is that correct? Am I reading this right? (A) That would be my interpretation of this; by looking at the legend. 'X' equals open, it means the item is 'open' which in normal semantics means 'not accomplished'. It has an 'X' on that line.

(Q) Your company issued Service Bulletin 52–40 on the 27 October 1972. So this would have come out after 52–37. I would ask that that be included in the record at this point and that you be given a copy of 52–40 to look at the first two pages. This Service Bulletin provides for a further modification of the aft cargo door assemblies. The notation on the first page states that the modification provided therein must be accomplished in conjunction with previously issued Service Bulletins 52–35, 52–37 and 52–38 which is another one we have not discussed. On the second page of the Bulletin, the reason for issuing 52–40 is set forth in reference to the cargo door problem in the Detroit accident.

Now, at the time Service Bulletin 52–40 was issued, the Turkish aircraft was in the custody of your company was it not? (A) That is correct, to the best of my knowledge.

(Q) So it was still at the Douglas factory somewhere. Our information – I think the records show it was not delivered to the Turkish Airlines until 17 December 1972. Can you explain why this modification was apparently not accomplished on this plane? (A) I think so. This particular modification put in a load limiting spring in lieu of a solid link between that handle that actuates the lock-pins and the vent door. It made it impossible to put more than 50 lbs. I guess, pressure on this. If you did, the handle would go down but nothing would happen. The door would not close and so on.

(Q) So that too was a safety feature? (A) Not necessarily. It was a refinement of what we had already done.

(Q) Are you saying that it is not a safety feature? Isn't the whole purpose of putting this spring on so that, if the guy is going to close the door improperly, the spring will release the pressure and he can just push the handle in without closing the vent door? And with that vent door not closed, the plane can't be pressurized or the pilot can't take off? This is a safety thing, isn't it? (A) No, sir. Not in our interpretation.

(Q) Well, what is the purpose of putting a tension spring on the handle? (A) It was a compression spring.

(Q) What was the purpose of putting that on in any event? Let me just put that question again. I thought you said before that if the guy leans on the handle he can push the handle 'in', but the vent door is not going to close? (A) I think that is it. Let me consult with my technical expert here.

Thus Brizendine in the above testimony contradicted portions of his earlier sworn statement to the effect that the Turkish aircraft had been modified in accord with all operable Service Bulletins prior to its delivery on the 17 December 1972, which would have meant that modification 52–40, issued on the 27 October 1972, was supposed to have been done.

In addition one can only conclude that Brizendine was trying to say that the 52–40 modification was *not* a safety feature, simply because Air Transport Association rules by which McDonnell Douglas were supposed to abide, says that *all* safety modifications *must* be issued as Alert Service Bulletins if not Airworthiness Directives, on special blue coloured paper to attract greater attention, and with the word 'Alert' stamped across each page. Therefore since Douglas had issued this vital Service Bulletin in an ordinary fashion without the 'Alert' notification, Brizendine therefore claimed that it was *not* a modification to ensure greater safety.

So now we learn that the *main two* modifications to the badly designed DC–10 cargo doors, had *not* been incorporated on the Turkish Airlines machine that crashed outside Paris, even though *both of these* were supposed to have been carried out by McDonnell Douglas prior to delivery.

Another valid point is that notices plastered all over the outside of the door and its surrounding area were, in the Detroit case, in English – the door closer's mother tongue. At Paris, the operator could read and write both French and Arabic, but the notices were in English and Turkish.

However there was a similarity between past occurrences experienced by both of the DC–10s involved in these accidents. In the lead-up time periods before the disasters, both American Airlines' and Turkish Airlines' door closing operators had experienced a large number of problems in shutting these things. On many occasions, flights were delayed because they could not get the doors closed.

But there is still no excuse for the Douglas Aircraft Company, a division of the McDonnell Douglas Corporation, that permitted or even employed persons whose lack of responsibility was shown in that they were prepared to, and actually did, sign off safety modifications as having been completed, when in actual fact they had not been done.

Now to take this whole matter one stage further. However, firstly it must be explained that since the very first days of flying machines and internal combustion engines, designers realized, from both hard experience and unnecessary deaths, that if a part could be installed the wrong way around, someone, somewhere, some day, would install it in such a fashion, and deaths and destruction could be caused as a result.

The whole of the aircraft industry knew of this possibility. They knew, because over the years, many crashes with much suffering and bloodshed were caused by designers building parts that were capable of being installed incorrectly or being mis-rigged. This is history – this is not a new discovery.

Now how good is McDonnell Douglas design?

At the products' liability trial in Los Angeles, it has been revealed that the internal workings of the crashed Turkish DC–10's aft cargo door, found in the forest near *Saint Pathus* had been mis-rigged.

One is not planning to enter the argument as to who mis-rigged it – either McDonnell Douglas before delivery, or Turkish Airlines afterwards – but the fact that the internal workings' design seems to be such that it can be mis-rigged, is clear proof of design failure.

CHAPTER 47

On 7 March 1974, smallish stories appeared in many of the British newspapers of great importance to potential DC–10 passengers.

AIRLINE WILL SEAL ALL CARGO DOORS AFTER DC-10 DISASTER

That was the London *Daily Mail's* headline piece.

Laker Airways was the only British operator of the DC–10 (British Airways had undertaken a 'study' of the machine, but had ordered in preference, the Lockheed TriStars). They now announced to the press that because of the adverse publicity given to the DC–10 with regards to design faults in the cargo door's latching system as well as the inadequate cabin floor strength and the disgraceful control cable placement, they were going to 'secure' these doors on their two machines, making them impossible to open, until such a time as 'the air had cleared'.

Mr. Freddie Laker, the company's managing director, had ordered this to be done off his own bat. The British Civil Aviation Authority, at this stage, had not ordered any added modifications.

The question now was, why did not the British Civil Aviation Authority take the most logical and necessary steps of *de-certificating* the DC–10?

An approach was made to the chairman of the Authority, to put these and more pertinent questions to him. But due to: ". . . pressures in his diary and the situation in which we find ourselves in regard to this particular case, he still feels that he cannot find the opportunity right now–in the foreseeable future that is – to arrange a face-to-face encounter." So was the excuse given by the chairman's private secretary, Michael E. Tredray.

In an aggrieved half-hour telephone conversation, Tredray admitted that the British had issued certificates of airworthiness for the two Laker Airways' DC–10s solely because the F.A.A. had themselves issued such certificates. His actual words were:

> We have close bearings with the F.A.A. for acceptance of their airworthiness standards. There is, after all, an understanding

between the U.K. and the U.S.A. on these subjects, and this means we obviously have the right to note differences which we would wish to be introduced into American types coming into this country. But basically it is expected that we will accept American Standards.

And what if a British designed and built aircraft does not come up to American standards? Do they have to accept it?

Well, obviously, it is entirely reciprocal as far as I understand it. I mean obviously I am not in the American administration, and I'm not too sure about the way in which they imported our BAC 1–11s. But I understand these things to be arranged in an entirely reciprocal way. They therefore would accept our B.C.A.R.s (British Civil Air Regulations), subject to any small refinements, modifications, which they feel are for good reason, necessary.

One presumes he means here, modifications, for instance, that were found necessary in the BAC 1–11s after a plane load of people had died when a fire – unbeknown to the pilots – caused the tail to fall off. One supposes that he means that these are the types of design modifications that they may require. But then, why could the British not sell one single Trident in America? Because of the F.A.A.'s certification standards? Or for some other reasons? What if they found a fault in the aircraft, or did not like the standards laid down in the British Civil Air Regulations?

Ah, well, obviously, they would quickly inform us of these things in order not only to ensure that the aircraft going to their country didn't include this defect, but that the defect was put right for all others on the register.

Yet the Federal Aviation Administration did not reciprocate in like fashion to the Civil Aviation Authority?

Yes, well, er, this is a thing that they said at the inquiry and other inquisitions into the DC–10s history. Because after all, the modification was 'suggested'; it wasn't perhaps suggested strongly enough – this is the door modifications to the DC–10. It is purely a matter of priorities which it was given, and then the manufacturer's arrangements for ensuring that these modifications are introduced in that way.

During the conversation, it was revealed that the British 'Authority' had mandatorily required the two DC–10s being delivered to Laker Airways to be modified to all Service Bulletin standards.

They did not demand modifications to the floor to strengthen it. They did not require modifications to the control cable placement *through* the floor beams. But they did require the whole 'gamut' of 'minor adjustments' – a notice here, a screw there etc., – to be made on the aircraft *before they were delivered.*

Then why was it suddenly discovered on the morning of 6 March 1974, that Laker Airways DC–10, fuselage number 47, which had been delivered to the airline on 16 October 1972, was legally unairworthy because McDonnell Douglas had *not* incorporated the addition of a support plate 'recommended' in their Service Bulletin 52–37, dated 3 July 1972, even though they had signed the aircraft's log book off as if the modification had been completed?

> Ah, yes, well, because we don't, er, ourselves, er, take every aircraft to pieces when we give them a certificate of airworthiness.

Then under what circumstances did the British Civil Aviation Authority issue a certificate of airworthiness to an aircraft that was *supposed* to be modified, but which, in fact, was not?

> Ah, well, er, that's something on which I'm not in a position to advise on conclusively. But if you would put that question to us, I am sure we will be able to answer it – and in correspondence for quoting.

Having already asked a large number of passenger-safety questions of the chairman of the British "Authority", it was realized that they, who have had just under 200 passenger deaths on their hands in the past couple of years, were again hedging around subjects which could mean life or death to passengers. From the conversation, it was quite evident that the British "Authority" had "rubber-stamped" the F.A.A.'s airworthiness certificates for the DC–10 with possibly – as no proof of their actions was forthcoming – the smallish addition that British certificated DC–10s (both of them) *must* include all the modifications "recommended" in the McDonnell Douglas Service Bulletins.

Whether these modifications were effective in correcting the faults, of course, was another matter. And so it was quite evident that these machines were imported *carte blanche* without any of the authority's inspectors actually making sure that the modifications had been done.

When the Civil Aviation Authority learned that I had given

up my search for answers to these questions and planned to publish details from the telephone conversations (which were, one might add, recorded), they quickly had pen to paper:

He (the chairman) has also noted that in the part of your analysis which concerns the C.A.A., and which you have already written, you propose to quote information I gave you in the course of one or other of the telephone conversations we had. If it would be helpful to you, we should be only too pleased and willing to arrange for your draft to be checked for accuracy and for any misunderstandings or errors to be pointed out.

This now turns out to be an interesting series of international 'cover-up' situations. There are thousands of civil aircraft in the air, most of them imported from one country to another. How many of these, one wonders, also suffer from DC–10 mania?

And what of the rest? What of the countries that knew about the DC–10 design faults having airline members of the KUSS group; the Dutch, the Swiss, the Scandinavians and the French? What of the remainder; the Spanish, the Germans, the Zairois, the Gabonese, the Belgians, the New Zealanders, the Pakistanis, and of course, the Turks? Did they all permit their airlines to import these DC–10s practically sight-unseen, as they must have done if they had not woken up to the door/floor/cables problems, and all the others listed elsewhere?

With all these countries, other than the Dutch and Germans, there could not be any bi-lateral agreements of the nature of "you certificate my aircraft and I'll certificate yours". None of the others build civil airliners. How do these exchanges operate under such circumstances?

Firstly, the Dutch *Rijksluchtvaartdienst Luchthaven*, after three months, had not replied to correspondence. On telephoning their London Embassy and speaking to the civil air attaché's secretary, she suggested the inquiry be put to the British Civil Aviation Authority, which of course, defeated the purpose of trying to find out why the Dutch certificated the DC–10s purchased by K.L.M., when it was known to have design faults. And remembering that the Dutch had realized this way back in 1970.

Neither Turkish Airlines nor the Turkish Civil Aviation department in Ankara cared to reply to correspondence, so the same telephoned inquiry was made to their London Embassy. After speaking to someone who did not know whether they employed an air attaché, let alone a civil air attaché, the telephone was answered by a young English-accented woman who said that the air attaché could not be spoken to, but she could, and what was the purpose of the call? It was then explained that the reason was to find out under what circumstances the Turkish authorities awarded certificates of airworthiness to foreign manufactured aircraft, and her answer was as follows:

> I am afraid you will have to write to us. We couldn't give you this information over the telephone. . . . If you write in, then the air attaché will look at your letter, and it's up to him whether he wishes to answer.

The Turkish could have been very touchy because of their direct involvement in the 346 deaths near Paris. The Turkish authority should have been well aware of the fact that Turkish Airlines had had difficulties in the past, in opening and closing the aft cargo door on the crashed airplane, and it therefore must solely have been for these reasons that the very simple information sought after: ". . . could not be given over the telephone . . ."

The same reticence was shown by the Swiss and Finnish attachés. The only difference between the Swiss and the Dutch was that the former suggested contacting the British Embassy in Bern. (You see, they are not as efficient as their publicity makes out.)

From the refusals in one way or the other, it seems that these countries had simply "rubber-stamped" the original F.A.A. certification, but without being involved in any "gentlemen's agreements" between the certificators and the president of the manufacturing company. They evidently, other than the Dutch, had accepted the aircraft, more than likely without even inspecting it, and now, since all the deaths in the *Forêt d'Ermenonville*, realized that they *too* could be implicated in the many writs and court actions that were developing. Certainly, if another accident were to occur to any of these aircraft that they had certificated, they would have absolutely no defence against ensuing litigation.

They were in an awkward situation: they were morally obliged to cancel the DC–10s' certificates of airworthiness until the design faults had been corrected – but this would have thrown their airlines' schedules into chaos. On the other hand, if they did not have these machines taken out of service, there was the possibility of a few hundred more passenger deaths. This is the present situation as it seems to stand today.

CHAPTER 48

THE National Transportation Safety Board, in referring back to their correspondence files, noted on 12 March 1974, that the F.A.A. had been deathly quiet on the subject of the DC–10 problems for just on one year.

However, one should give credit where credit is due. In America, unlike all other countries which certificated the DC–10 – places like Britain, Switzerland, France, Germany and even Turkey – there is a Federal Air Regulation that says that the pilots and airlines *must report all incidents* to the F.A.A. An "incident" is any untoward event that occurs during a flight, which does not result in injury or worse to any person on board, and does not result in substantial damage or worse to the aircraft.

If a civil air regulatory body was supposed to be conscientious in its determination to stamp out accidents, it is essential that such a department knows about these, because most accident causes vividly show themselves in incidents, months before the deaths and injuries occur.

Incidents related to the causes, can be traced back as happening long before the major disasters. British European Airways (now with its new, "cleaner" name – British Airways), had evidence but did not act on it, years before the 118 people had to die near London Heathrow Airport on 18 June 1972. Boeing, other airlines, and the civil air authorities, had evidence of incidents of exactly the same nature as the causes (a raging fire that started in a lavatory, racing through the interior of the passenger cabin with the aircraft in flight), years before the 123 people had to die near Paris, *Orly* Airport, on 11 July 1973, when a Varig Boeing 707 was trying to land.

As we can vividly see in our DC–10 story, there were hundreds of incidents *and* a non-fatal accident all caused by problems in the DC–10's cargo door design and related difficulties with the floor and control cables, long before the 346 people had to die near Paris on 3 March 1974.

The terrible part, is that the F.A.A. were well aware of the situation – the Douglas Aircraft Company and the "rubber-stamping" civil aviation departments of the other countries which imported DC–10s, were well aware of the situation, but neither Shaffer, nor Butterfield, nor McGowen, nor Brizendine, nor anyone else did sufficient about it. And at the time of writing, *the position is exactly the same today because only a small part of the overall design problems has been corrected.*

But are these two prior major events with the doors the only ones?

During a recent trip to Washington D.C., I walked in to the F.A.A.'s third floor information filing department. It took about two hours for them to realize that a penetration of their artificial barriers had been made, but in that time, a copy of their DC–10 incident report summaries had been written out. As these only offer the barest of facts, an official request was then lodged for photocopies of eighteen of the full reports. The information, one might add, comes under the 1966 Act that says that information must be given on request, and must be made available to anyone who specifically asks for it.

Yet the F.A.A. has refused to hand over these photocopies. Now, why? Because here, we would have proof of seventeen other instances of possible near disasters caused by the design of these doors? Is it because, if the F.A.A. were to hand over the documents, there would be even further proof of gross negligence in letting the DC–10 design faults perpetuate to the stage of the Paris disaster?

Before presenting the list of incidents obtained from the F.A.A.'s files in the short time they were available, it must be said in all honesty that the number of reports is nothing unusual. With the introduction of a new aircraft into passenger services, because of the lackadaisical methods of testing the designs concerned, there always are large numbers of incidents reported in the first couple of years of operations.

However, this list is not complete because of the lax attitudes of the other civil air departments which have also "rubber-stamped" this aircraft's design certificates. Of the whole lot of them, only the F.A.A. insists on the mandatory reporting of incidents. And therefore it is more than likely that many more cases have occurred, but no one knows officially about them. For instance, there are a number of known Turkish Airlines incidents, but the "finer details" are not available.

In order that the reports be located in the F.A.A.'s files, they are listed under the date the incident occurred and then given a code number for the report received concerning that day. Further, the F.A.A. uses the terminology "UL" to indicate that the aircraft was involved in an "unscheduled landing", as a result of the incident. Here is the list:

14 Jan 1972	1	United Air Lines	Wing slat did not retract – UL.
23 Jan 1972	1	American Airlines	Unruly passenger on board – UL.
31 Jan 1972	1	National Airlines	Electric power loss – UL.
4 Feb 1972	1	National Airlines	Landing gear door jammed – UL.
8 Feb 1972	3	American Airlines	Fuel weight problems – UL.
17 Feb 1972	3	American Airlines	Engine failure.
3 Mar 1972	3	United Air Lines	Ran off the taxiway.
5 Mar 1972	2	National Airlines	Fuselage skin damaged in airport accident.
5 Mar 1972	3	United Air Lines	Ran off taxiway.
6 Mar 1972	3	American Airlines	Landing gear unsafe – UL.
8 Apl 1972	1	United Air Lines	Nose landing gear remained down – UL.
23 Apl 1972	1	National Airlines	Loss of hydraulic pressure – UL.
26 Apl 1972	1	United Air Lines	Fire in number 3 engine.
1 May 1972	6	United Air Lines	Collision at airport with aircraft – being started.
2 May 1972	1	Continental Airlines	Failure in number 2 engine.
7 May 1972	3	United Air Lines	Slats 'out of position' light remained 'on'.
14 May 1972	3	American Airlines	Ran off taxiway while turning.
31 May 1972	4	Continental Airlines	Brake failure while taxiing.
6 Jun 1972	8	National Airlines	Slat struck a loading cart while taxiing.
11 Jun 1972	1	National Airlines	Severe turbulence *en route* LAX – MIA.
18 Jun 1972	3	United Air Lines	Nose landing gear door vibration – UL.
23 Jun 1972	3	American Airlines	Bird ingestion No. 1 engine on take-off – UL.
6 Jul 1972	2	United Air Lines	Hydraulic fluid loss No. 3 and No. 2 systems.
12 Jul 1972	5	American Airlines	Low oil pressure – No. 2 engine shut down.
20 Jul 1972	4	American Airlines	False stick-shake – equipment failure.
29 Jul 1972	1	American Airlines	Aircraft pushed back into jetway gate.
3 Aug 1972	1	United Air Lines	No. 3 engine disintegrated on take-off.

10 Aug 1972	1	American Airlines	No. 3 engine shut down – severe vibrations.
13 Aug 1972	1	American Airlines	Hydraulic leak in No. 2 hydraulic system.
26 Aug 1972	4	American Airlines	Hydraulic leak in No. 3 pump seal.
30 Aug 1972	8	American Airlines	Fire warning in No. 3 engine – UL.
2 Sep 1972	3	American Airlines	Hot brake system on No. 8 wheel.
17 Sep 1972	7	American Airlines	Moderate turbulence – 1 stewardess injured.
20 Sep 1972	3	American Airlines	Pressurization loss – UL.
21 Sep 1972	1	United Air Lines	Unable to retract nose landing gear – UL.
22 Sep 1972	1	United Air Lines	No. 3 engine shut down, vibration – UL.
23 Sep 1972	3	National Airlines	Fire warning in No. 3 engine – shut down.
2 Oct 1972	1	American Airlines	No 3 engine shut down because of oil loss.
5 Oct 1972	4	American Airlines	No. 2 engine shut down – oil pressure loss.
6 Oct 1972	1	American Airlines	Reported fight between two passengers.
10 Oct 1972	2	United Air Lines	Fire warning in No. 1 engine; stuck in mud on landing.
13 Oct 1972	2	United Air Lines	Taxied onto grass off taxiway.
26 Oct 1972	5	United Air Lines	Multiple No. 1 and No. 3 engine fire warnings.
29 Oct 1972	5	Continental Airlines	No. 3 engine compressor stall – UL.
29 Oct 1972	8	United Air Lines	Fuel leak – turbine failure No. 3 engine.
30 Oct 1972	2	American Airlines	Bird strike.
2 Nov 1972	1	McDonnell Douglas	Ran off taxiway and sank into asphalt.
4 Nov 1972	1	United Air Lines	No. 1 engine failure.
16 Nov 1972	1	United Air Lines	Ran off taxiway and was bogged down in mud.
30 Nov 1972	4	American Airlines	Fire warning in No. 3 engine – UL.
13 Dec 1972	2	National Airlines	No. 2 engine high fuel consumption – stopped – UL.
13 Dec 1972	6	United Air Lines	No. 2 engine shut down – vibrations in No. 3 – UL.
15 Dec 1972	3	American Airlines	Slid off end of icy runway landing.

22 Dec 1972	1 United Air Lines	No. 1 engine cowl loose – UL.
24 Dec 1972	1 National Airlines	No. 3 engine shut down, oil loss – UL.
26 Dec 1972	1 American Airlines	Unsafe nose landing gear indication.
27 Dec 1972	4 United Air Lines	AC electrical failure – UL.
30 Dec 1972	6 American Airlines	No. 2 hydraulic system failure, flaps not extended.
3 Jan 1973	1 United Air Lines	Oil loss in No. 1 engine – shut down.
4 Jan 1973	4 National Airlines	Low on oil in No. 1 engine.
9 Jan 1973	1 American Airlines	Left MLG door would not close after take-off.
9 Jan 1973	5 Continental Airlines	Unable to pressurize aircraft – UL.
15 Jan 1973	2 American Airlines	Two unruly passengers *en route*.
18 Jan 1973	3 National Airlines	Right wing flap and slat, stuck.
23 Jan 1973	2 National Airlines	Fire warning in No. 1 engine – shut down.
25 Jan 1973	1 American Airlines	Low oil pressure – No. 2 engine shut down.
4 Feb 1973	1 American Airlines	Fire warning in No. 1 engine – shut down.
7 Feb 1973	1 United Air Lines	Unable to pressurize aircraft – UL.
9 Feb 1973	10 American Airlines	Unable to retract right landing gear – UL.
10 Feb 1973	4 American Airlines	Vibration in No. 3 engine – shut down – UL.
11 Feb 1973	1 American Airlines	No. 1 thrust reverser deployed in flight.
11 Feb 1973	2 American Airlines	No. 2 throttle stuck at 88.9% N_2.
22 Feb 1973	3 American Airlines	Ran off runway – bogged down in mud.
27 Feb 1973	2 American Airlines	Oil pressure failure in No. 3 engine.
28 Feb 1973	4 Continental Airlines	No. 3 oil pressure failure – take-off aborted.
3 Mar 1973	4 American Airlines	Oil loss in No. 3 engine – shut down.
4 Mar 1973	3 American Airlines	Unsafe landing gear indication.
7 Mar 1973	1 Delta Air Lines	No. 1 hydraulic pump failure – UL.
10 Mar 1973	3 American Airlines	No. 1 and No. 2 hydraulics systems failures.
11 Mar 1973	3 United Air Lines	Flame out and fire in No. 3 engine.

18 Mar 1973	6	United Air Lines	No. 3 hydraulics system failure.
19 Mar 1973	1	American Airlines	Slid off runway on landing.
27 Mar 1973	6	American Airlines	Oil loss in No. 3 engine – shut down – UL.
3 Apl 1973	4	National Airlines	ILS control head malfunction.
5 Apl 1973	2	Northwest Airlines	No. 3 engine 1st stage fan failure – UL.
15 Apl 1973	4	United Air Lines	Co-pilot's sliding window opened after take-off – UL.
16 Apl 1973	3	National Airlines	No. 1 and No. 3 engines ingested seagulls.
18 Apl 1973	4	American Airlines	Loose asphalt on runway damaged flap on landing.
23 Apl 1973	3	United Air Lines	Smoke in cabin – 1 passenger fainted – UL.
23 Apl 1973	4	National Airlines	Take-off aborted and taxied off taxiway.
25 Apl 1973	3	American Airlines	Oil loss in No. 3 engine – shut down.
2 May 1973	3	American Airlines	Unable to pressurize aircraft after take-off.
3 May 1973	4	American Airlines	Cargo door warning light "on" – UL.
5 May 1973	6	American Airlines	Fire warning in No. 1 engine – shut down.
7 May 1973	3	American Airlines	Access door lost in flight – No. 3 access door.
8 May 1973	6	National Airlines	No. 2 engine translating cowl separated in flight.
9 May 1973	1	American Airlines	No. 3 hydraulic system failure.
12 May 1973	1	American Airlines	Stewardess cut her hand on serving cart.
19 May 1973	2	American Airlines	Oil loss in No. 1 engine – shut down – UL.
21 May 1973	5	Delta Air Lines	Hit tug cart on taxiing to Gate.
27 May 1973	2	American Airlines	Pressurization uncontrollable – UL.
27 May 1973	6	United Air Lines	Dragged tail on runway on landing.
5 Jun 1973	5	American Airlines	Fire warning in No. 2 engine – shut down – UL.
7 Jun 1973	4	Northwest Airlines	1R escape chute deployed on taking off.
12 Jun 1973	4	Northwest Airlines	Taxied over "ramp" sign and on to grass.
14 Jun 1973	3	Trans International	Tyre blows out on take-off from Malaga, Spain.
14 Jun 1973	6	United Air Lines	Unable to retract slat – UL.

18 Jun 1973	3	United Air Lines	No. 1 and No. 2 tyres blew out on take-off.
19 Jun 1973	1	Northwest Airlines	Split flap indication/blew tyres on landing.
19 Jun 1973	6	United Air Lines	Stewardess injured in turbulence.
21 Jun 1973	2	American Airlines	Fire warning in No. 3 engine–shut down – UL.
21 Jun 1973	3	Iberia Air Lines	Hit TWA 707 on push back at Boston, Mass.
22 Jun 1973	2	American Airlines	Take-off aborted – escape slide deployment.
29 Jun 1973	6	United Air Lines	Dragged tail on landing.
2 Jul 1973	4	United Air Lines	High vibrations No. 1 engine – shut down – UL.
4 Jul 1973	4	Delta Air Lines	Slats would not retract – UL.
4 Jul 1973	3	American Airlines	Oil loss in No. 3 engine – shut down – UL.
7 Jul 1973	6	Northwest Airlines	No. 4 passenger door opened in flight – UL.
8 Jul 1973	3	American Airlines	Med-oxygen duo-pack burnt through seat – extinguisher failure.
8 Jul 1973	8	National Airlines	Translating engine cowl separated on landing.
14 Jul 1973	1	Delta Air Lines	Extensive vibration No. 1 engine – shut down – UL.
14 Jul 1973	4	National Airlines	Jumped chocks, hit work stand injuring mechanic.
18 Jul 1973	3	Continental Airlines	Slat would not retract – UL.
18 Jul 1973	6	United Air Lines	Dragged tail on landing.
18 Jul 1973	8	National Airlines	During tow, right wing struck rudder of EAL DC–9.
19 Jul 1973	6	American Airlines	Tail cone work platform door came open in flight.
22 Jul 1973	4	Continental Airlines	During taxi, right gear ran over obstruction light – tyre blew – fire – emergency evacuation.
25 Jul 1973	5	American Airlines	Left landing gear failed to retract – emergency evacuation.
26 Jul 1973	3	Northwest Airlines	Unable to retract slats – stall warning in flight.
29 Jul 1973	2	KLM Dutch Airlines	Pallets in forward cargo compt. not locked down.
2 Aug 1973	4	American Airlines	Brakes locked – excessive air in system.
6 Aug 1973	2	American Airlines	Fire warning in No. 2 engine – abort take-off.

7 Aug 1973	1	Delta Air Lines	Right wing refuelling door fell off in flight.
9 Aug 1973	1	United Air Lines	No. 2 hydraulics system failure.
10 Aug 1973	6	Northwest Airlines	Fire warning in No. 3 engine – shut down.
10 Aug 1973	9	National Airlines	Stewardess injured by food service cart.
15 Aug 1973	2	Continental Airlines	Radioactive package damaged *en route*.
21 Aug 1973	4	United Air Lines	No. 1 engine failure – take-off aborted.
23 Aug 1973	4	United Air Lines	Lost hydraulic fluid and pressure.
25 Aug 1973	4	Northwest Airlines	Lost tyre recap on take-off – blew on landing.
27 Aug 1973	5	Continental Airlines	No. 3 engine failure – take-off aborted.
4 Sep 1973	3	Northwest Airlines	Unable to maintain cabin pressurization.
6 Sep 1973	6	National Airlines	Ran off taxiway after landing doing high speed turn.
14 Sep 1973	4	Delta Air Lines	Compressor stall in No. 3 engine – UL.
15 Sep 1973	1	United Air Lines	No. 2 engine caught fire before take-off.
18 Sep 1973	5	American Airlines	Made landing with nose gear "unsafe" light flicker.
21 Sep 1973	2	American Airlines	Vibration in No. 1 engine – power reduced.
3 Oct 1973	5	American Airlines	Stewardess struck by food cart – back injury.
13 Oct 1973	1	American Airlines	High oil consumption in No. 2 engine – shut down.
16 Oct 1973	1	American Airlines	No. 2 engine exploded on take-off.
16 Oct 1973	6	United Air Lines	Right wing spoiler inoperative.
30 Oct 1973	3	American Airlines	Overhead panel fell on passenger's head.
31 Oct 1973	4	American Airlines	No. 3 engine would not de-reverse on landing.
1 Nov 1973	3	United Air Lines	No. 3 engine exploded during descent.
5 Nov 1973	2	Trans International	No. 1 engine vibrated after take-off – bird ingested.
5 Nov 1973	3	National Airlines	Oil pressure zero in No. 3 engine – shut down – UL.
6 Nov 1973	2	United Air Lines	Low oil – No. 3 engine shut down.

6 Nov 1973	5 Delta Air Lines	Right landing gear did not retract after take-off.
7 Nov 1973	3 Northwest Airlines	Forward cargo door seals damaged – loss of pressurization – UL.
7 Nov 1973	4 National Airlines	No. 2 engine failure on climb out.
7 Nov 1973	6 United Air Lines	Due to bad weather – UL.
8 Nov 1973	3 United Air Lines	Complete loss of pressure in No. 3 hydraulics system.
11 Nov 1973	3 United Air Lines	No. 1 brake system hydraulic valves defective.
11 Nov 1973	6 Continental Airlines	Arrived with tail cowl access door missing.
13 Nov 1973	5 Northwest Airlines	Jet blast hit tail section of Braniff B–727.
15 Nov 1973	8 United Air Lines	Open vent door in aft cargo compartment prevents pressurization after take-off.
25 Nov 1973	2 National Airlines	Fire warning in No. 1 engine – shut down.
26 Nov 1973	3 American Airlines	Reverse thrust unlock light came on during climb.
27 Nov 1973	2 Delta Air Lines	Right hydraulic pump shaft sheered inside.
3 Dec 1973	1 Delta Air Lines	High temperature in No. 3 hydraulics system.
9 Dec 1973	1 American Airlines	Vibration in No. 2 engine – shut down.
11 Dec 1973	2 American Airlines	No. 3 engine power loss – overtemperatured – blades bent.
13 Dec 1973	1 American Airlines	High speed compressor stall No. 1 engine – take-off aborted.
14 Dec 1973	3 Continental Airlines	No. 3 engine did not develop power – take-off aborted.
16 Dec 1973	1 Continental Airlines	Loss of oil quantity and pressure in No. 3 engine – shut down.
16 Dec 1973	4 American Airlines	Engine fan cowling damaged during taxiing – hit a conveyor.
17 Dec 1973	3 United Air Lines	Ice ingestion in No. 2 engine – shut down.
20 Dec 1973	5 Northwest Airlines	On taxiing, APU fire warning – extinguisher used.
29 Dec 1973	3 American Airlines	Loss of oil quantity and pressure in No. 3 engine.
29 Dec 1973	4 American Airlines	Unsafe nose landing gear extension warning.

6 Jan 1974	1 American Airlines	Unable to reduce power to No. 2 engine – shut down.
6 Jan 1974	6 American Airlines	Woman passenger hysterical.
8 Jan 1974	7 Continental Airlines	Unsafe nose landing gear indication.
10 Jan 1974	9 National Airlines	Fire warning on starting No. 3 engine.
10 Jan 1974	11 Northwest Airlines	On taxiing, slid into snow and sank.
11 Jan 1974	13 Continental Airlines	Elevator damaged in flight by work platform piece.
14 Jan 1974	3 Northwest Airlines	No. 3 engine reverse thrust light on – take-off aborted.
14 Jan 1974	4 National Airlines	Oil leak in No. 1 engine – shut down.
18 Jan 1974	1 Trans International	Left main gear failed to retract after take-off.
18 Jan 1974	7 American Airlines	After landing, ran off runway.
18 Jan 1974	8 National Airlines	Engine failure light on – abort take-off.
20 Jan 1974	1 American Airlines	No. 3 engine reverse thrusted in flight.
23 Jan 1974	2 Continental Airlines	Hydraulic and control problems *en route*.
4 Feb 1974	4 American Airlines	Right landing gear ran off taxiway – sank into mud.
15 Feb 1974	1 Continental Airlines	Yaw and strong vibration in aft of aircraft during climb out.
11 Mar 1974	3 American Airlines	On landing, No. 3 engine stuck in reverse thrust.

It must be pointed out that the airlines have 90 days in which to send in these reports of incidents to the F.A.A., and this is the reason for the "tailing-off" of information for the months of January, February and March 1974. It is not as though DC–10 aircraft suddenly improved.

Most of the incidents – well, at least some of them – seem to have happened as a direct result of bad pilot training and operations. For instance, it will be noted that all of the reports of DC–10 tails being scraped along the runways during landings, happened to United Air Lines aircraft.

At around the time of the DC–10 disaster near Paris, United were paying for advertisements, seemingly shown hundreds of times each week on all the major American television networks. United stresses "The Friendly Skies of Your Land" in its publicity; and the DC–10 according to McDonnell Douglas propaganda, is the "Friendly Neighbour" aircraft.

Here are the words of the beautifully sung advertisements:

Friendship service,
The Friendly Skies Of Your Land,
Ever seen the other side . . .
It's people, making sure your trip goes smoooooth and easy, taking care of your things like we take care of our own . . .
. . . And so much more to give.
Mmmmmmmmother country.
. . . Looking after your comfort every step of the way . . .
Don't let yer convan pass yer by . . .
Pass you by – eye – eye – eye.
Friendship . . . good reasons why people choose The Friendly Skies more than any other airline in the land . . .
Wild beneath the skies.
Come along, sing the song,
People, now's the time.
The Friendly Skies Of Your Land,
United Air Lines.

The "league table" of incidents is another interesting point, though one cannot gain any hard conclusions from it simply because, for a correct analysis, it would be necessary to do a day-by-day survey, taking into account other factors such as the number of flights, numbers of aircraft, and so on. However, a simple league table comes up with these results:

Airline	*No. of aircraft*	*Incidents*	*Incidents/ Aircraft*
American Airlines	25	79	3.2
National Airlines	11	27	2.6
United Air Lines	20	46	2.3
Continental Airlines	8	16	2.0
Delta Air Lines	5	9	1.8
Northwest Orient Airlines	15	14	0.9

But then one must also remember that there have been uncovered over 100 incidents that these carriers have simply not reported to the F.A.A. which solely involved troubles closing the doors. However, it is still interesting to see that American Airlines have the worst record in this field – remembering, of course, that it was American Airlines that presented the DC–10 Paper to the Air Transport Association on 10 December 1971, proudly pointing out how much involvement they had had in the machine's design and concept.

The incident reports mentioned earlier, copies of which were requested from the Federal Aviation Administration and refused, are as follows:

20 Sep	1972	3	American Airlines	Pressurization loss – UL.
26 Oct	1972	5	United Air Lines	Multiple No. 1 and 3 engine fire warnings.
9 Jan	1973	5	Continental Airlines	Unable to pressurize aircraft – UL.
7 Feb	1973	1	United Air Lines	Unable to pressurize aircraft – UL.
23 Apl	1973	3	United Air Lines	Smoke in Cabin – 1 passenger fainted – UL.
2 May	1973	3	American Airlines	Unable to pressurize after take-off.
3 May	1973	4	American Airlines	Cargo door warning light "on" – UL.
7 May	1973	3	American Airlines	Access door lost in flight.
27 May	1973	2	American Airlines	Pressurization uncontrollable – UL.
7 Jun	1973	4	Northwest Airlines	1R escape chute deployed on take-off – UL.
22 Jun	1973	2	American Airlines	Take-off aborted due to escape slide deploying.
7 Jul	1973	6	Northwest Airlines	No. 4 passenger door opened in flight – UL.
19 Jul	1973	6	American Airlines	Tail cone work platform door came open in flight.
7 Aug	1973	1	Delta Air Lines	Right wing refuelling door fell off in flight.
4 Sep	1973	3	Northwest Airlines	Unable to maintain cabin pressurization.
7 Nov	1973	3	Northwest Airlines	Forward cargo door seals damaged – lost pressurization – UL.

11 Nov 1973	6	Continental Airlines	Arrived with tail cowl access door missing.
15 Nov 1973	8	United Air Lines	Open vent door in aft cargo compartment prevents pressurization – UL.
15 Feb 1974	1	Continental Airlines	Yaw and strong vibration in aft of aircraft during climbout.

As you can see, eight of these incident reports concern cases of problems in pressurizing the aircraft; six others refer to doors coming open or falling off; two more seem to deal with fires or major fire potentials; and a further two state that passenger escape slides, which are loaded into the bases of the doors, inexplicably inflated themselves.

The eight cases of pressurization problems are the ones that are more interesting. The United Air Lines incident of 15 November 1973, states already that the aft cargo compartment vent door was OPEN when the plane took off. At the time of this occurrence, Service Bulletins making many recommendations concerning the operation of the cargo door and how to "remedy the problems for once and for all" were supposedly passed around to the airlines operating these machines, which, supposedly, also included United Air Lines.

Yet if all the modifications had been made, according to propaganda from both McDonnell Douglas and the Federal Aviation Administration, (1) it would have been impossible to lock the door without the "vent door" being closed, (2) it would have been impossible to stow the locking handle without the "vent door" being closed, (3) the pilots' door "open" light should have been illuminated if any part of the door was not completely locked, and (4) the person who closed the door should have inspected the locking mechanism via the "recommended" viewing peephole, to make sure that the door was properly locked.

No thank you, Mr. United. I'll look after my things – you have fun with yours.

CHAPTER 49

BUTTERFIELD's letter to the National Transportation Safety Board, dated 15 March 1973, had been couched in such terms as anyone reading it would consider the F.A.A. a terribly conscientious organization; rapidly looking into the problems and coming up with decisive answers. He now had a further letter from the N.T.S.B., this one dated 12 March 1974, asking exactly the same questions about what he was planning to do to correct the design problems in the DC–10. Only this time, the 346 deaths at Paris had already occurred.

On 19 March 1974, Butterfield attempted to handle this situation in the following correspondence:

Dear Mr. Chairman,

On 3 March 1974, a U.S. manufactured McDonnell Douglas DC–10 airplane was involved in a tragic accident near Paris in which the aft cargo compartment door appears to have failed in flight for reasons still under investigation. As of 1 March 1974, McDonnell Douglas Corporation has delivered 133 DC–10s – 95 to 9 U.S. airlines, and 38 to 14 foreign operators. Knowing of your interest in this accident, and in the developmental history of the DC–10, I want to review briefly for you the certification status of the airplane and the steps taken by the F.A.A. to ensure that the objectives of U.S. aircraft design standards are met.

The DC–10 was type certificated by the F.A.A. on 29 July 1971. The certificating authority was the F.A.A. Western Region, Aircraft Engineering Division, Los Angeles, California. Although the type certification was predicated on a "fail-safe" structure of all aircraft exterior doors, a DC–10 accident near Windsor, Ontario, on 12 June 1972, demonstrated that there were peculiar design deficiencies in the aft cargo door actuating system which required correction. The official accident report found that: "the probable cause of this accident was the improper engagement of the latching mechanism for the aft bulk cargo compartment door during the preparation of the airplane for flight. The design characteristics of the door latching mechanism permitted the door to be apparently closed when, in fact, the latches were not fully engaged, and the latch lockpins were not in place." In addition to the design modifications which followed the re-evaluation by the

F.A.A. and the manufacturer, the N.T.S.B., as a result of its investigation of that accident, addressed certain recommended modifications to the F.A.A.

Shortly after the Windsor accident, the F.A.A. Western Region drafted what was to have been the first of a series of Airworthiness Directives seeking the modification of DC–10 cargo compartment doors in order to ensure the integrity of fuselage pressure containment capability. Upon appeal to the Administrator by the manufacturer, a decision was made to forego the issuance of Airworthiness Directives and, instead, to seek the modification of the cargo compartment door latching mechanism through voluntary compliance by operators with McDonnell Douglas Service Bulletins. On 16 June 1972, telegraphic confirmation of the agreement between the F.A.A., the manufacturer, and operators was transmitted by the F.A.A. to airlines then operating the DC–10. Thereafter, a series of Service Bulletins was developed by the manufacturer and approved by the F.A.A. for the further modification of the cargo compartment door. Modifications related to the integrity of the type design were to have been accomplished on all aircraft then on the production line, and retroactively in accordance with arrangements to be agreed by the operators and the manufacturer. The most critical of these modifications were to have been accomplished during the calendar year 1973.

On 5 March 1974, within a few hours after having received notification of possible aft cargo door involvement in the Turkish Airlines accident near Paris, I formed a task force under the leadership of Mr. Oscar Bakke (the recently appointed Associate Administrator for Aviation Safety), and requested a thorough review of the status of cargo door modifications on all DC–10 aircraft. On 6 March 1974, when I learned that certain of these modifications had not been carried out, and that there were some uncertainties in the records available to us with regard to the status of DC–10 aircraft, I directed that the door modification program be formalized through Airworthiness Directives and made the subject of an immediate F.A.A. regulatory surveillance program. An Airworthiness Directive was issued that day. It meant, of course, that no DC–10 could be flown by a U.S. operator except in compliance with its terms. Copies of the Airworthiness Directive were transmitted to all foreign airlines concerned, to the Washington Embassies of the States in which these aircraft were registered and, additionally, to the F.A.A. regional offices in Brussels and Honolulu for distribution to appropriate aviation authorities overseas. This multiple communication was initiated to effect maximum practicable coverage of all DC–10 aircraft in the hands of foreign operators. Then, on the following day (7 March 1974), a fleet-wide inspection by F.A.A. inspectors checked Airworthiness Directive compliance by DC–10 operators.

In order to ensure adequate latching of DC–10 cargo doors, it was required that a fleet crewmember inspect each cargo door before the plane's departure. Provision was made for other equivalent safety inspection arrangements provided explicit approval was given by the

F.A.A. Specific approval was also required of the cargo door training portion of the air carrier training program to ensure that all flight crewmembers and other maintenance and inspection personnel are thoroughly familiar with the cargo door and its operation.

On the weekend of 15 and 16 March 1974, I directed a second intensive fleet inspection to be certain that all carriers were complying in full with instructions and approved procedures. A total of 130 inspections were carried out, and findings indicated that there is a high degree of discipline and attentiveness in compliance.

There remains a number of additional cargo door modifications which, while not critical to the integrity of the type design, are nevertheless desirable. These have been the subject of numerous Service Bulletins developed by McDonnell Douglas as part of its continuing voluntary compliance understanding with the F.A.A. These modifications were to be accomplished during the next two years. I have directed, however, that necessary prospective modifications also be included in Airworthiness Directive form. The first of these follow-on Airworthiness Directives will be issued this week. It will require prompt completion of the aforementioned modifications.

There have been other characteristics of the aircraft design which have been under serious reconsideration by the F.A.A. since the American Airlines accident near Windsor, Ontario, including, for instance, strengthening of the passenger compartment floor, venting into the aft cargo compartment and isolation or other protection of critical systems passing through the passenger compartment floor. The N.T.S.B. as well as other interested persons, has made specific recommendations concerning certain of these characteristics. Consistent with the decision to seek voluntary compliance in lieu of explicit regulatory action, the F.A.A., in February 1973, raised several questions with McDonnell Douglas concerning these design characteristics. Specifically, we asked that we be informed as to the manufacturer's intentions when the necessary design reanalyses were completed.

In February 1974, McDonnell Douglas responded to the effect that the design characteristics to which the F.A.A. had referred, involved so comprehensive a reconsideration of present aircraft design standards – and the possible modification resulting therefrom would prove so burdensome to the manufacturer – that action instead should be initiated by the F.A.A. on an industry-wide and government financed basis. I sought clarification of the McDonnell Douglas position in this matter during a recent personal visit to the manufacturer's plant in Long Beach, California. After lengthy discussions with Mr. John Brizendine and his senior technical staff, I was unable to accept that portion of their proposition relating to the motion that specific design improvements (floor strengthening, intercompartmental venting and isolation of critical systems) should await industry-wide rulemaking actions. I directed the Western Region to develop as soon as practicable an analysis of all DC–10 design changes which might reasonably be required in the near future.

Finally, we are isolating these far-reaching regulatory issues which we believe might impact on all wide-bodied jet aircraft. I intend to deal with these matters by way of appropriate rulemaking action. This latter effort will require extensive consultation with all interested persons in accordance with the Administrative Procedure Act.

Sincerely, (Alex Butterfield).

Even though these statements have been penned by Butterfield, practically nothing has, up to the time of writing, happened as a result. The F.A.A., one must remember, has written this kind of letter before with no follow-up actions at all.

Butterfield in this correspondence, admits that the airlines have not been incorporating design modifications, handed around as "pleading requests" in McDonnell Douglas Service Bulletins. The proof may be seen in his paragraph: "There remains a number of additional cargo door modifications which . . . have been the subject of numerous Service Bulletins . . . and . . . were to be accomplished during the next two years".

He publicly admits that Douglas made, and the F.A.A. certificated, (a) a badly designed door latching mechanism, (b) an understrength, weak floor, and (c) a system of cables and controls placement *through* such a floor, knowing that it would be highly probable that the aircraft would become unflyable were anything to happen to the weak design.

But one must remember that Butterfield should have given such a reply to the letter dated 23 February *1973*; long before the Paris disaster – long before the pain and suffering. Surely the F.A.A. should have acted months earlier?

But curiouser and curiouser – in his letter, Butterfield for the first time would have us believe that Jackson McGowen: ". . . appealed to the administrator . . . to forego the issuance of Airworthiness Directives . . ." which was never stated by either McGowen, Shaffer or Basnight.

Butterfield then goes on to say: "On 16 June 1972, telegraphic confirmation of the agreement (to use Service Bulletins instead of Airworthiness Directives) between the F.A.A., the manufacturer and the operators, was transmitted by the F.A.A. to airlines then operating the DC–10". There could have been no such agreement with the operators, since they had not been contacted to be asked whether they agreed. And besides, Shaffer, who was then administrator, has said nothing about the airlines agreeing to incorporate these Service Bulletins on a "compulsory/voluntary" basis.

And further, if there was such an agreement between the F.A.A.,

McDonnell Douglas and the airlines, why did Douglas send out pieces of paper called Service Bulletins, asking the airlines to fill in the forms, note all the numbers, and post them back, *if* they wanted the parts, in order to make the modifications? There are simply no signs of anyone agreeing to anything, other than the telephone conversation between Shaffer of the F.A.A. and McGowen of the Douglas Aircraft Company.

However, even worse than all the errors, delays and inconsistencies, nothing has been done to correct what is now admitted to be a highly dangerous situation that is *still* permitted to exist in these machines – namely, the understrengthened floor, and the control cable placement through the beams. If Butterfield had meant sincerely and seriously that he was worried about the situation, why did he not have the DC–10 taken completely out of service until such a time as the faults, which he readily admits are still in the design and for which he says he will try and have some corrective action performed within the next two years, are corrected? The N.T.S.B. replied to his long diatribe, three days after they received it, on 22 March 1974:

Dear Mr. Butterfield,

The National Transportation Safety Board has received your letter of 19 March 1974, concerning the actions taken by the Federal Aviation Administration intended to assure the integrity of the aft cargo door and its actuating system on the McDonnell Douglas DC–10 series aircraft.

Concerning implementation of our Safety Recommendation A–72–97, we have reviewed the requirements of your Airworthiness Directive of 6 March 1974. We further note in your letter that this door modification program is the: ". . . subject of an immediate F.A.A. regulatory surveillance program . . ."; the F.A.A. is: ". . . carrying out DC–10 fleet inspections to ascertain that all carriers are complying in full with instructions and approved procedures . . ." We understand that additional Airworthiness Directives will be issued in the near future.

The telegraphic Airworthiness Directive of 6 March 1974, requires immediate modification to all DC–10 cargo doors in accordance with McDonnell Douglas Service Bulletins 52–27 of 30 May 1972 – A52–35 of 19 June 1972 – and 52–37 of 3 July 1972, except for those aircraft which have been modified in accordance with McDonnell Douglas Service Bulletin 52–49 of 25 October 1973. Additionally, the Airworthiness Directive requires prior to each take-off, that a flight crew member shall check each cargo door for proper security.

Service Bulletins 52–27, A52–35 and 52–37 were issued and in existence prior to the release date of 6 July 1972, of the Board's Safety Recommendation A–72–97. In making our recommendation at that time we stated:

We are aware of the inspection procedures currently in effect to ensure safe operations of the DC–10 as well as the existing safety features of the door design. Nevertheless, in order to preclude the recurrence of similar accidents, the Safety Board recommends that the Federal Aviation Administration:

> Require a modification to the DC–10 cargo door locking system to make it physically impossible to position the external locking handle and the vent door to their normal door locked positions unless the locking pins are fully engaged.

The Airworthiness Directive of 6 March 1974, represents the first mandatory action taken by the F.A.A. to treat this cargo door problem. The provisions of that Airworthiness Directive will substantially reduce the risks of cargo door failures. Nevertheless, the Board finds that the Airworthiness Directive fails to meet the full intent of our Safety Recommendation A–72–97, which was to make it physically impossible to position the locking handle and vent door unless the locking pins are properly engaged.

We have reviewed the provisions of Service Bulletin 52–49 which provide for the installation of a revised "closed loop" cargo door locking mechanism to eliminate the possibility of the human factor error by ensuring prevention of vent door closure unless the cargo door latches are engaged and locked. The Safety Board believes that modification of the DC–10 cargo doors in accordance with Service Bulletin 52–49 would carry out the intent of our Recommendation A–72–97, by eliminating the human error factor and making it physically impossible to pressurize the aircraft without the door's locking pins being fully engaged.

Our Safety Recommendation A–72–98, proposed that the F.A.A.:

> . . . require the installation of relief vents between the passenger cabin and aft cargo compartment to minimize the pressure loading on the cabin flooring in the event of sudden depressurization of the cargo compartment.

In the Board's follow-up letter to the F.A.A. of 23 February 1973, we stated that:

> . . . if complete venting is not possible, partial venting would be beneficial. Such venting could prevent the complete collapse of the aft cabin floor, or it could reduce the amount of floor deflection, and attendant control cable damage in a DC–10.

In this connection we note in your letter that you have had discussion on this subject with McDonnell Douglas and that the F.A.A. is developing as soon as practicable, an analysis of all DC–10 design changes which might reasonably be required in the near future.

As the result of our inspection of several DC–10 aircraft recently modified in accordance with the Airworthiness Directive of 6 March, we noted that the cargo door viewing window used for the inspection of lockpin engagement does not provide for visual observation of the area. The positioning of the lockpin is difficult to see, even with the aid of a flashlight. We are of the opinion that the viewing window should be improved and illuminated so as to allow full visual observation of the lockpin area in order to assure positive verification of proper locking pin engagement.

In view of the above, the Safety Board recommends that the Federal Aviation Administration:

1. Require that the provision of the McDonnell Douglas Service Bulletin 52–49 entitled: "Doors – Cargo – Install Revised 'Closed Loop' Cargo Door Locking Mechanism", be made mandatory by the immediate issuance of an Airworthiness Directive.
2. Expedite your resolution of the cabin floor venting problem so as to incorporate the flooring modification by Airworthiness Directives at the earliest possible date.
3. Amend the Airworthiness Directive of 6 March 1974, to require an improved viewing window and illumination of the lockpin area in order to permit full visual observation of the locking mechanism so that positive verification can be made of proper locking pin engagement.

There is further and full proof that a disaster on the scale of the Paris DC–10 crash could happen again at any moment – any time of any day or any night – until the F.A.A. takes some positive action in correcting the design problems in the door, the floor and control cable placement.

You will notice that the above letter was written by the N.T.S.B. on 22 March 1974. Up to 1 November 1974 – the F.A.A. has implemented only the first of these recommendations.

CHAPTER 50

On 22 March 1974, the F.A.A. stirred itself into action – or gentle movement – once again. Following on their Airworthiness Directive of 7 March 1974, they issued another. But this time, it was related to the Iberia Air Lines crash at Boston Logan Airport on 17 December 1973. That means it took them just over three months to effect action, whereas the earlier Airworthiness Directive, partially correcting some of the problems in the DC–10's cargo doors, took them just on 20.5 months to issue.

> The Agency has received a report, that during an emergency evacuation after a DC–10 landing accident, a forward right-hand passenger door failed to operate. In another case, during a routine pneumatic function check of the left-hand forward passenger door, the door handle could not be moved into the emergency position, preventing operation of the door in the emergency mode. Deformation of the forward passenger door emergency air bottle actuating crank stop bracket caused the failure of the door. This condition allowed the air bottle actuating crank to move past its normal stop position, allowing a malfunction in the door's operation.
>
> Since the condition is likely to exist or develop in other airplanes of the same type design, an Airworthiness Directive is being issued to require certain inspections and/or reworks, tests and/or replacement of defective parts.
>
> Since a situation exists that requires immediate adoption of this regulation, it is found that notice and public procedure hereon are impracticable and good cause exists for making this amendment effective in less than 30 days.
>
> In consideration of the foregoing, and persuant to the authority delegated to me by the Administrators, §39.13 of Part 39 of the Federal Aviation Regulations is amended . . .

It is pleasing to see that the F.A.A. then said that the above situation needed immediate adoption. However, there are a large number of other design faults in the DC–10 that also need "immediate action". Why no grandiose words on these?

CHAPTER 51

In Europe, the word "compensation", when associated with air crash victims, is a dirty one. You are egged on to travel by air with large advertisements, such as the one that appeared in the London *Evening Standard* on 19 March 1974, headed:

FLY DAN-AIR – LONDON TO BERNE – IT'S THE ONLY WAY TO GO.

Besides being unable to spell Bern correctly, Dan-Air, a British charter and part-time scheduled airline, was the carrier when 112 persons died in an accident near Barcelona, Spain, a few years earlier, and the "normal" payout for compensation for a victim without dependants in that case, was £500 ($1,200) – a disgustingly low figure in anyone's language.

Daily in Europe, the newspapers detail claims for compensation in road accidents involving sums anywhere from £25,000 to £75,000 ($60,000 to $180,000), yet air crash victims, who are more likely to be victims of manslaughter as nearly all crashes involve cases of gross negligence in one way or the other, are "paid off" with these paltry sums.

The type of thinking that has perpetuated such disgracefully low figures for compensation, is that shown by the British government for instance. In 1961, they passed a law called The Carriage By Air Act; a mean piece of legislation which aligned British levels of compensation with those of the Warsaw Convention of 1929, as amended at the Hague in 1955. The Warsaw Convention stipulates for a maximum compensation figure of about £3,886 ($9,326), but under the circumstances, this amount does not apply if the airline is guilty of negligence. And moreover, the Convention states that it *is* guilty until such a time as it proves itself innocent. If it cannot prove innocence of negligence, the maximum figure for compensation under the Warsaw Convention becomes "negotiable", without limitations, dependent upon each individual case at hand.

Countries like Britain, France and some others, together, amazingly enough, *with most of the communist countries*, have ratified the 1955 amendment at the Hague. This calls for cases of compensation under these circumstances to have a maximum figure of *double* that of the Warsaw Convention – still as paltry as it is – *but*, instead of the airline being guilty of negligence until *it proves itself innocent*, the aggrieved persons – the suddenly involved passengers' relatives – have themselves to prove the airline guilty of negligence, and in such a specific fashion that it would be nigh impossible to do.

It should be pointed out here that the United States has most certainly not ratified such rubbish as this Hague Amendment. They consider both of these earlier agreements as disgustingly low levels of compensation, and therefore retain the regulatory effects of the original 1929 Warsaw Convention because, realizing that most air accidents are caused by negligence in one form or the other, this at least gives passengers' relatives a fair chance of obtaining adequate sums of compensation.

Britain, France and all the rest – in the main, the countries that have nationalized or State owned airlines, together with the communist countries (so you can gain an idea of the thinking behind those responsible for ratifying the Hague Amendment), which have agreed upon the terms discussed at the Hague, have done so purely to "cover-up" negligence on the parts of their own government departments responsible for civil air activities, and not to cause their nationalized airways boards to be involved in large compensatory payouts when and as the accidents occur. There can be no other reasons for any country in the world ratifying the terms of the Hague Protocol of 1955.

In the case of Turkey being interested in carrying on civil air activities, it is a member country of the International Civil Aviation Organization (a division of the United Nations – a wasteful, highly expensive "community of friends" association that is supposed to lay down the rules for international air travel, but which can do nothing other than "make recommendations" in similar fashion, you will note, to the McDonnell Douglas Service Bulletins!). Yet Turkey is permitted to participate in international air transportation without being a signatory to either the Warsaw Convention, the Hague Protocol, or the Montreal Agreement.

Turkish Airlines is a member of the fare price-fixing cartel known as the International Air Transport Association (I.A.T.A.) (an association of airline managements that fixes trade practices and passenger fares). But Turkish Airlines has not signed the special contracts for increased amounts of compensation, with a maximum figure of $75,000 – including legal fees – which, even in itself is still a paltry sum.

How a country and an airline has been permitted access to other countries' landing spaces, and has been permitted to schedule flights in to and out of other countries' air space without signing any of these international agreements, one does not quite understand.

Anyway, that is the situation as it is today. It is a complete mess as far as passenger compensation is concerned because of the Turkish attitudes to life and the inefficiencies of other governments such as France and Britain, that permitted Turkish owned aircraft landing facilities.

Because the matter of compensation in this Turkish Airlines' Paris accident is so terribly complicated and indeed in such a mess, naturally it is advisable for passengers' relatives to seek to institute legal action in *any* country that could offer them what must simply be called "adequate amounts of compensation".

The choices before them in this case are: *Turkey*, the country which owned the aircraft (where no international agreements are in force, evidently because the Turks consider a life to be of no value). *France*, being the country from which the aircraft took off on the fatal flight and also being the territory on to which it crashed, (but France has ratified the Hague Protocol purely for her own government's private interests in "covering-up" its own inefficiencies). And *Britain*, the country to which the aircraft was flying when it crashed, (but which values a person's life at £500 – three lives being worth as much as a family saloon car – and which also has ratified the Hague Protocol).

The only other possibility for the victims' relatives, and luckily it is a case in which the defendants will have a lot of questions to answer, is to take legal action against the aircraft's manufacturer and certificator, McDonnell Douglas and the Federal Aviation Administration, in *America*. Under these circumstances, the Warsaw Convention, the Hague Protocol, and all the other international agreements are not involved, and so in reality it seems that all that has to happen is for the lawyers to sit around a table and work out the amounts of compensation applicable in terms of manufacturer's liability and current monetary values.

Like British justice, there is a lot to say for American justice. But at the same time, there is a lot to be said "against" it. One thing that fits into both categories is the system of fees payable to lawyers in America who are involved in litigation. For instance, in a case such as the one just described, the aggrieved parties would pay nothing to the lawyers acting on their behalf, but the lawyers would be recompensed by removing 15% or so out of whatever moneys they were able to extract from the defendants. As will be realized, this has both its good and bad sides.

Naturally, it invites lawyers to search out aggrieved persons to encourage them to permit them to start an action on these above-mentioned grounds. It also means, on the other hand, that the genuine lawyer will work exceedingly hard on the case because he realizes that the more evidence he can produce against the defendants, the harder, more hitting his case will become; the harder his case, the higher the compensation he is likely to get for his clients; the higher the compensation, the greater his 15% cut will be.

In cases such as the Turkish Airlines DC–10 crash, the American legal system permits "class actions" – that is to say, where a number of persons are involved in the same incident to the same degree (such as the passengers who died near Paris on 3 March 1974), instead of 340-odd different cases going to different Courts at different times, the whole lot may be bundled into the one case at the one Court. Naturally, application has to be made to the Court to show that there are grounds for such a "class action"; that indeed there are more than just a couple of people involved in the same event to the same extent.

On 24 March 1974, the British Sunday *Observer* revealed to the world that the wife of their recent European correspondent, Mary Hope, whose husband Francis Hope had died in the Turkish Airlines disaster, initiated an action, later to be turned into a "class action", against the McDonnell Douglas Corporation for the nominal compensation figure of £54,000,000 ($130,000,000) on behalf of the relatives and those who died in the crash.

This was quite a legal, and moral thing to do. It is about the only way that passengers' relatives involved in this catastrophe, stand any chance of receiving even slightly adequate amounts of compensation. And the only way to go about this degrading business is for action to be taken against the aircraft's designers, builders and certificators in a country that understands better matters of such nature as compensation for manslaughter.

The Observer article said in part:

Mrs. Hope wishes the lawsuit to be tried in California where high damages are known to be available. If the California Court allows the suit to proceed there, it will give all other dependants of victims the opportunity to join in the action . . .

But if the competency of the Californian Court is settled in Mrs. Hope's favour, the resulting action, if successful, will almost certainly undermine the Warsaw Convention as a basis for the settlement of aviation compensation. The Convention, with its attendant protocols and caveats, is rapidly coming to be seen as an anachronism which has served to suit the airlines, but not the passengers, by limiting the amounts of damages payable . . .

Speiser & Krause, the New York lawyers handling Mrs. Hope's action, are advising dependants wanting to join in the suit against McDonnell Douglas that any prior compensation offered by an airline should be worded so as not to exclude dependant's rights to sue in the United States.

The costs of the American lawyers in this case, will be met out of the final settlement, which means that the only expenses immediately incurred will be those of lawyers who act for them in their own country.

All this, together with the headline mentions of the figure of £54,000,000 would sound like a godsend to poor aggrieved passengers' relatives. It would sound as though they have suddenly "struck it rich" – all they had to do was to "jump on the bandwagon". *The Observer* had printed the name and city of address of the lawyers concerned – Speiser & Krause, New York – and all one had to do being a relative of a victim of the disaster, was simpy to write to them saying that you too, would like to join the class action.

Besides, with the information given in the newspapers, it would not have taken a person of even below average intelligence too long to realize that they could obtain the actual address of the firm of Speiser & Krause, either from the local American Embassy's Information Library, or even from the telephone service's directory inquiries. The two essential pieces of information needed were the name and the city of address of the lawyers concerned – both of which were supplied.

The British *Sunday Times* was also featuring this story and were about to venture right outside their ranks of newspaper journalism by not only writing a book on the subject, (a matter *The Observer* also allegedly decided upon much later), but then producing a television documentary. Nothing would stop them, and journalists were taken off regular newspaper duties for a year and more, with two of them assigned to staying in touch with the firm of Speiser & Krause, so that all concentration could be given to these outside activities. Around £10,000 as a starter was allocated to the television expedition, and miles upon miles of colour film was shot in *Ermenonville* forest – the scene of the disaster (after most of the debris had been cleared away).

The *Sunday Times* decided that it too, like *The Observer*, would keep the Mary Hope compensation attempts to the forefront of the public's gaze, and so on the 24 March 1974, they too covered in length, the story of the issuance of the writ in the California Courts at 1259 hours on 18 March 1974:

Mrs. Mary Hope, whose husband was killed in the crash, cut through many of these international complexities by filing a lawsuit against McDonnell Douglas in the Central District of California, where the DC-10 was built.

Mrs. Hope's suit alleges that McDonnell Douglas were negligent, demanded damages for her two year-old daughter and herself – and claims that the firm must be liable for at least £54,000,000 if all the other afflicted families should follow her. This they could easily do, regardless of nationality, if the Court grants Mrs. Hope's application for a "class action", that is to treat her as a representative plaintiff.

Mrs. Hope issued a statement saying that although she herself was not in immediate need (Francis Hope, Paris Correspondent for *The Observer* and a *New Statesmen* columnist, came from a well-to-do-family, and was insured), she was bringing the suit to discover the truth about the accident, to help obtain adequate compensation for all the people affected, and to make it less likely that such a thing could happen again.

Two formidable groups of U.S. lawyers are acting in the case. The suit was filed through Butler, Jefferson & Fry of Los Angeles, specialists in personal injury cases, who obtained $750,000 damages for one child (relatively lightly injured) in the only thalidomide case yet fought to a conclusion. Combined in the action are the New York and Washington firms of Stuart M. Speiser, a former Liberator pilot and aviation law-specialist, who last week was settling cases resulting from the Lockheed TriStar crash in Florida . . .

One of the people killed in the Florida TriStar crash was a Cuban citizen, Mr. Jesus Lorenzo, who was earning about £5,000 a year as a building supervisor in New York. Last week, Speiser & Krause settled on behalf of Mr. Lorenzo's widow and two children for £274, 957.

The B.B.C. and the commercial news bulletins featured stories, and actually showed the speaking Mrs. Hope making her plea for others to join her.

On the 22 March 1974, Mary Hope was interviewed on the B.B.C. television programme *Nationwide* with regard to her plans for litigation:

HOPE: I've been advised by my lawyers, American and English lawyers, that I and all other people concerned in the crash, can get more compensation by suing the manufacturers of this aircraft in America than by going through the British Courts. They have to be sued in California.

REPORTER: Are you taking this action in the American Courts entirely on your own behalf?

HOPE: In the first instance, I am. My name is first on the list. But the point is that anybody can join me in this from this minute onwards. I want as many other names in this with me as possible. Anybody involved can add their name to mine as soon as possible, and I am anxious that they should do so. We must all act together on this.

REPORTER: So it's a matter of principle as far as you are concerned?

HOPE: Absolutely. Absolutely. I just feel that we can get some sort of a meaning out of a disaster like this by trying to first of all stop such a thing ever happening again, which maybe we can by taking this action, and also by bringing both financial and some kind of spiritual comfort to all those who, like me, are bereft.

May it be made adequately clear right here and now that I am *not* against such form of litigation taking place – but am amazed that such wide publicity had not been afforded many other similar cases in Europe. I would like to see this action in America fully succeed, but not at the expense of possibly raising unfortunate people's hopes by citing examples of huge compensatory payments that need not necessarily be applicable in the DC–10 affair.

Press outlets that were featuring this general subject were in contact with Speiser, and therefore his office addresses, telephone numbers and even cable codes were readily available to victim's relatives who may have wanted to reach him.

And then, Mrs. Hope, composed a letter which one assumes was then posted to *all* of the British victims' relatives known at that time, repeating her suggestion that they join her in the fight for compensation.

The letter was dated 4 April 1974. Both Speiser and his Washington office partner Don Madole, travelled from New York to London on 1 April 1974, and set up their headquarters in the offices of a Robert Gurland, who was identified to me on the telephone as an 'international law consultant'.

Here is her letter, which had been run off on a duplicating machine:

Dear Sir or Madam,

I am the widow of Francis Hope, who was killed in the Turkish Airlines crash in France last month. I am writing to you because I feel very strongly about the legal and the financial positions of the families who suffered losses in this accident. I have come to the conclusion that I should like to to (sic) try to help these families – and to discover the truth about the causes of the crash in any way I can.

At the moment I feel that the best course of action is to share with you what I have learned about the legal and financial aspects of compensation for victims of this disaster. The compensation situation is extremely complicated, but I have been advised by my American lawyers that there is an excellent chance of recovering very substantial damages, free of any treaty limitations, by suing in the United States. On March 18, 1974, I filed a suit in California. My American lawyers are handling the case on a "contingency fee" basis, which means that there is no cost to me unless compensation is collected.

Therefor(e), I am writing to you to suggest that you instruct your solicitors to ask for information regarding the legal situation from my American lawyers. You or your solicitors can obtain this information in London from the office of Robert Gurland, Esq., Suite 5, Eaton House, 39/40 Upper Grosvenor Street, London W1X 9PF, telephone 01-409 1903.

My American lawyers are Butler, Jefferson & Fry, 628 Wilshire Boulevard, Los Angeles, California – 90017; Speiser & Krause, 200 Park Avenue, New York, N.Y. 10017; and Speiser, Krause & Madole, 851 National Press Building, Washington D.C. – 20004.

Several of my American lawyers are now in London, meeting with solicitors at Mr. Gurland's offices. These meetings will continue until about April 11th. Information on the meetings is available from Mr. Gurland's offices.

I should add that my American lawyers have advised me that under Californian law any claimants must sue within one year of the accident.

Yours sincerely,
(signed: Mary Hope).

On the receptionist's table at the entrance to Gurland's office, copies of this letter, already in envelopes, were stacked. During a meeting between Speiser and solicitors that started at 1500 hours on the 4 April 1974, one of the lawyers present was called outside to answer a telephone call received for him, and he espied, on the top of this pile of envelopes, one that was ready to be sent out to one of his clients.

At about the same time as Mrs. Hope's letter was being written, the story of the crash was repeated on B.B.C. television's *Nationwide* programme on the evening of 3 April 1974.

From a recording made of the broadcast, here is what was said:

COMMENTATOR: A month ago today there occurred the worst disaster in aviation history when a Turkish Airlines DC–10 crashed near Paris. We've already spoken about the plight of the dependents of the 200-plus British dead and of their need for financial support. But the daily revelations about the causes of the crash must have made appalling reading for them.

For those who don't know the details – the story goes back to 1971 when the McDonnell Douglas DC–10 emerged onto the market. But in flight, it later turned out, the aircraft had one crucial problem.

At the rear of the aircraft is a cargo door. The locking device on it was not completely safe. If the door comes off in flight, this is what happens. The cargo hold depressurizes causing the floor above to collapse. In doing so, it cuts vital controls to the tail and the rear engine.

Many believe this is what caused the Paris crash. They believe this because the evidence points to it and because it has happened before. On 12 June 1972, over Canada, the cargo door fell off another DC–10. By what amounts to a miracle, the pilot was able to get the plane back to an airport. But there was no miracle for the passengers on the Turkish Airlines DC–10.

But why was the lesson of the original Canadian incident not learned? It was thoroughly investigated and the Federal Aviation Administration who looks after aircraft safety in the United States were about to issue a safety directive to McDonnell Douglas. This never happened. Instead a 'gentlemen's agreement' about safety modifications was made between the two men seen here. Mr. Jackson McGowen, then President of McDonnell Douglas, and Mr. John Shaffer, then head of the Federal Aviation Administration. The so-called 'gentlemen's agreement' to make the cargo door safe, was apparently not honoured.

But worse was to follow. An inspection failure in the McDonnell Douglas factory meant that one vital modification was missed out on the airplane that was later sold to Turkish Airlines. The new president of the company said last week, 'it could have been an oversight.'

If all this is true – if the accident was the result of a series of mistakes, the families of those who died are entitled to justice. Some, like widow Mary Hope, will take their case to America where the aircraft was made.

HOPE: I just feel that we can get some sort of meaning out of a disaster like this, by trying to first of all, stop such a thing ever happening again which maybe we can by taking this action.

And also by bringing both financial and some kind of spiritual comfort to all those who are, like me, bereft.

INTERVIEWER: Well, since Mrs. Hope first spoke of her plight on *Nationwide*, some thirty other British families have said they're interested in joining her. But they face a difficult dilemma. Compensation laws are more generous in the United States, but to take advantage of them, British families will have to put their futures in the hands of foreign lawyers.

If they decide to settle in Britain, the most they'll get from Turkish Airlines is £10,000 ($24,000) – and that might sound a lot but in fact of course, it wouldn't even buy a three-bed-roomed home.

Now, Don Madole, why can't these people do both? Why can't they accept a settlement here and then sue in America?

MADOLE: Well its possible that they can. But it must be done in precisely the right manner and by that I mean that very careful attention of the solicitor to the form of the agreement of settlement would be required so that it wouldn't be prohibited from the laws in the United States for settlement.

INTERVIEWER: Now, legal action of course, costs a great deal of money and one gets the impression that it costs even more in the States than it does over here. Isn't this going to be far beyond the means of the British dependants concerned?

MADOLE: No. Our system, particularly in aviation litigation, is one which is based on a contingency fee system. The purpose of that system is to allow a person in any economic circumstances to be able to hire, and it does result in the hiring, of some of the best legal talent in the United States for their representation. The contingency fee system is one where the attorney participates if, but only if, a recovery is made.

INTERVIEWER: Now you, as I understand it, are an aviation lawyer. I'm not quite sure what that is. How do you work?

MADOLE: Well, I specialize in aviation cases and my background with the government, I was with the F.A.A. and an airline, and the Civil Aeronautics Board, and for the past eight years I have been practising, handling aircrash litigation cases.

INTERVIEWER: In these sorts of cases presumably, one needs to employ a great number of technical experts who, I would have thought, are very expensive. What happens if a case is brought and is lost? Who bears the costs then?

MADOLE: Well, as to the technical experts first, if I may.

Most of our counsel that we associate with have a great deal of technical expertise themselves. That's first. As to the answer of what happens if the case is lost; so far we haven't lost one. So I don't know the answer to that. But the technical answer is usually . . . You see, all of our clients are generally other lawyers who hire us to try a large number of cases at the same time. Usually, it would be spread out amongst a number of claimants in case there was a loss, which we have not experienced as yet.

INTERVIEWER: Is this a very long, drawn-out process? How long would people be likely to have to wait for a decision?

MADOLE: An average time in a major air crash litigation case is 2 to 3 years. However, I just finished handling the Lockheed 1011 crash in the Everglades in Florida. That litigation from the date of accident until the date where the defendants admitted liability, which concluded the liability phase, was less than 1 year.

INTERVIEWER: And what sort of figure should, or do people claim in this kind of suit?

MADOLE: Well, of course it depends on the person. The system is not a system of bonanzas or rewards, it is simply a system that is designed to bring people back to the position that they were financially . . . that they would have been but for the wrongful death, generally the wage earner or a member of the family.

INTERVIEWER: So it could be £100,000?

MADOLE: Could . . .

INTERVIEWER: In certain circumstances?

MADOLE: Oh yes. Or much more than that. *(See note on page 286.)*

Copies of the Martindale-Hubbell Law Directory entries on Speiser & Krause, and Speiser, Krause & Madole, were given to various lawyers who represented relatives of victims, and attended the meetings, which besides giving them a rundown on the different partners' backgrounds, also included a 'rating' for each firm.

These are explained in the 'foreword':

The ratings are based upon confidential recommendations from lawyers and judges in the city or area where the lawyer practices and at times from such sources elsewhere. Martindale–Hubbell does not undertake to make an evaluation of a lawyer's scholastic background, the type of clients represented, the number and kinds of cases handled, or participation in professional or community activities. The rating action reflects the confidential recommendations received.

Legal Ability: No arbitrary rule for determining legal ability has been formulated, but ten years' admission is the minimum required for the legal ability rating of "a" ("very high"), five years for "b" ("high") and three years for the "c" ("fair").

Those to whom inquiries are directed are requested to make their recommendations on the basis of a standard of ability for the place or area where the lawyer practices, taking into consideration age, practical experience, nature and length of practice and other relevant qualifications.

General Recommendations Rating: The rating in this category is "v" ("very high") and embraces faithful adherence to ethical standards; professional reliability and diligence; and other relevant factors.

Speiser & Krause a v

So according to this reference book these lawyers hold the "very high" ratings in both categories. However, it is interesting to note that (a) there are no "bad" categories as such; (b) the ratings appear to be more aligned to length of time the firm has been in operation than to abilities; and (c) it will be noted that under the General Recommendation, there is only one rating – "very high".

Then in the edition of *The Observer* of the 7 April 1974, under the headline:

DC–10 CRASH TAPES SUPPORT CONTROL FAILURE THEORY

the following extract may be seen:

Turkish Airlines, British Airways and other airline companies involved in the DC–10 disaster may soon be asked to agree compensation terms for dependants of people who died in the crash.

Any money awarded to relatives of the passengers would be accepted only on condition that families were not prevented from going ahead with a much bigger claim against the firm that built the plane, McDonnell Douglas, in the American Courts.

This was one of the preliminary strategies raised last week at meetings in London between groups of British solicitors and a team of New York lawyers. These gatherings (there were two last week and more will probably be held tomorrow and on Tuesday) are without precedent in British law.

Each day, between 20 and 25 solicitors drawn from all parts of the United Kingdom have met the American lawyers in the Mayfair offices of a firm of international law consultants. They are briefed about procedures necessary under American law, what technical evidence is available against Douglas and what civil remedies are available to non-American nationals in the United States.

Mr. Stuart Speiser, a member of the American law firm handling a £54 million claim for damages on behalf of Mrs. Mary Hope . . . reckons that solicitors representing the majority of the 200 English victims have now attended these informal 'briefing sessions'.

Each solicitor is afterwards handed an agreement form which his client must complete before becoming a party to Mrs. Hope's action before the Californian Courts. Mr. Speiser said he was impressed by the quality of the British lawyers. One London solicitor described Mr. Speiser as 'extremely high powered', adding 'I don't think any of us have attended anything quite like this before. I am sure it couldn't happen inside the British legal system'.

Speiser emphasises that he is not there to influence the English lawyers to join in the Hope case. 'I only present the facts and advise on the American law, and they must go away and decide what is best in the interests of their clients. They have the choice of opting out and persuing their own claims.

'Of course, much of this is fresh ground for an English lawyer. No action against an aircraft manufacturer has ever been taken in Britain and indeed the cost of such litigation would be staggering'.

By the time the article appeared in *The Observer*, the Hope letter would have been received by most of the relatives.

With respect to class actions, the following selection is taken from the *Manual For Complex Litigation* – a quasi-official series of suggested standards and practices to be upheld by American lawyers, created by a committee of Judges of the American Federal Courts.

1.41 Suggested Local Rule No. 7 – Prevention of Potential Abuses of Class Actions.

In every potential and actual action under Rule 23 (the one covering 'class actions' as such), all parties thereto and their counsel are hereby forbidden, directly or indirectly, orally or in writing, to communicate concerning such actions with any potential or actual class member not a formal party to the action without the consent of or approval of the communication by order of the Court. Any such proposed communication shall be presented to the Court in writing with a designation of or description of all addressees and with a motion and a proposed order for prior approval by the court of the proposed communication and proposed addressees. The communications forbidden by this rule include, but are not limited to: (a) solicitation directly or indirectly of legal representation of potential and actual class members who are not formal parties to the class actions; (b) solicitation of fees and expenses and arrangements to pay fees and expenses, from potential and actual class members who are not formal parties to the class actions . . .

At the time Mary Hope sent her letter to British subjects who were potential class members, she was a party to a potential class action, since on that date her attorneys had pending before the Court in California a request to have the case declared a class action.

A month later, however, on 5 May 1974, the following appeared in the British *Sunday Times:*

CHANGE IN TACTICS IN DC–10 CASE.

Lawyers acting for Mrs. Hope, who were suing the makers of the DC–10 airliner that crashed in March, have withdrawn their application to have the case made a 'class action'.

The firms involved – Speiser & Krause of New York, and Butler, Jefferson & Fry of Los Angeles – now act for a number of American, British, French and Turkish families whose relatives were killed in the crash.

Because of arguments over jurisdiction which McDonnell Douglas has raised, the lawyers believe that individual actions will now be able to go ahead more rapidly. Time is critical, because under the statute of limitations in California – where the cases have been launched – action must be got under way within one year of the crash.

Another problem with the 'class action', according to the lawyers involved, is that it led non-American families – who are the majority – to believe that they might be forced into going along with Mrs. Hope's case. In fact, the type of 'class action' asked for by Mrs. Hope's lawyers would not have had that effect, but it nonetheless appears to be a device only American lawyers are comfortable with.

In addition to the reasons for withdrawing the request for a class action reported in this above article in the *Sunday Times*, Speiser & Krause have also given as a third reason the fact that the Judicial Panel on Multidistrict Litigation had by that time taken jurisdiction over the cases, thus assuring their consolidation in the Central District of California and giving access to many of the favorable features of a class action without actually being a class action. However, in light of the advantages to unnamed plaintiffs in a class action, it is still somewhat surprising that the request for a class action was withdrawn.

In any event, although it is possible that the case may, at some time in the future, be declared a class action through the efforts of other attorneys representing other plantiffs, as of the writing of this book it is not a class action, but rather an action or series of individual actions on behalf of named plaintiffs. Speiser & Krause originally representing two named plaintiffs are now, I gather, acting for over 130 of them.

On 23 May 1974, it was determined by the Grievance Committee of the Bar Association of the City of New York that the Mary Hope letter did not involve any professional misconduct on the part of her attorneys. A copy of that letter had been sent anonymously to the Committee from an unknown source, and upon receipt of the letter the Committee gave Speiser & Krause an opportunity to explain in writing the circumstances under which the Mary Hope letter had been sent. Stuart Speiser submitted a statement of certain facts to the Committee, which then made its favorable determination. The Grievance Committee did not, however, address itself to the propriety of the letter as a communication from a party to a potential class action, and indeed by the time the Grievance Committee made its determination, the request for a class action has been deleted by Mrs. Hope's attorneys.

This writer is admittedly not an expert either in American law or associated forms of practice and ethics. But I was so thoroughly surprised by the events on this particular legal front – events which were so sharply in contrast with legal activities associated with some of the many airline crashes and incidents that I have studied and written about over the last ten years – that I was moved to consult what I take to be some of the relevant guidelines for American legal practices, and these are quoted before and below. For instance one of the documents given to the British lawyers who attended the meetings with Speiser in London, was a retainer agreement designed for signature by the clients – a contract saying that the relatives would retain the three American law firms to act on their behalfs, with most of it setting out the financial aspects. For example:

It is understood that the services of the above law firms shall be rendered on a contingent fee basis, so that if no recovery is made no fees will be payable, and if recovery is made, the total fee of all of the above law firms and any other attorneys rendering service on these claims in the United States, shall be a sum equal to fifteen per cent (15%) of the amount recovered, by suit, settlement, or otherwise; provided, however, that the first eight thousand three hundred dollars ($8,300) of any recovery shall be excluded from any fee calculation, there being no fees payable thereon to the above firms. . . .

It is understood that the said United States attorneys shall advance all out of pocket expenses required for the prosecution of these claims subject to reimbursement by the undersigned out of the proceeds of settlement; provided, however, that the said attorneys shall apportion the common litigation expenses equally among all cases in which they are retained arising out of the Turkish Airlines accident of March 3, 1974.

On quite another aspect, the American Bar Association's *Code of Professional Responsibility and Code of Judicial Conduct*, as adopted by the New York Bar Association, states:

DR 2–101 – Publicity in General.

(A) A lawyer shall not prepare, cause to be prepared, use or participate in the use of, any form of public communication that contains professionally self-lauditory statements calculated to attract lay clients: as used herein 'public communications' includes, but is not limited to, communications by means of television, radio, motion picture, newspaper, magazine or book.

DR 2–103 – Recommendations Of Professional Employment.

(C) A lawyer shall not request a person or organization to recommend employment, as a private practitioner, or himself, his partner, or associates, except that he may request referrals from a lawyer referral service operated, sponsored or approved by a bar association representative of the general bar of the geographical area in which the association exists and may pay its fees incident thereto.

(E) A lawyer shall not accept employment when he knows or it is obvious that a person who seeks his services does so as a result of conduct prohibited under this Disciplinary Rule.

But then, what of Speiser & Krause's London associates? Under what circumstances is such a group of international law consultants permitted to operate in England?

The British Home Office advised that anybody, qualified in his own home country, could set up and operate as an international law consultant within the shores of the United Kingdom, so long as a few preliminaries are undertaken. Firstly, they must apply for and be issued with both a residency and a working permit. They must supply statements from four British referees as to their abilities, standing and conduct, and they must sign a declaration stating that they will not offer advice or act in any advisory capacity on British law; they will uphold the standards and professional ethics of the British Law Society and this latter point of course, includes the stringent rules that no advertising what-so-ever will be permitted; and they must, in no way, shape or form, solicit business. The Home Office also advised that were a person, permitted to enter Britain and set up business as an international consultant, to breach, for example, any of the Law Society's rules, his residency and working permits would be immediately revoked which would result in his having to get out of the country.

Robert Gurland, whose office address was given on the top of the Mary Hope letter, and on whose receptionist's table a visiting solicitor had seen a stack of typed and sealed envelopes containing same, had applied to the Home Office to act as an international law consultant and had duly signed the statement that he would abide by the Law Society's rules.

However, in my opinion, the Law Society, when it comes to cases involving product liability and the injuries and deaths of airline passengers, has in the past, shown itself to be more selective than wholly abiding by its own standards of doing the utmost for clients of members of the British legal profession. The amounts of passenger compensation extracted from potentially and actually guilty airlines, aircraft manufacturers, and civil air government departments, have been negligible even in cases where proof of negligence has been readily cited.

For example, take the case of a recent major disaster to a British owned, built and operated airliner. The machine crashed, falling apart in the air, because of lack of proper maintenance, negligent maintenance, negligent design, and negligent certification. Everyone on board died when the machine crashed to the earth. There were no Court actions for product liability; there were no Court appearances by officials of the British Civil Aviation Authority, the certificators of both the aircraft and airline; there were no Court cases being defended by the aircraft's operators. Yet, just on one year prior to this major disaster, officials of the British accidents investigation branch of the Department of Trade and Industry allegedly had warned officials of the Civil Aviation Authority that these aircraft were not being properly maintained in the very area that caused the crash, in such a fashion that would be condusive to safe flight.

And even though all this was ignored, the airline itself was in the knowledge that, a few months prior to the deaths and destruction, the particular aircraft involved was reported as 'having flight difficulties', but that somehow, and for some reason, the pilots regained control and the machine landed safely.

With all this knowledge, it now seems quite evident that certain members of the British Law Society whose clients were involved in this accident, could themselves be culpable of gross inefficiency in their attempts to serve their clients to the best of their abilities. The Law Society itself could also be cited as being a party to such malpractice for at least two reasons: firstly, some influential members of this Society have direct personal and business contacts with the aviation industry, and therefore could be biased against passenger claims realising that they more than likely would gain more prestige, influence and *money* from being on the industry's side in such matters; and secondly, because of their biasedness on the first count, they have done absolutely nothing to rectify a ridiculous situation by acting as a 'co-ordinator' in past cases of claims for litigation made by clients of their members, and therefore, because there was not a "force" behind the passenger cases, and because most British lawyers are hopeless anyway at matters dealing with passenger compensation for negligence and product liability, the low levels of compensation have been permitted to persist.

Another British accident which was placed in the hands of lawyers and of which they made a relative mess, was the British Airways Trident crash near Staines, London, on 18 June 1972. At the Public Inquiry, which was an all-lawyers-get-together, most of them did not have the faintest idea of what they were talking about when it came down to reliability and product liability with regards to the aircraft or the way it was certificated or operated. This complete lack of knowledge of a subject about which they were supposed to be presenting a legal case, is again, the fault of the Law Society. (It is discussed in much detail in my book *Papa India – The Trident Tragedy.*) In such matters where a large number of widely scattered people are involved, and therefore a large number of widely scattered, non-specialist lawyers, it is the Law Society's fault that these people's affairs have not been channelled into specialist groups, with specialist advisors, who know and understand the situation.

Europeans do not want, and have no need, to take their grievances to America in order to receive basic justice.

Actions of a similar nature to those conducted in America, can also be effected in Britain, and in France, and in Japan. This would spread the overall excessive costs of such litigation amongst everyone involved, reducing it to fairly reasonable levels. If large sums of money can be awarded for wrongful death in road accidents, then equally large, if not greater sums should be available to innocent victims of manufacturers of faulty products, and/or their certification and operation through gross negligence.

Note: When Mr. Madole, a member of one of the firms representing Mrs. Hope, had appeared on the B.B.C. programme *Nationwide* (described in detail on pages 277-280) on the evening before the Mary Hope letter was posted, he was identified only as an attorney specializing in aviation cases. Although on the programme a film clip of an earlier interview with Mrs. Hope was shown, the interviewer on the programme did not indicate that Mr. Madole was in any way connected with the pending case in California or with Mrs. Hope, and in fact there is no indication that at that time he had been identified in any of the media as one of Mrs. Hope's attorneys.

CHAPTER 52

It was a sunny, but cool Tuesday morning in Washington D.C. on 26 March 1974, the day set down for the start of the Senate Commerce Committee's inquiry into the whole sad and sorry mess, with particular emphasis on the parts played by McDonnell Douglas and the Federal Aviation Administration.

The inquiry was set up under the auspices of the Senate Committee on Commerce, and was called the subcommittee on aviation. Its chairman was Senator Howard W. Cannon of Nevada, and besides he and Senator John V. Tunney of California, the remainder put in a very poor show, only two of them turning up for short periods (mainly those when the television cameras were operating).

The Senators on this committee who did not make an appearance were: Warren G. Magnuson, Washington; Philip A. Hart, Michigan; Vance Hartke, Indiana; Ernest F. Hollings, South Carolina; Daniel K. Inouye, Hawaii; Frank E. Moss, Utah; Norris Cotton, New Hampshire; James B. Pearson, Kansas; Howard H. Baker Jr., Tennessee; Robert P. Griffin, Michigan; Marlow W. Cook, Kentucky; and Ted Stevens, Alaska. 'Five minute appearances' were made by Adlai E. Stevenson III, Illinois; and J. Glenn Beall Jr., of Maryland.

The Inquiry was spread over two days and of the number of people called to give evidence, the main characters everyone wanted to hear from were of course, John Hixon Shaffer the former F.A.A. administrator, and Jackson McGowen the former president of the Douglas Aircraft Company. The third person people wanted to see questioned by Senators Cannon and Tunney, was Shaffer's deputy Kenneth Myers Smith. But Smith wrote that he could not attend and he was not interrogated.

Shaffer was interviewed by the press before he was called to give evidence.

When finally called before the committee, he was sworn in by Senator Cannon, who then started the questions:

(Q) Mr. Basnight (the director of the F.A.A.'s Western Region office) testified as to a 'gentlemen's agreement' between you and the president of the Douglas Aircraft Company with regard to modifications of the DC–10 and the cargo door problems. Would you tell the committee just how you reached the decision to make that agreement and forego the issuance of an Airworthiness Directive, if that is a fact? (A) Well, the first event that happened after the Windsor, Ontario incident or accident involving the DC–10 . . . I called Mr. Wardy who was the president of American Airlines . . . to call George to find out what he thought went wrong with the airplane . . . and then after that I initiated calls to other people who also would have had some information through their channels, and eventually I got to Mr. McGowen and *we* discussed how we were going to fix it. As a result of that discussion, when we had determined that the thing that was needed was a simple plate added in the latching mechanism of the cargo door, and when we learned how simple the 'fix' was really, and how quick we could get the parts, and how easy it would be for Douglas personnel who were familiar with the airplane to campaign the fleet – and at that time I think there were about 35 airplanes in the system – we agreed that we could do it by Service Bulletins principally because the Service Bulletin gives you the instructions; it gives you the parts; and it gives you the trained personnel to accomplish the work. And we were under the impression, or rather I was under the impression, that we could virtually do the job over-night.

This testimony should be considered in the light of the following facts:

The Windsor, Ontario accident happened on 12 June 1972. The conversation with McGowen took place (according to Shaffer) on the 13 June 1972. A possible further conversation took place between Shaffer and McGowen on 15 June 1972. On 16 June 1972, McDonnell Douglas telegrammed all DC–10 operators, stressing the importance of implementing the modifications contained in the Service Bulletin A52–35 which would be issued the following Monday, the 19 June 1972. Service Bulletin A52–35, as we have earlier seen, was the 'recommendation' that a small viewing window be installed in the lower cargo door section so that the operator could see whether the lock-pin was in place, and it also contained a series of notices to be stuck all over the outside of the fuselage in the vicinity of the door. It contained NOTHING like: ". . . a small, simple plate . . ."

Shaffer then went on to tell the committee: ". . . when we learned how simple the fix really was, and how quick we could get the parts . . ." However, the Service Bulletin that did 'recommend' the 'simple, quick plate', was not issued until 3 July 1972, in standard Service Bulletin form – 22 days after the Windsor, Ontario accident; and 21 days after Shaffer, according to his statement under oath, learned "how simple the fix was really". Three weeks is certainly not a *quick period of time*. Three weeks is certainly not ". . . over-night . . ." There were 44 DC–10s in service at that time, and imagining that each was involved in a conservative five landings and take-offs each day, simple mathematics tells us that there were 4,620 plane loads of passengers flying on DC–10s prior to the notification to the airlines of this 'simple, quick' supposed remedy.

When the Service Bulletin finally came out on 3 July 1972, it contained instructions to the airlines of *how* they had to go about securing the modification kits: filling in the forms, posting the forms back to Douglas, awaiting their place in the line for the parts to be made, having the parts then shipped to their maintenance factories, and all this was covered by the specification that unless their order was received within 90 days of the date the Service Bulletin was issued, McDonnell Douglas would make them pay.

Shaffer then went on to testify: ". . . and how easy it would be for the Douglas personnel who were familiar with the airplane, to campaign the fleet . . ." The Service Bulletin says nothing about Douglas effecting the modifications, but it does give the airlines full details of how long it will take *their engineers* to do the job. Moreover, the Service Bulletin quite clearly states: "Parts to be modified by the operator".

Thus contrary to Shaffer's statement, McDonnell-Douglas would *not* provide personnel to do the modification.

(Q) Now when you said: "We agreed . . .", who were you referring to? (A) I meant that I agreed with Mr. McGowen that we would do the job with a Service Bulletin. It was not really a request of his, because of my confidence in our system; confidence in him and his company; and again, trust in the – or faith and trust, I guess – in the people to follow through with what they agreed to do.

(Q) Were you aware at that time of the fact that your own personnel in the F.A.A. had recommended an Airworthiness

Directive? (A) Mr. Chairman, I am not sure that I think that the sequence of events were quite like that. The record is already replete with dates and telegrams, G-notes, Service Bulletin information, that sort of thing, and I would say the events were going concurrently. We were doing in the headquarters things that we thought were appropriate, and the region was doing, of course, what it believed to be the appropriate way to go.

(Q) Well the decision on whether to issue the Service Bulletin or the Airworthiness Directive was not reached, apparently, until early on 3 July, which was about seventeen days after the incident. Was this an investigative period to determine what had actually taken place? Or what did the F.A.A. actually do during that period? (A) Well, we did several things during that period. First of all, Douglas alerted all their operators by telegram on the 13 June, the day following the incident. We published our first message, I believe, on the 14 June, and if I recall it was signed by Mr. Melugin. I'm not certain of that, but I believe either Mr. Melugin or Mr. Rudolph. Then the G–note (General Notification) went to our people, I believe on the 18 June. That's the general notice to F.A.A. Personnel – it's an internal communication. And I believe prior to that there were communications addressed to and signed by Mr. Smith, on either the 16 or 18 June, to all of the presidents and chief operating officers of the airlines.

(Q) I think, from our records here, that the Airworthiness Directive was prepared on the 16 June: however it was not issued. Now did you have any discussion with your top people about whether there should be an Airworthiness Directive or a Service Bulletin, in this case? (A) I am not sure there was a . . . well, when you say discussion, I'm not sure that we had a definite debate. I think I would have remembered it. I'm positive we discussed how to accomplish the work and I determined that it would be done by Service Bulletin. And I think I have a couple of reasons for thinking that way. One, I am totally familiar with Service Bulletins and how they are accomplished. I believe Mr. Smith was at one time in his career, a service representative for General Dynamics – at that time, Convair. So the Service Bulletin method of accomplishing work in the field, is something I believed in. Secondly, I would suggest that since there were no foreign airplanes involved at that time, the Airworthiness Directive because of its wide-spread distribution to the Embassies and so on of the foreign countries operating this airplane, would not have been germane – because there were none of them operating it at the time. The question of getting the word to them either by Service Bulletin or Airworthiness Directive doesn't, I don't think, contain (sic) in a situation like this.

Shaffer's argument about the unnecessary advice being sent to foreign Embassies in his attempts to justify the use of Service Bulletins, is hardly a justification for eschewing the compulsory device of an Airworthiness Directive.

(Q) Well did Mr. McGowen or any other official of Douglas pressure you not to issue an Airworthiness Directive in regard to the DC–10 but rather to issue a Service Bulletin? (A) No, sir. There was no pressure involved in this. I think the thing . . . I'll accept the total responsibility in that I think the thing was so simple to do, that I couldn't believe that a Service Bulletin wouldn't accomplish it in the shortest possible time.

(Q) Well now, you apparently had a discussion with Mr. McGowen on the 15 June 1972, in relation to this matter. Can you relate the substance of that discussion? (A) Yes, sir. Well I was furnished with a copy of Mr. Basnight's memorandum to the file and that refreshes my memory on the fact of the date of the discussion. I'm sure that it was immediately after the incident or accident over Windsor, and I don't remember whether it was the 13, 14 or 15 June, but it was in that time period. I wanted to know what Douglas thought the problem was. What they developed was what we already had decided was the problem, and they had already developed the 'fix'. They knew how to do the job. And the part that was to be furnished to accomplish the initial part of the job is a very simple thing and I'm sorry there isn't one around here to show. It's a piece of hogged out aluminium about three-eighths of an inch thick, or something like that, and it is about the size in total area of about two silver dollars. That's the part. Three bolts, and a drill and a couple of ratches and a qualified mechanic, and you could get that job done right now.

If the whole answer to the door problems was as he had just described, why did it take well over 22 days for the initial advice of this small piece of: ". . . hogged out aluminium . . ." and the notification of: ". . . such a simple remedy . . .", to be sent around to the airlines? With the importance of the modification to the safety and well-being of passengers, and with Shaffer being sure that the job could be done: " . . . right now . . .", why the 22 days delay in initially advising the airlines of such a modification? And you will notice that Shaffer has now said that the part would be fitted by: ". . . a qualified mechanic . . .", whereas earlier he said that McDonnell Douglas engineers would be doing the installation: " . . . over-night".

(Q) Were you involved in the solicitation of campaign funds for the president, from McDonnell Douglas at or about that time, or during 1972? (A) No, sir. I was never involved in campaign funds solicitation.

(Q) Were you aware of the fact, or were you appraised of the fact that McDonnell Douglas officers and executives contributed more than $90,000 to President Nixon's campaign during 1972? (A) I'm not aware of that fact, sir.

(Q) Were you aware of the fact that Mr. Herbert Kalmbach personally met with Mr. McGowen in the summer of 1972 to solicit funds for the president's re-election campaign? (A) Absolutely not. I couldn't believe anyone in a regulatory relationship could ever get himself involved in it. And I wasn't involved.

(Q) Now, pursuant to your agreement with Mr. McGowen, did you instruct your deputy Mr. Smith to conduct a telephone conversation with all U.S. operators of the aircraft, notifying them of your decision to proceed with modifications to the rear cargo door on a voluntary basis, that is, in accordance with a Service Bulletin? (A) Well, yes, sir. I agree with all that except the word "voluntary". I did discuss it with Mr. Smith. He, I would say, is a kind of electronic "nut"; and that is not considered to be a criticism of Mr. Smith. He had all of the electronic equipment and all of the communications gear in his office and all he had to do was to "flip a switch" and he would get anybody he wanted. And that is why he was given the duty.

(Q) Well now, did you at that time have a discussion with him concerning the recommendations that an Airworthiness Directive be issued rather than a Service Bulletin? (A) I really don't believe that it was ever really a topic of conversation. Certainly, it wasn't a debate. And again, I think, you're totally familiar, Mr. Chairman, with your Air Force background of what a Service Bulletin can accomplish, and how we won the War in Europe with Service Bulletins. I survived, I think, because of Service Bulletins. So I believe in them. And I think you can get the job done with them.

(Q) After your discussions with Mr. McGowen, did you check with your technical people in the F.A.A. to ascertain the opinion on the nature and effectiveness of the "fixes" Douglas proposed as part of the "gentlemen's agreement"? (A) I believe we concurred in the work, or the re-design work, and as the flood of paper indicates there was a continuing design review and we added improvements as we went along. And I think our instructions so stated that we do this and in the event that there was something else to accomplish we superseded or supplemented it with successive messages, and we did. I think we had five messages within the period between the Windsor incident and the date of the memorandum for the record.

(Q) You heard both Mr. Basnight and Mr. Rudolph testify that they recommended that Airworthiness Directives be issued. Both of them believed that. They were the two foremost agency experts in that particular field, were they not? (A) That is correct. They . . . one is a director of flight standards and the other is the director of the Western Region. Both are career employees. I have no problem with that at all except that an Airworthiness Directive would, by reference, include, or be included. or would include, the Service Bulletins. So the information that I would suggest is that the Service Bulletin information is developed before the Airworthiness Directive information is developed. In other words, it prefaces and includes the same information. As a matter of fact, in substance they are virtually identical, and I would like the record, if you please, or if it's agreeable Mr. Chairman, to actually compare the contents of those two pieces of paper which we generated, to the piece of paper, the Airworthiness . . . er, draft included. I would suggest that there aren't eight words that are different, and I've never seen them. I mean, I've never seen the Airworthiness draft. Because I believe Mr. Rudolph would have used the Airworthiness draft to develop the piece of which . . . er, I guess Mr. Slaughter developed, according to my copy. But I think they used the Airworthiness draft to develop that piece of paper.

Shaffer was of course, still defending the approach he had taken. He is glossing over the fact – that there is an enormous difference between an Airworthiness Directive and a Service

Bulletin – in nearly every respect. But then, his conversation was outside the more serious aspects of the differences between the two ways of affecting modifications. One is mandatory upon every single aircraft owner and also the manufacturer. The other is purely a "recommendation", a "pleading request", a "suggestion", without any form of compulsion attached to it whatsoever. The Service Bulletin is a complete hit-or-miss affair as we have seen, whereas the Airworthiness Directive is an order in law.

(Q) The testimony was that no foreign carriers had DC–10s. There were some, I believe, on order at that time. Is that correct? (A) Oh yes, sir.

(Q) Did you take any steps to notify the foreign carriers that had DC–10s on order, of this Service Bulletin? (A) I don't recall that I did. I'm sure that the company would have. And in this case, I believe that every airplane to be delivered to foreign carriers, I know they were still with the Douglas Company in the process of manufacture or on their ramp. So they would have been responsible for informing the foreign carriers of the content. But I am sure that the foreign carriers are all alert to it simply because every one of them have resident here in Washington – civil aviation attachés – who stay in touch with the Aviation Agency.

(Q) Did you take steps to see that the F.A.A. inspectors in the field took any action with relation to the Service Bulletins, to see if they had in fact been complied with? (A) Well . . . the answer, is, yes, sir. We . . . and I believe this is one that either Melugin or Rudolph signed. The last paragraph says: "Your chief maintenance inspector will be assigned this responsibility". That is, each regional director has a maintenance inspector who maintains surveillance on the air carriers whose certificates reside with him. And he was the man to ensure that any discrepancies were corrected before further flight – I think that was the words they used.

(Q) Are you aware that inspections in March 1974, just a few weeks ago, found instances where not all the carriers were not following all the provisions of Service Bulletins, and in some cases the called-for modifications had not been performed? (A) I am now aware of it. Yes, sir. I wasn't aware of it before yesterday.

(Q) Do you think that the failure to comply with a Service Bulletin would have been possible to occur if an Airworthiness

Directive had been issued, rather than a Service Bulletin? (A) Well, here I would suggest that someone made the statement, I believe they used the expression: "... the fact that if an Airworthiness Directive were to have been issued instead of the Service Bulletin, it would have not necessarily precluded this accident ..." I think there were words to that effect in somebody's testimony. I think that's a fact. I really believe that the record of an airplane itself would indicate that if you follow the instructions that the field had and closed the door properly, that the airplane is safe. And 500,000 hours, and in this case about 21 months of operation with this specific airplane without that part in it, would lead me to believe that if you close the door you have a safe airplane. Er ... close the door properly.

(Q) Following the Safety Board's two recommendations (after the Windsor accident), under which the F.A.A. would have required the manufacturer to impose two mandatory modifications to the DC–10, what actions did you take with respect to these two recommendations of the N.T.S.B.'s? (A) I think the first recommendation, you'd have to refresh my memory on this, we did accept that one. We probably ... er ... denied the second one on the basis that it was ... er ... in that compartment ... I think its a Class "D" compartment – venting is a big problem. I don't recall which of the two recommendations you are talking about, sir.

(Q) Well, the number one was the recommendation that you apparently implemented by way of the Service Bulletin, and the number two then was to require the installation of relief vents. (A) Yes, sir. Well that will be ... er ... I know you are totally familiar with it, but those that aren't, there's a ... er ... hermetically sealed, I guess is the word, compartment, and that's the problem there was that venting. They don't want fresh air, or they don't want air introduced into the compartment which would support combustion in the event there was a fire in the baggage hold. And I think there's ... er ... you can find this information from better experts than me now – I'm trying to forget – not this incident, I want the record to show – I've got my own career to pursue and it doesn't involve redesigning an airplane at this time.

Shaffer's statements about the rear baggage compartment supposedly being: "... hermetically sealed ...", are not strictly accurate. He is wrong when he says: "... they don't want the air that would support combustion . . ." in the rear cargo compartment.

For simple proof, if they did not want air in the aft baggage compartment, why not have the compartment unpressurized in the first place? Why pressurize a compartment – that is, why pump air into the area – when they do not want it in there? Besides, if there was no pressurization in this compartment, we would have seen a greater strength built in to the passenger cabin floor and a much stronger fuselage construction in this area, both of which would have reduced even further Douglas's attempts to make this aircraft "super economical". If there was no pressure in the aft cargo compartment, there would not have been 346 deaths when the door came off. The trouble is, the economic/strength ratios are so critical that even with the cabin floor strengthened, the machine probably would become economically uninviting to potential airline customers – and if that is not correct, then why are there difficulties in strengthening the floor?

If Douglas did not want a fire to sustain itself in the rear baggage hold because there are no fire installations in there, they should at least realize that fire needs three things to start and to keep it going. It needs a combustible material; it needs an ignition source; and it needs oxygen. An unpressurized cargo hold at 35,000 feet altitude would offer such a fire, only a quarter of the amount of oxygen available to it at ground level. Therefore, if what Shaffer said he wished to be believed, in order to stop combustion the hold would need to preferably have the air sucked out of it as in a vacuum. If that was not possible, the next best thing would be *not* to pressurize the hold. And if neither of these are possible, the only other thing is to fill the hold with an inert gas, thus removing the oxygen from the area.

To maintain low-altitude amounts of oxygen is to invite combustion, which is exactly what happens when the hold is pressurized.

(Q) In connection with the certification of the DC–10, did it come to your attention as administrator of the F.A.A. that a DC–10 at the Douglas plant had a failure of the cargo door leading to a pressurization blow-out? (A) I am certainly aware that I heard that yesterday, and I think I also was well aware of it during my tenure in the job. Yes, sir.

(Q) Well, did that raise any red flag as far as you were

concerned with respect to the certification of the aircraft? (A) Yes, we made . . . I'm sure the contractor was required to demonstrate the integrity of the airplane . . . more than once in the pressure vessel testing. So I believe the certificating authority has already been established as the Western Region. They not only certified this Dash–10 airplane, they certified another model of the airplane later, after the . . . and I think the Windsor, Ontario incident intervened. So we not only certified it once, we reconfirmed it after.

(Q) Without the modification specified in Service Bulletin 52–37, it was stated that the company investigation had found that the Turkish airplane that crashed apparently did not contain this modification. Had you issued an Airworthiness Directive rather than the so-called "gentlemen's agreement", do you think that it would have been reasonable to assume that F.A.A. inspectors at the Douglas plant would have been checking more thoroughly to see that necessary modifications were actually performed? (A) Oh, I think there's reason to believe that an Airworthiness Directive would have gotten more attention, but I rather doubt it. And I don't really think the question of Airworthiness Directive versus G-note, versus Service Bulletin affects production aircraft. These airplanes are fixed by master change records and the parts are installed either in the production line when the plane is being built, or on the ramp after, in what we would call the mod-operation, or modification operation. And so the Airworthiness Directive – I don't want to offer opinions – I just believe the Airworthiness Directive isn't germane to this accident, because that airplane was still there and Service Bulletin, G-note, Airworthiness Directive, would not have changed it. The thing that's needed . . . the thing that's missing is, why wasn't that part in the airplane? And the reasons for getting the part in the airplane are, obviously . . . well, there are procedures for getting that part in the airplane. The master change records on that airplane should have incorporated it, and I believe the documents delivered with the airplane would show if the part's installed. I think the thing that really . . .

(Q) So is what you're saying that whether it was a Service Bulletin or an Airworthiness Directive, if it had been certified by the company that the work had been performed that that would indicate that it had been performed and a "fix" had been accomplished. Is that it? (A) I think that's the way I feel about it, and the fact that the inspection hole, the window, is in that airplane might have caused everyone to believe that the whole thing was there because that was later in the sequence of modifications. And if the window were in the airplane, then you would assume . . . I guess you could assume, but I don't think that's the way to do it, that all the parts were there.

(Q) Were there other occasions during your tenure as Administrator when you made a decision not to issue an Airworthiness Directive, but proceeded with a Service Bulletin instead, as against the recommendations of your staff? (A) I believe . . . against the recommendations of my staff. I don't think I made this against the recommendations of my staff. We used Service Bulletins frequently, and we used Airworthiness Bulletins frequently, and I would guess one is – I think – . . . Here I have this problem of, if you call it a problem, I think people think of Airworthiness Directives being 24 carat gold, and a Service Bulletin is a damn piece of paper. I don't think that's the way it is. I think they both get the job done. And this is the . . . I think the emphasis is wrong – Service Bulletins versus Airworthiness Directives. I really think you can get this job done either way, if you get the people to do the work. And you get your inspectors to certify or to inspect it and if you have our follow up system that . . . and I think that's the key. The follow up system didn't catch it.

Shaffer's questioning was then taken up by another Senator, who turned up solely for a few minutes, evidently of the opinion that press coverage would be extensive, and some of the old ground was re-covered. Finally, Senator Cannon wanted to make one thing adequately clear:

(Q) Just one further question by way of summary and I think you've already answered this, but let me just ask you again: Do you believe that had an Airworthiness Directive been used in this instance rather than a Service Bulletin, that it could have made or would have made any difference with respect to the crash of the Turkish Aircraft? (A) I do not believe so, Mr. Chairman, because the airplane was still at the factory when the Service Bulletins, G-notes, and Airworthiness . . . well, no Airworthiness Directive, but, the point was that the airplane was still in the manufacturer's custody; still required him to perform that work; and I believe – and they could correct the record if I'm wrong – I believe that the documents that went with the airplane indicate that the work

> had been accomplished in the . . . in that specific aircraft fuselage number.

Of course, of all the things, Shaffer said, this last statement was one of the very few that was 100% correct. The Turkish Aircraft was under the control of McDonnell Douglas until the end of 1972; and it was during the latter half of that year that the whole gaggle of Service Bulletins was issued, each one with the assurance that *this* was the answer, folks, to the cargo compartment door problems. Therefore it certainly was the full responsibility of McDonnell Douglas to ensure that *all* modifications, no matter how small and stupid they appeared to be, were carried out. There is no truth in the excuse that because a small viewing window had been installed (which should have been there before the machine was even certificated, according to the Federal Air Regulations), that that was sufficient evidence that the other modifications in the same series had also been performed. McDonnell Douglas must explain and fully, the reason for this evident complete lack of quality control in their factories. And of course, the ever present question now remains; if three separate cases of this lack of quality control were found in the DC–10 design, how many other McDonnell Douglas products also suffer from the same fate?

This question possibly could have been answered by either or both the past and present presidents of the Douglas Aircraft Company, who were next to testify.

> (Q) Do you solemnly swear that the testimony that you are about to give in this matter will be the truth, the whole truth and nothing but the truth, so help you God? (A) We do.

Brizendine the present president, then went ahead and read a prepared statement adding here and there throughout it as he thought fit, some ad libbed comments. One of these, pointed out that Shaffer was not completely correct during his earlier questioning under oath – either that, or Brizendine was not telling the truth in his ad libbed comment. It has to be one or the other.

Brizendine was discussing the matter of how and when the Service Bulletins and drafted Airworthiness Directive derived. As will be recalled, Shaffer said that McGowen, the past president of the Douglas Aircraft Company, had told him that all that was needed to correct the door problems was: "a small, simple plate". The conversation in this regards (according to Shaffer) took place on 13 June 1972. The first Service Bulletin issued after the Windsor accident was number A52–35 on 19 June 1972, which was a "recommendation" to install the small viewing window in the door and an assortment of instruction notices on and around it. The Service Bulletin that included: "a small, simple plate", was number 52–37, issued on 3 July 1972 – 22 days after the accident, and 21 days after Shaffer, according to him, was told that the answer to this problem was such a small plate, which, again according to him, could be installed by McDonnell Douglas engineers: "over-night".

Now Brizendine was commenting on this exact same topic:

> There was a time separation between Service Bulletin A52–35 and the issuance of 52–37. Much of the conversation centred around whether 52–35 should have been an Airworthiness Directive or not. Fifty-two–37 did not exist at the time 52–35 was issued. We were in fact, still conducting engineering tests based on information learned from talking to the mechanic who had actually closed the door (to the American Airlines DC–10 at Detroit International Airport, on 12 June 1972), to understand what he had done, and then what we could do to prevent him from doing that again. There was a time-span, and I repeat that 52–37 did not exist at the time the 52–35 was issued.

So whom out of these three were telling the truth, and which were trying to "cover-up" a desperate situation that now involved 346 deaths outside Paris? Did McGowen tell Shaffer that:". . . the simple plate . . .", was all that was needed, and that the modification would be done: ". . . over-night . . ."? If he did not, would this mean that Shaffer was making this story up? And is Brizendine telling the truth that on 19 June 1972, when Bulletin A52–35 was issued: ". . . 52–37 did not exist . . .", which would certainly mean that at the most, only the very initial thoughts of such a "plate" were being contemplated, and certainly no solid information could have been given at that time, detailing how, when and where such an iota of an idea could or would be incorporated.

Or was Shaffer correct in saying that they did in fact discuss: ". . . a small, simple plate . . ." as being the answer to their problems, but because of a complete disarray in Douglas's manufacturing and design plant, combined with gross incompetence, it took three weeks to work out the method of installation and then a further period of time to make the parts, whereas McGowen, according to Shaffer, had incorrectly indicated that the whole thing would be ready in a very short period of time.

One of these stories must be false.

After the long, and sometimes boring public relations statement, the questioning of Brizendine began from Senator Cannon:

(Q) With regards to pressure relief venting. Venting ranges from 430 square inches in the aft bulk cargo compartment to 970 square inches in the enlarged forward compartment, without damage to the passenger floor. Was that in the initial design, or is that something that had been accomplished later? (A) In the initial design.

(Q) Well then, how does that square with the recommendation of the N.T.S.B. when they say: "... require the installation of relief vents between the cabin and the aft cargo compartment to minimize the pressure loading on the cabin flooring in the event of sudden decompression of the cargo compartment ..." Are you talking about two different things here? (A) The venting area provided, provides for an opening of 430 square inches in the fuselage shell, and the differential pressure thereby caused on either side of the floor, will not damage the floor. That is a 430 square inch sized opening.

(Q) Well aren't you saying that you've got what the N.T.S.B. is recommending that you do? If that isn't what you are saying, then explain it to me. (A) In part, Mr. Chairman. I'm saying that we had provisions for a hole 430 square inches in size. Now a cargo door, I believe, is about 22 square feet in size. Four hundred and thirty square inches is approximately three square feet in area. So we had provisions for a coverage of say, a bomb blast or something like that.

(Q) Well is there a difference in the blow-out area in venting and the terms you use? (A) No, sir.

(Q) Well the recommendation of the N.T.S.B. doesn't say a venting to take care of the relief in the event of a blow-out of the cargo door size. It just simply says: "Require the installation of relief vents between the cabin and the aft cargo compartment to minimize the pressure loading on the cabin flooring in the event of sudden depressurization of the cargo compartment." (A) I think what the ... er ... I'm only speculating because I've not been privileged with discussions with them. I think they are saying they would like us to go farther. If they understand the design of the airplane, that is what they would have to be saying, that they want us to go farther than the venting provided.

(Q) You were the program manager for the DC–10 during the development of the aircraft at Douglas, were you not? (A) Yes, sir.

(Q) Did you encounter problems concerning the design and certification of the airplane with respect to the pressurization problem in May of 1970? (A) May of 1970 was prior to the first flight of the airplane. The airplane was still in the construction stage, and that is the incident (of the door coming off during pressure vessel tests) that I reported upon in my prepared statement.

(Q) Is that the incident that resulted in what you described as, the re-designing of the door latching mechanism, prior to it being put into service? (A) Yes, sir.

(Q) The Committee is having trouble with the information that we've received, that the Turkish DC–10 and at least one other, that being the Laker's plane, left the Douglas plant in December of 1972, for delivery to the operators, with Douglas certifying that its own Service Bulletin 52–37 had been accomplished when in fact it had not. Now I am sure you have investigated this matter thoroughly and you referred to it somewhat. Could you tell us how such an incident could happen? (A) We're having a lot of trouble with that question, too, Mr. Chairman. Our system is so arranged and organized as to prevent this. We have a rigid system of work authorization – work compliance – and work verification. And how this can happen I do not know yet, but it is one of the things that absolutely must be determined and will be determined. Our investigation is not complete in this regard, sir.

(Q) Well, if the investigation of the Turkish accident does determine ultimately that the loss of the door and the ultimate decompression was the cause of that accident, then it would appear that something in the quality control with regards to compliance of that aircraft, could be the guilty factor in that occurrence. Wouldn't you say? (A) No, sir; not necessarily. That door would have been perfectly safe had it been properly operated.

(Q) If it had been closed and properly latched, as Mr. Shaffer says, the door would have been perfectly safe? (A) That is correct.

(Q) But if 52–37 had been complied with in full, it could not have been closed and properly latched without 440 pounds of pressure (being exerted on to the "close and lock" handle). Is that correct? (A) That is correct, sir.

(Q) And so then a 200 lb. (weight) service operator on the ground, particularly at the angle at which he would have to be, would find it very difficult trying to put that amount of pressure on, to really close the door. Isn't that correct? (A) That was the purpose of the design. Yes, sir.

(Q) That's why you designed it that way? (A) Yes.

(Q) But, you find now in your inspection of the parts that have been retrieved so far, that the modification of the inspection window was there, so you could see whether the pins were in fact locked, and your instruction placard, I presume, was intact. Was it? (A) To the best of my knowledge. I do not know.

(Q) But some part of that 52–37 "fix", you've been unable to recover from the accident. Is that correct? (A) Yes, sir.

(Q) And that part that is missing. Do I understand your testimony to be that the part that is missing would have permitted that door to be closed with less than 440 lb., of pressure? (A) — (long pause) — Will you excuse me for just a moment, sir. — (another long pause) — Well, the question is the inverse. The purpose of the part is to make the force required 440 pounds, greater than if it were missing. The force would have been less than that. If the rigging portion of the service change or Service Bulletin had been accomplished, then the force would have been about 250 pounds. And we have reason to believe that the rigging change was accomplished. The Turkish airplane and the Laker airplane were side by side, not in production as has been indicated, but in our pre-delivery modification. And the work was accomplished about the same time, by the same people, according to the inspection stamps on our records. Now the Laker airplane, the rigging had been accomplished, thereby making the force required to close that door about 250 pounds. It is logical to speculate at least, that the same thing was the condition of that on the Turkish airplane.

(Q) And the modification work on both these aircraft would have occurred by the same crew and at about the same point in time? (A) The manufacturing records are stamped with the stamps of the same individuals on both aircraft. Yes, sir.

(Q) Well, it would appear then that this was a problem of quality control, if in fact that is what happened. What steps have you taken at the plant to avoid a recurrence of this type of incident? (A) The particular people involved have been removed from their positions, having any inspection or work accomplishment verification responsibility. An investigation is underway by our security organization, and I have a special investigatory team has been formed that is also working on the problem. I will personally be involved in the problem, as the investigation continues. Mr. Chairman, I might add for the record that normally, operating that door handle requires about 30 pounds (of pressure) if the door is properly latched. It only takes about 30 pounds to close that handle and push the pins in. So anyone who is at all familiar with the door would know that if the force was high, or greater than approximately 30 pounds, something is wrong – it should be investigated.

(Q) So that any operator who was familiar with the normal operation, if they found difficulty in forcing the door to lock, should have been briefed or should have known that there was something wrong with it? (A) Absolutely. It's a built-in warning. The man's got it in his hand. He normally would close it with 30 pounds. If it takes more than that, he must be alerted to that.

At this stage in the proceedings, Jackson McGowen was asked to join Brizendine at the witness' table.

(Q) Now after the Windsor, Ontario incident concerning the American Airlines DC–10 on 12 June 1972, did you, or did Mr. McGowen, call anyone at the F.A.A. to discuss the substitution of a Service Bulletin for an Airworthiness Directive? (A) (Brizendine) I did not. (McGowen) I don't, er, I certainly made no call for that purpose, if that's what you mean.

(Q) Well did you talk to Mr. Shaffer? (A) (McGowen) I talked to Mr. Shaffer, and as I recollect the situation, he called me shortly after this incident and asked what we knew about it. And if I'm not mistaken, we had several conversations over a period of a few days about what had happened; what was being done about it. And the last conversation which I think everyone refers to, has to do with the fact that he told me, as I remember it, that if we would get the job done, he would get the airlines to put these changes on. And we were then talking about two changes; the main one being the inspection holes. And if we were to do our part of the job and get the parts out, he would personally call the airlines and see that they were done.

(Q) Now was that what later became known as – well – er – which Service Bulletin? (A) (McGowen) The only one existing then as I recall, was 52–35. The other one didn't exist.

(Q) The 52–37 according to Mr. Brizendine did not come into effect until a later point in time? (A) (McGowen) That is correct but the . . .

(Q) Now the F.A.A. . . . (McGowen continues) . . . discussion with regards to the Airworthiness Directive versus not, was referred to as 52–35. It was not referring to the later one of course.

McGowen was now saying that Shaffer's earlier information was not correct. Shaffer had said that they discussed one thing only, the installation of: "a simple, small plate". Now, McGowen said that their discussions were mainly centred on: "the small viewing window". The Service Bulletin detailing the addition of the viewing window was circulated a couple of days after these discussions, whereas the "simple, small plate" was not "recommended" until three weeks later.

(Q) And the 52–35, required which modifications? (A) (McGowen) The inspection holes and the wire, I think.

What he means is the inspection holes and all the notices.

(Q) And so the 52–35 did not require the pressure "fix", which would have required the 440 pounds pressure to close? (A) (McGowen) No. That did not exist at that time.

(Q) So at the time the proposed Airworthiness Directive was written up by the Western Region of the F.A.A., that would not have, in and of itself, have required the pressure "fix" that was later imposed by the 52–37? (A) (McGowen) My understanding is that the Airworthiness Directive, which I am told would have been issued, would only cover those two items, with a statement that other items may be added later.

(Q) Now did you have any discussion with the F.A.A. people at the Western Region regarding the fix and what ought to be done? (A) (McGowen) I had very little conversation personally. My people had talked to them in length, but I didn't. They apparently talked to the head of the region out there and relayed a message from Mr. Shaffer, as I recall, that, er, we would proceed without an Airworthiness Directive and get the job done. And we did get the job, I think, in less than two weeks. Is that correct? (Brizendine) It was completed on all aircraft. (McGowen) In less than two weeks, at this time.

(Q) You heard Mr. Basnight (the head of the F.A.A Western Region office) testify that you entered into a "gentlemen's agreement" or "understanding" on what was to be done to modify the DC–10 aircraft. Is that correct? (A) (McGowen) I wasn't here then, but I understand he said something of that nature.

(Q) Oh, I'm sorry. Well. Then let me ask you, did you enter into a "gentlemen's agreement" with Mr. Shaffer or anyone else at the F.A.A. as to what modifications would be accomplished, and how they would be accomplished on the DC–10 aircraft. And particularly, would they be accomplished by a Service Bulletin rather than an Airworthiness Directive? (A) (McGowen) I don't know what you call a "gentlemen's agreement" or not. But what I made an agreement with Mr. Shaffer on, was that we would work 24 hours a day to get those parts to the airlines so that they could immediately install them. And that was agreed to; and that was what the Airworthiness Directive was all about.

(Q) And those were the parts that were covered in the 52–35? (A) (McGowen) Yes. That's my understanding.

(Q) And did you in fact get those out to the airlines in a short period of time thereafter? (A) (McGowen) As Mr. Brizendine has testified, all of those changes were done in less than two weeks from that time. (Brizendine) That is correct. (McGowen) On all airplanes.

(Q) Now what would have been the difference between a Service Bulletin and an Airworthiness Directive? You would simply have made the part kits available and shipped them out to the carriers as you did in this instance? (A) (McGowen) We'd have done exactly what we did. To my knowledge. The Airworthiness Directive was not significant as to whether we did them this way or that.

(Q) And did you know whether or not the carriers did or did not comply with the modifications in that 52–35? (A) (McGowen) Yes. They complied immediately. Under great pressure, both from us and from the F.A.A. in their own interests.

(Q) There was an inspection made by the F.A.A. on 6 and 7 March 1974, which uncovered the fact that one aircraft, even then, was not in compliance. (A) (McGowen) I don't believe – er – is that correct? Not of this modification, no, sir. (Brizendine) Not with respect to 52–35, Mr. Chairman. (McGowen) 52–37 was not in existence at this time.

(Q) Mr. McGowen, were you aware that an Airworthiness

Directive had been drafted by the F.A.A.'s Western Region and that they were recommending that it be put into effect? (A) (McGowen) I am not sure whether I was or not. I am sure I had heard that they were discussing it.

(Q) Did you have an occasion to talk to anyone in the F.A.A. concerning the problems of the cargo door at this time? (A) (McGowen) I talked to Mr. Shaffer several times as I remember, about what we knew about what was going on.

(Q) Are either of you aware of complaints by Mr. Sliff (the chief of the Aircraft Engineering Division of the F.A.A.'s Western Region) that Douglas was unco-operative in providing the F.A.A. with information regarding complaints received from airlines on the cargo door problems? (A) (McGowen) I recall that that came to my attention, but I don't remember when it was. Do you remember the timing?

(Q) Well it was roughly in the 12 June 1972 period. (A) (McGowen) Well in any event, I discussed it with our management that this had to stop if it was going on, and as far as I know that was the end of it. They were to give full co-operation to the F.A.A.

(Q) As far as you were concerned, you instructed your people to give full co-operation to the F.A.A.? (A) (McGowen) Absolutely. They . . . my people . . . had denied that they had been reluctant to give them information.

(Q) Was there any attempt on the part of Douglas, or anyone within Douglas, to try to 'cover-up' this cargo door problem? (A) (McGowen) Not to my knowledge. (Brizendine) I think, absolutely not, within my knowledge.

(Q) In as much as the Turkish airplane and the one sold to Laker's were not delivered until several months after 52–35 was issued, would these operators have had any way of knowing that the necessary modifications had not been accomplished? (A) (Brizendine) If they'd inspected the airplane for that specifically. They would have found it, if that were the case.

(Q) Well, is the normal procedure to simply go to the aircraft records and find out if the modifications have been made or have been indicated that they have been made? (A) (McGowen) Well that depends somewhat on the airlines. Some airlines are very meticulous in their inspection of the whole airplane. Others are less so. (Brizendine) Some ask us for verification.

Both McGowen and Brizendine, in their attempts to set the record straight, were implying that Laker Airways and Turkish

Airlines – both of whom had taken delivery of aircraft that were supposed to have been modified by Douglas – were lax and flippant in inspecting their new machines. These two were saying, firstly, that these airlines should have thoroughly inspected their new acquisitions in order to insure that the correct number of rivets had been used during the construction, as well as checking to see that Douglas had done the work properly. Secondly, if they had not done that, then the carriers were sloppy, inefficient, uncaring outfits. Yet *of course*, not so sloppy; not so inefficient; and not so uncaring as to have ordered a Douglas product in the first place.

(Q) What is the company's position regarding the modifications recommended by the N.T.S.B. for the venting or partial venting of the cargo holds, thereby lessening the danger of the collapse of the floor? (A) (Brizendine) Well . . . the . . . er . . . the subject is under study to decide if, we believe, design changes should occur. And if so, how they should occur. This involves a great many considerations in the design of the airplane. For example, the control systems are already divided . . . separated, for safety reasons. Part of the control cables go down the left side of the airplane, through the floor beams, and part of them go down the right side. The left-hand elevator controls are on the left side; the right-hand elevator controls are on the right side, so that if either is damaged you can still fly the airplane with the other side. Now we are considering things such as, what if your controls are damaged and jammed? And this sort of thing. We're also looking at the problem of venting.

(Q) Do you know whether or not in this Turkish accident, at this point in time, whether the vent panel that you designed into the aircraft, whether that panel gave way? (A) (Brizendine) We would have no way of knowing, Mr. Chairman. I assume they did . . .

(Q) The purpose of that was to relieve the pressure, so that you didn't have an undue amount of explosive decompression in that floor section? (A) (Brizendine) It's designed to . . . The panels are designed to 'blow-out' at one half pound per square inch, so undoubtedly they did.

(Q) Do you know, did it blow in the American Airlines incident? (A) (Brizendine) I believe they did, but I do not have personal knowledge of that.

(Q) Will you supply that information for the record as to whether the panel actually did blow in that accident? (A) (Brizendine) Yes, sir.

Something strange seems to have happened to that last question and answer. Firstly, it *was said* at the inquiry – it *is* recorded on sound tape. However, it has been omitted from the official committee on commerce's transcript altogether, being replaced by a section: "(McGowen) I do not remember precisely" – on the top of page 170. Neither anywhere else in this transcript, may the answer to that question be found.

Secondly, there should be clarification of this whole venting business. The aft cargo compartment itself as such, is a plastic-type box within a larger type box. The larger one, for want of explanation is made up of the lower section of the passenger cabin floor; the inner extremities of the fuselage skin on either side; and the dividing walls between this compartment and those ahead and behind it. The smaller type box is in effect, a plastic lining, turning the aft cargo compartment into a room, with walls, floor and ceiling. The floor of the 'plastic lining', rests directly on the floor construction beams underneath it. The walls and the ceiling are attached at various points to the beams incorporated in the larger type box outside. But because of the rounded angulation of the aircraft's lower outer skin, the plastic walls are against the skin construction at the bottom of the inner box. However above the floor level there is a space between the inner plastic box and the aircraft's fuselage. (The box inside a box-type construction may be seen in the drawing on Page 311).

The point is, and no matter what Brizendine says, there is *no adequate venting at all* between the interior of the aft cargo compartment, or the inner box as we have been calling it, and the space surrounding this compartment. Pressurized air is blown in to the compartment through two air vents at the forward end of the inner box. However, there are three very small pressure relief valves between this outside area and the passenger cabin above, built in to the passenger cabin floor.

As Brizendine had said, these three small vents between the space immediately outside the aft cargo compartment and the passenger cabin, are only designed to be sufficient to offer pressure equalization in the case of a 'slow leak' of air occurring. They are hoplessly inadequate should the pressure be lost through a larger hole. He further admitted that the size of the hole through which pressure could be lost and for which the present venting system would be able to cope, was something in the order of 3 square feet. The aft cargo compartment door, for example, is 14.66 square feet, nearly 5 times the size catered for.

Now why would they have supplied the venting between the

space and the passenger cabin above to a standard of 3 square feet pressurization loss? Brizendine admitted that it was not in case a mid-air collision occurred, because, as he so rightly said, a collision would involve a hole much greater than 3 square feet. What else would cause a hole in the fuselage? Well, there is metal fatigue, but in most cases in the past, a hole greater than 3 square feet has been involved; so it does not seem to be metal fatigue. A Fire? Yes, should a fire develop in the aft cargo compartment, because there are no heat and smoke detection probes, no one would know about it, and possibly it could burn a hole in the fuselage being fed by the nice plastic walls and ceiling of the compartment, of around 3 square feet in size. But it would also burn its way up through the passenger cabin floor thus achieving the venting required by the National Transportation Safety Board's recommendation. The only other way this assessment of a hole 3 square feet in size could have been landed upon, would have been a pure guess that possibly something of this size would have been blasted through the aircraft's skin if a bomb happened to explode in this compartment. But then, even this is illogical, because in two recent cases of sabotage explosions occurring in cargo compartments, both have resulted in fuselage damage far in excess of 3 square feet.

So you can now see that there is in reality, no logical reason for choosing this size of 3 square feet through which pressure would be lost, other than in the case of the fire. But then if that did occur, the aircraft and its contents would more than likely be a write-off because of the lack of fire detection and extinguishing equipment and access to this hold.

Douglas were well aware why they did not and cannot increase this pressure equalization venting, and it is simply because the floor of the aircraft is weak. The National Transportation Safety Board engineers, following the Windsor, Ontario accident, determined that the floor was strengthened to something UNDER 4.5 lbs. per square inch pressure. From other sources, the figure of 3 lbs. per square inch was mentioned. So let us imagine the actual strength is somewhere between 3.5 and 4.0 lbs. per square inch in the DC–10, and then let us further recall that in comparison with the DC-8, the DC-10's floor is about half-strength. But the DC-8 takes six passengers in a row across the fuselage, whereas the DC-10 takes eight – an average of one quarter more in weight, on a half-strength floor.

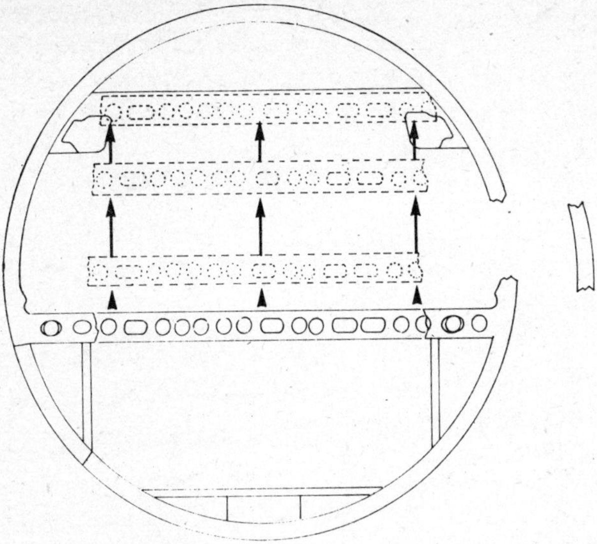

A mid-air collision, affecting the passenger cabin fuselage, could mean the floor collapsing upwards.

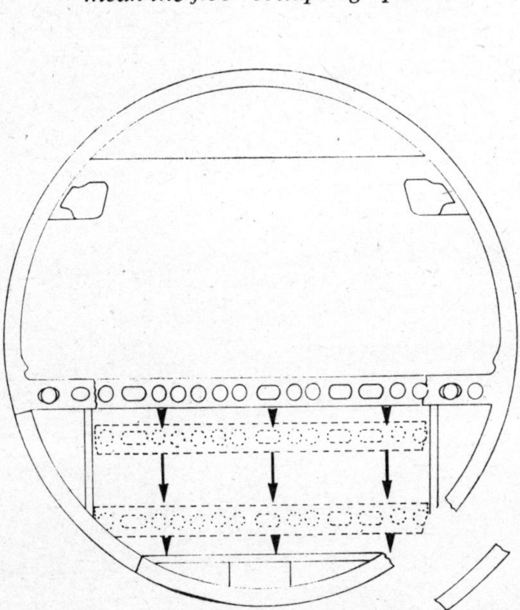

A bomb explosion in the aft cargo compartment, could mean the floor collapsing downwards.

During the Court action in Los Angeles, with regard to passenger compensation for products' liability, a further horrific revelation came to light. It was discovered that the Turkish DC-10's aft cargo compartment door's operating mechanism had been misrigged. This conclusively proves that the DC-10's door design is *still* one of the major faults in this machine.

These three major differences; the floor strength, the positioning of the control cables, and the lack of venting in the DC–10's aft cargo compartment, therefore combine to give us the accident risk that we know so well today.

Correcting these design discrepancies in the door, is only a partial remedy. It nowhere near fixes the combination of problems so lovingly placed in the fuselage, seemingly by the combined efforts of the McDonnell Douglas Corporation, in assistance with American Airlines and others who were also meddling with it since inception, making sure that it would be nice and cheap to maintain, and nice and easy to operate.

But even though, at the time of writing, the door problems have been partially remedied, the main overall dangers are still there. The floor will still collapse should there be a major pressure loss because of the lack of venting or for other reasons. And when this happens, the control cables running through the beams will still be jammed or cut, and the aircraft will still finish up in a similar fashion to the American Airlines DC–10 near Windsor, Ontario if luck happens to be on everyone's side – and if not, another Paris catastrophe will certainly occur.

Everyone is, of course, aware of the major problem – an economic one for McDonnell Douglas and the airlines that bounded in to the sales rooms, thinking they were securing a bargain, evidently, without checking up on the machine's design to any great extent – other than the Dutch Civil Aviation Authority which was well aware of the problems, as we have seen, but were evidently not too worried about them because they still issued certificates to the DC–10 so that K.L.M. – The Royal Dutch Airline – could operate it.

The economic problems are that if the floor had to be strengthened in order to include adequate venting into this aft cargo compartment, then in ratio, so would the other main construction parts of the airplane. It is no good strengthening the floor only to have the side fuselage collapse because it was not designed in the first place to withstand *that* amount of weight; so it too would have to be strengthened. And all this, of course, means greater fuselage weight, less payload, greater fuel usage, and therefore the machine will cost more to operate, making the airlines' bargains suddenly turn into very expensive exercises.

However, until this strengthening and this re-routing of the control cables has been done, passengers should certainly avoid travelling in these aircraft, unless of course, they, seemingly like the rest of the industry, do not really care whether the disastrous consequences repeat themselves, and they too end up as statistics.

And even though McDonnell Douglas announced, early in November 1974, that they were going to strengthen the DC–10's floors, they have *no plans* of correcting this problem in the present fleet of DC–10s, already in the air, but they are purely planning to do *some* floor strengthening work on any other DC–10s they may happen to sell. Naturally, with all the truth being made public about the DC–10 design problems, it possibly could be highly unlikely that any further DC–10s would be sold, which would mean that none of them would be "strengthened" – meaning that *nothing* would be done. Besides, any airline ordering DC–10s *now*, must be of the knowledge that this aircraft does have major design problems, and therefore their standards also could be questioned.

It will be interesting to see any further developments and any further press statements that either McDonnell Douglas or the Federal Aviation Administration may make, in these regards.

CHAPTER 53

On 28 March 1974, evidence appeared that the Federal Aviation Administration had acted more out of panic than thoughtfulness on the 6 March 1974, when they issued their first Airworthiness Directive.

They now had to issue another, fairly similar in nature, but one that corrected the faults in the original.

Butterfield had admitted that he had ordered the immediate issuance of the Directive on the 6 March, soon after it was known that the Paris disaster was pointing to a further failure of the aft cargo door, and therefore it seems quite evident that he acted long before he thought. You may recall that he had also said that he knew nothing of the DC–10 problems until the 5 March 1974, when he was enlightened 'for the first time' by one of his minions, although he had signed the letter dated the 13 March 1973, which stated that the F.A.A. was working on the DC–10 problem.

The amended directive reads in part:

Since the issuance of the telegram dated the 7 March 1974, the F.A.A. has determined that the requirement contained in paragraph 3 of that telegram, that pertains to depressurizing the airplane, requires clarification. That paragraph, in pertinent part, requires immediate depressurization of the airplane if any abnormalities in pressurization of the airplane are observed during flight. Literal interpretation of that requirement could result in the crew depressurizing the airplane in any phase of flight, including high altitude cruise conditions, regardless of the reason for the abnormality.

Since the F.A.A. has determined that an improperly secured cargo door would manifest itself during the initial phases of pressurization following take-off, this need not be required by the Airworthiness Directive. Accordingly, the Airworthiness Directive is revised to clearly state that the Airworthiness Directive requires depressurization of the airplane only if an abnormality in pressurization is observed during the initial phases of pressurization.

The F.A.A. here admitted that they had made a mistake. The Airworthiness Directive had a lot to say about what must be done in the case of any problems noticed in the pressurization soon after taking off. They are presuming that *all* problems of this nature *must be* connected with an improperly closed cargo door. The one thing that has not been stated by anyone to date, is that during the initial phases of flight of both of the accident aircraft (that is, the American Airlines DC–10 which had just departed Detroit, Michigan, on the 12 June 1972, and the Turkish Airlines DC–10 which was lifting off from Paris, *Orly* on the 3 March 1974), neither pilots reported *any problems at all* with pressurization. There was not a single word said on either cockpit voice recording indicating the slightest difficulty with the pressure systems.

Does this mean that the F.A.A. are working on the basis that an accident will occur once in every five depressurizations, as this is what this part of the Airworthiness Directive seems to be aimed at? With a failure rate of one in five, and a disaster rate of one in ten, what are the F.A.A. trying to get at, especially in the light of both accidents having occurred to aircraft that were NOT experiencing pressurization problems?

Their 're-thinks' continued:

> Furthermore, the F.A.A. has determined that the requirements contained in paragraph 2, of the 7 March 1974 telegram that pertain to checking of the cargo doors for security before each flight, are unnecessarily restrictive. Compliance with Paragraph 2 requires the opening of cargo doors that need not otherwise be opened and that may have previously been checked for security in compliance with this Airworthiness Directive. Accordingly, Paragraph 2 of this Airworthiness Directive is amended to require that doors known NOT to have been opened since last being checked need only be checked from the outside before subsequent flights.

So now we have the let's-pretend-it-hasn't-been-opened risk factor added. Can you imagine all the ground engineers and baggage handlers gathering around in a little circle asking one-another if anyone had opened this door, or that? Can you not imagine the eventuality occuring that on one occasion, or possibly more, with the flight running late and the airline trying to catch up time, that someone, somewhere, sometime, will *forget* to make a final check of a door that has been opened?

The only way to correct this wholly unacceptable situation is for the aircraft to be taken out of service.

The doors need to be changed to an INWARD OPENING

design. The floor in the whole aircraft needs to be strengthened. There needs to be complete inter-venting of air pressure between all of the compartments. There needs to be fire detecting and extinguishing equipment in all compartments, and especially those inaccessible in flight, such as the aft cargo hold. And the control cables to the tail and the rear engine need to be re-routed in the passenger cabin ceiling.

If all this were to be insisted upon, the result might well be "good-bye" DC–10, and then possibly a few years later, "hello" DC–11.

Finally, a number of non-substantive changes have been made. Since these amendments clarify requirements or relax unintended restriction notice and public procedure thereon are unnecessary and good cause exists for making them effective in less than 30 days.

Since it was found that immediate corrective action was required, notice and public procedure thereon was impracticable and contrary to the public interest and good cause existed for making the Airworthiness Directive and the amendment to the Airworthiness Directive effective immediately as to all known U.S. operators of McDonnell Douglas DC–10 series airplanes by individual telegrams dated 7 March 1974, and 22 March 1974. These conditions still exist and the Airworthiness Directive as amended, is hereby published in the Federal Register as an amendment to § 39.13 of the Federal Aviation Regulations to make effective as to all persons.

They say that the original Airworthiness Directive was issued of: ". . . public interest and good cause . . ." Public interest and good cause was needed after the American Airlines mess of the 12 June 1972. But it was evidently *not in the public interest* at that time to issue such an order.

If they really meant what they said, it would have also been in the public interest to have the DC–10s grounded until such a time as *all of the problems* were corrected.

CHAPTER 54

On 3 May 1974, after just two months of what was called 'intensive investigations', John Brizendine wrote to Senator Cannon, the chairman of the Senate subcommittee which was looking into the 'Oversight Hearings On The DC–10 Aircraft'.

During the hearings, it may be recalled, Senator Cannon asked for a full explanation as to how Douglas's quality control could have permitted two aircraft to be signed off as being 'modified', when in fact only a part of the overall changes had been made.

Now Brizendine offered his excuses:

During the hearings conducted by your Subcommittee on the 27 March 1974, I testified that the Douglas Aircraft Company was investigating the question of how our records could indicate that a support plate was installed on the aft cargo door of the Turkish DC–10 aircraft which crashed near Paris on the 3 March 1974, when physical examination of that door indicated that the support plate was not installed. Our investigation regarding this discrepancy commenced on 11 March 1974, and is still continuing. However the investigation has progressed to a point where some pertinent findings have been made and certain actions have been taken by the Douglas Aircraft Company. At this time, I would like to submit to you information regarding these findings and the actions we have taken to date.

We have been able to identify the Quality and Reliability Assurance Inspector and the Manufacturing Assistant Foreman whose stamps appeared on the paper work indicating that the support plate had been installed on the Turkish DC–10 aircraft. We have been unable to identify the Manufacturing Mechanic who was assigned to install the support plate on the aircraft. From our investigation to date, it appears that the absence of the support plate was not discovered in the inspection of the modification work by either the Manufacturing Assistant Foreman or the Quality and Reliability Assurance Inspector. These two individuals have been removed from any work involving inspection and they have been given reprimands.

Also, we have had our manufacturing and inspection control system audited by independent agencies and while we are satisfied it is basically sound, we are taking steps to further improve the system. For example, we have reviewed the number of outstanding stamps in both Manufacturing and Quality and Reliability Assurance and as a result, have reduced the number of outstanding stamps and have restricted the usage of such stamps. Each individual in either Manufacturing or Quality and Reliability Assurance who possesses a stamp has signed an accountability record which identifies the stamp issued to him and sets forth the conditions under which the employee agrees to use the stamps. . . . In addition, we have revised our manufacturing and inspection work instructions as they relate to 'out-of-position' work to be more consistent with the similar type instructions utilized in our production line. All of our supervisory personnel have been reinstructed in the application of our work procedures.

At the present time, the F.A.A. is conducting its regular, periodic QASAR audit of our manufacturing and quality control system, which action our company certainly welcomes.

We want to be responsible and co-operative to the requests of your Subcommittee and trust that the foregoing information will be helpful to you in your investigation.

One would have thought that the sole reason for issuing inspection stamps, as they are called, was so that the inspector's jobs would be speeded up. That is to say that as the inspections are made, the stamp is applied to the official construction document, thus saving the inspector's time having to tick and sign each individual part as he inspects it. Therefore, surely it would have been logical in the first instance, to make these stamps identifiable as to which inspector did what. However, from Brizendine's reply it seems that Douglas simply ordered a few dozen copies of the same stamp and then handed them about.

Following a visit to the Douglas plant and actually seeing what seemed to be super efficiency all round, noticing time clocks, work in progress and work completed cards liberally spread around the walls of the different factory sections, it seems incredible that the persons investigating the reasons why both the Turkish Airlines and the Laker Airways DC–10s were not modified prior to delivery, were unable to trace the mechanic that was supposed to have carried these adjustments out.

There seems to be no reason why the two quality and manufacturing control inspectors 'stamped off' as being seen to have been done, the work that was supposed to have made it impossible for a door closing operator to force the handle into the locked position.

The partial excuses in this letter in no way elucidate the full reasons for how these aircraft could have been released to their purchasers without these 'recommended' modifications being incorporated. It certainly does point to faults within McDonnell Douglas's own inspection processes – faults of a basic nature that really should not have been there in the first instance. And then, possibly in an attempt to take some of the blame off the Corporation's shoulders, it adds that the F.A.A.: "is conducting its regular, periodic (check) of our manufacturing and quality control systems." The F.A.A. had, long before this modification oversight had taken place, made a thorough examination of the methods being used, and therefore they had agreed with Douglas that their systems were capable of maintaining 'adequate' quality control. Taking the inference one step further, they could be saying that the F.A.A., having found their methods 'acceptable' could now be held as equally as responsible in the oversight as the Douglas Aircraft Company. Of course this could be a possibility – no one really knows. But it still does not alter the fact that these two machines were released without the modifications being incorporated, and so much play had been given to the effectiveness of these very modifications in making it impossible for the door to be 'mis-handled', that it seems not only is the Douglas quality control at fault, but the whole system that would permit such an oversight to slip through.

Possibly the fault could be referred back to Douglas, as, as we have seen, the original Service Bulletin that recommended the installation of this strengthening plate, was published and disseminated to the airlines in such a way as it would appear to them to be 'just another normal piece of paper'. The Bulletin contained no mention of the word 'safety'; it did not have the word 'Alert' stamped on it; it was not preceded or followed up by either a telephone call or a telegram; yet, after the 346 persons had died near Paris, this ordinary piece of paper was suddenly realised by everybody to be of far more importance to the safety of the aircraft than had ever been given it before.

Besides, if the airlines themselves relatively ignore Service Bulletins as such, presumably because they do not have the 'pressure of Law' behind them (whereas the Airworthiness Directive does), why should McDonnell Douglas do more than simply issue the notice.

To back up this point of lackadaisical attitudes shown to Service Bulletins, by airlines, let us look at a chronological survey, taken for the American House of Representatives inquiry into the whole DC – 10 business, with specific relationship to Service Bulletin 52–37, the one, part of which was found 'missing' from both the Turkish and Laker aircraft:

3 July 1972	– Service Bulletin 52–37 issued by Douglas.
1 August 1972	– United Air Lines modify fuselage No. 35
12 August 1972	– United Air Lines modify fuselage No. 45
14 August 1972	– United Air Lines modify fuselage No. 42
18 September 1972	– United Air Lines modify fuselage No. 15
2 October 1972	– United Air Lines modify fuselage No. 11

At this point in time, three months had passed from the date of original issuance of the Bulletin. As stated on it, these airlines were in possession of the detailed aircraft:

American Airlines: 3, 5, 7, 9, 12, 13, 20, 21, 22, 23, 24, 30, 31 and 37
Continental Airlines: 34, 40, 41, 43 and 44
National Airlines: 14, 16, 18, 19 and 38
United Air Lines: 4, 6, 8, 10, 11, 15, 17, 25, 26, 27, 32, 35, 39, 42 and 45.

In all, 39 machines needed to be modified. Three months after the Service Bulletin had been issued, which, as Shaffer had testified before the Senate Subcommittee hearing as being what both he and Jackson McGowen had determined to be the answer to the DC–10s' door problems, only 5 modifications had been made – and all of them by the one airline.

Continuing down the list:

6 October 1972	– United Air Lines modify fuselage No. 32 (95 days)
9 October 1972	– United Air Lines modify fuselage No. 17 (98 days)
11 October 1972	– United Air Lines modify fuselage No. 8 (100 days)
15 October 1972	– United Air Lines modify fuselage No. 10 (104 days)
16 October 1972	– United Air Lines modify fuselage No. 27 (105 days)
17 October 1972	– United Air Lines modify fuselage No. 6 (106 days)
20 October 1972	– United Air Lines modify fuselage No. 39 (109 days)
24 October 1972	– United Air Lines modify fuselage No. 4 (113 days)
30 October 1972	– United Air Lines modify fuselage No. 25 (119 days)
2 November 1972	– National Airlines modify fuselage No. 18 (122 days)

9 November 1972	– United Air Lines modify fuselage No. 26 completing all modifications.
9 November 1972	– National Airlines modify fuselage No. 19 (129 days)
14 November 1972	– American Airlines modify fuselage No. 7 (134 days)
2 December 1972	– National Airlines modify fuselage No. 38 (152 days)
6 December 1972	– National Airlines modify fuselage No. 16 (156 days)

Now, six months have elapsed since McDonnell Douglas had issued Service Bulletin 52–37. Of the 39 machines to be modified, 20 only had been completed, and only one of the 4 operators had converted all his fleet. It will be noticed that American Airlines, who were so outspoken earlier on with regards to their nosey participating in the DC–10's design, and who were in possession of 14 aircraft at the time the Service Bulletin had been issued, had taken 134 days to modify just one machine.

11 January 1973	– American Airlines modify fuselage No. 12 (192 days)
23 January 1973	– American Airlines modify fuselage No. 21 (204 days)
2 February 1973	– Continental Airlines modify fuse. No. 34 (214 days)
17 February 1973	– American Airlines modify fuselage No. 9 (229 days)
5 March 1973	– American Airlines modify fuselage No. 22 (245 days)
15 March 1973	– American Airlines modify fuselage No. 5 (255 days)
18 March 1973	– American Airlines modify fuselage No. 23 (258 days)
28 March 1973	– Continental Airlines modify fuse. No. 41 (268 days)
12 April 1973	– American Airlines modify fuselage No. 20 (283 days)
2 May 1973	– American Airlines modify fuselage No. 31 (303 days)
7 May 1973	– Continental Airlines modify fuse. No. 40 (308 days)
11 May 1973	– American Airlines modify fuselage No. 3 (312 days)

14 May 1973	– Continental Airlines modify fuse. No. 43 (315 days)
23 May 1973	– American Airlines modify fuselage No. 24 (324 days)
24 May 1973	– American Airlines modify fuselage No. 37 (325 days)
29 May 1973	– Continental Airlines modify fuse. No. 44 completing all modifications.
2 June 1973	– American Airlines modify fuselage No. 13 (334 days)
18 June 1973	– American Airlines modify fuselage No. 30 completing all modifications.

Now, one year had passed after the date on which Service Bulletin 52–37 was issued. It was not until the 6 March 1974–3 days after the Turkish Airlines disaster – that it was suddenly 'discovered' that National Airlines, fuselage number 14 *still* had not been modified – meaning it took National 611 days to incorporate these modifications.

Airlines simply cannot afford to ignore Airworthiness Directives when these are issued, as the carrier could quite easily lose its license to operate if it did not incorporate the changes by the date specified in the Directive. However, taking McDonnell Douglas DC–10 Service Bulletin 52–37 as a typical example, and especially in the light of Administrator Shaffer admitting that one part of this Bulletin was the: "simple" answer to all the DC–10's catastrophic problems, we now find that it took each of the airline owners of this equipment the following average times to have these changes made:

United Air Lines – average of 90.5 days per modification.
American Airlines – average of 267.7 days per modification.
National Airlines – average of 285.0 days per modification.
Continental Airlines – average of 287.0 days per modification.

In the questioning of John Shaffer by members of the House of Representatives committee looking into this DC–10 matter, here are the series of questions and answers; Daniel J. Manelli, the committee's chief counsel was the questioner:

> (Q) Following the incident of June 1972 . . . the Western Region Administrator . . . Mr. Basnight was considering the issuing of an Airworthiness Directive. Did you consult with Mr. Basnight concerning the possible issuance of this? (A)

No. As I say, I was not asleep on the job. I want you to know that what we were doing in Washington was going on concurrently with what Douglas and they were doing. The Airworthiness Directive is no more than a document that incorporates a Service Bulletin. A Service Bulletin has to proceed it. The Airworthiness Directive is worthless without the parts. That is what a Service Bulletin does. . . .

(Q) You don't mean to say that you don't see any difference between a Service Bulletin and an Airworthiness Directive? (A) There is a principal difference. The Service Bulletin takes with it parts that have to be there, and it provides trained manpower to do the job.

(Q) And the Airworthiness Directive doesn't? (A) I didn't say that. The Airworthiness Directive will incorporate the Service Bulletin.

(Q) So the Airworthiness Directive has everything the Service Bulletin has, but it also has something else? (A) Yes, it has everything the Service Bulletin had and the communications that went out from my headquarters had everything that the Airworthiness Directive draft would have had. I would like to have them compared for the record.

(Q) I think we have the Airworthiness Draft in the record. (A) The three or four communications that went out from June 14 through 19, included everything that any of the Air worthiness Directives might have included. Every word.

(Q) Are you saying that the requirements, the enforcement requirements and the mandatory effect of what you did was exactly the same as an Airworthiness Directive? (A) I think it was just as good and perhaps more efficient. I gave it the person-to-person touch. These people are people I know. . . .

(Q) Did you discuss with either Mr. Rudolph or Mr. Smith that the Western Region was drafting or contemplating an Airworthiness Directive? Did that come up at all? (A) I don't believe that I was ever aware that there was any question about whether it would be an Airworthiness Directive or Service Bulletin. To my mind, a Service Bulletin works beautifully if people perform. A Service Bulletin can say anything that an Airworthiness Directive can, and say it just as emphatically and have the same deadlines. You keep telling me it is a mandatory thing and the other is voluntary. That is not the fact.

(Q) What is wrong with that statement? (A) The Airworthiness Directive is mandatory on all U.S. carriers. It is information to others. At the time of this incident, there were no others.

So there was not a question as to whether or not the people were going to be well informed. They were going to be just as well informed with a Service Bulletin, or as alerted to it, by an alert Bulletin, Alert Service Bulletin, as they would be with an Airworthiness Directive.

An Airworthiness Directive *only* affected American operated aircraft – but no other countries were operating the DC–10. An Alert Service Bulletin was just as effective as an Airworthiness Directive – yet the modifications in 52–37 were sent around as normal pieces of paper. Shaffer said he telephoned the airlines about the importance of this modification. "A Service Bulletin can say anything an Airworthiness Directive can" but it did not. Furthermore every single Service Bulletin that was issued was simply 'recommended', – not one was 'enforced'.

(Q) I will call your attention to the fact that the Service Bulletin 52–37 was not complied with, for in some cases, 1 year and 9 months. (A) You have to relate the papers to it. There was a communication with the airline presidents and vice presidents of operations. They had specific statements in it and say: 'if you find the foregoing discrepancies you will correct before further flight,' and things like that. This is all of the related documents.

(Q) We have to be specific here. Are you saying that in our estimation, Douglas Service Bulletin 52–37 was mandatory, and accomplished the same as an Airworthiness Directive? (A) No. I did not say it was mandatory I am going back to it again. I would like the record to be replete with this. This is a safe airplane if you perform the function properly – if you close the door properly – in accordance with instructions. We certificated, I keep reminding the record, we certificated the airplane because we thought it was a good airplane then, and we still do. . . .

(Q) Mr. Basnight testified . . . that he has designated authority which would allow him to consider an Airworthiness Directive, have it approved by his review board, and sign it, and it would be effective as an F.A.A. Airworthiness Directive. Where in that process would the administrator, or Mr. Smith, your deputy, where would you be involved? You have delegated that authority I understand? (A) You delegate the authority, but that doesn't mean that you forgive it or never have the same authority.

(Q) But you would not intervene . . . (A) I didn't intervene.

I think that Mr. Basnight can swear to that now. I didn't tell him not to issue it.

(Q) The question is, you would not intervene unless you felt the delegated authority was not being used properly? (A) I didn't intervene. I prefer to do it by Service Bulletin because it has the parts and the people with it.

(Q) The question was not quite answered, I think. I am saying you would not intervene in the issuance of an Airworthiness Directive unless you thought that the delegated authority is being misused? (A) The answer is, no. I would not intervene.

(Q) Mr. Basnight . . . (A) I don't think Mr. Basnight's memorandum says that.

(Q) Let me finish the question. Mr. Basnight testified to the effect that as far as he is concerned, he was on the verge or on the point of issuing an Airworthiness Directive, and he is of the opinion that his decision was countermanded. (A) I don't think it was countermanded. I would call it, not anticipated, but pre-empted. We beat him to it, so to speak.

When questioning Richard Sliff, the head of the aircraft engineering division of the F.A.A.'s Western Region, Daniel Manelli asked:

(Q) I would like now to have you review three letters, and I will ask that you be given copies of them. They should be in order. I will describe what they are supposed to be. The first is supposed to be a letter dated 12 July 1972, from you to McDonnell Douglas requesting biweekly summaries, by individual airplane, of modifications made after the Detroit incident. I presume these are the modifications as set forth in Service Bulletins 52–35 and 52–37. Do you have that letter? (A) Yes. I am aware of the letter.

(Q) Have I described it accurately as to what it called for? (A) Yes, sir.

(Q) The second letter is dated 7 August 1972, and that is the McDonnell Douglas reply. That reply states that modifications under Service Bulletins 52–27 and 52–35 have been completed and kits have been sent to carriers to perform 52–37. Is that a correct summary? (A) That is correct.

(Q) The third letter is dated 18 August 1972, in further reply from McDonnell Douglas, and it states that modifications of the entire fleet of DC–10s would now commence. You have the letter I believe? Are they referring here only to the United DC–10s or is this the entire fleet of DC–10s? (A) I don't think I could make that determination. From the way it is phrased here, I am assuming that they meant the entire fleet as being all the aircraft regardless of airline.

(Q) Did you continue to receive these summary reports about Service Bulletin 52–37? (A) To my knowledge, no.

(Q) Do you know whether anyone in the F.A.A. received a running tab on the completion of work on Service Bulletin 52–37 on the existing fleet? (A) I don't believe anyone did.

(Q) Should the F.A.A. have kept such records? (A) It could be assumed that they said it was commencing and would be accomplished on the entire fleet as closing the issue. However, it is not my opinion that this is the way it should have been closed.

(Q) How do you think it should have been closed? (A) There should have been biweekly reports that were originally agreed upon.

Of course we know why Douglas "lapsed" into forgetfulness – firstly the lateness for which the parts were delivered, and secondly, the fact that none of the airlines seemed to be enthused into action in getting the modifications effected.

Does one need many more reasons for the 346 deaths at Paris? Is it not a wonder that many more plane loads of passengers have not been killed in a like fashion when these "could not care less" attitudes seem to abound in an industry so intent on painting a public picture of itself as being so "safety conscious"?

CHAPTER 55

WHAT HAPPENED AT PARIS

. . . TURKISH Airlines, Flight 981 from Ankara to London, with intermediary stops at Istanbul and Paris, arrived in the French capital on time, on the morning of Sunday, 3 March 1974. It was due to depart at 1200 hours on the final leg. However, because of the British Airways engineers' strike in London, that company was placing people on any aircraft flying between the two capitals that had seats available.

On landing at Paris, and after disembarking those terminating their journeys there, the DC–10, with a capacity of 345 passengers, had 226 vacant seats. This, for British Airways, was too good an opportunity to miss, however because of the utter confusion in the passenger terminal at *Orly* Airport, only 216 of these seats were allotted to awaiting customers.

In order to facilitate the loading and unloading of baggage, Turkish Airlines had arranged for the suitcases of the people getting off at Paris, to be loaded into the aft cargo compartment. It did not matter where passengers boarded the flight, be it Ankara or Istanbul, if they were only flying as far as Paris, their baggage would have been stored in this rear hold. So at the same time as confusion reigned amongst passengers and airline staff in the terminal building, the baggage handlers were emptying the aft cargo compartment of all of its contents, and were loading into the forward cargo hold, the checked-in luggage of those joining the flight at *Orly*.

The chap in charge of the unloading operations was a tall, thin, tanned Algerian who had been employed for the past 5 years by the *Aeroport de Paris*, the government department responsible for the management and operation of all of the Paris airports. Contrary to nearly all of the press reports at the time, this man could fluently speak and write both French and Arabic though he was, possibly in its most correct form, called "illiterate" because even though he was proficient in his own languages, he did not know either of the two that were presented to him on the notices stuck all over the outside of the DC–10's cargo doors, namely Turkish and English. If one were to use that word "illiterate" in its correct meaning, it would be doubtful if there is a person anywhere on earth that could not be classed in that category – each and every one of us is "illiterate" to differing degrees.

However, there is one small consideration that should be given to this point. In international civil aviation circles, English has been adopted as the "international language" until something better comes along. The International Civil Aviation Organization (I.C.A.O.) determined that the English language will be used on all "international occasions", except in the circumstances where a pilot, proficient in the local language (if it is not English) happens to be overflying, entering, leaving or grounded on such territory. This means that if you can speak and understand French, you may converse with the air traffic controllers in that language while you are in radio contact with them. If, on the other hand, your mother tongue is Turkish and you enter France, if you are unable to speak French proficiently then English must be used. The French, by the way, have agreed to this I.C.A.O. recommended practice.

It will of course, also be remembered that on 5 March 1973, just two days under one year before the DC–10 holocaust, the French civil air traffic controllers were on strike. Because of this, the French government called in the military air traffic controllers to look after civilian aircraft – but they evidently omitted to realize that the methods of ATC-ing the civil and the military, differed one from the other.

On that day, two airliners, one an Iberian DC–9 heading north towards London; the other a Spantax Coronado CV–990 also heading north towards London, but from a slightly different angle, were being directed over the *Nantes* radio beacon in northwestern France, by two separate military controllers situated at two different military air traffic control centres. Both machines had been allotted the same flight altitude. Because of differences in winds and aircraft flying speeds, both were simultaneously arriving at the position overhead of the ground navigation beacon, and relatively at the last minute, the captain of the CV–990 realized that unless something was done and quickly, there was a ready possibility of a mid-air collision over the area of the beacon. After repeated requests to the military controller in charge of the CV–990's path, the captain decided that the situation was far too serious to be permitted to continue, and so he effected an "avoiding" right-hand turn in dense cloud, during which his outer left wing tip sliced into the DC–9's fuselage causing it to distintegrate in mid-air and crash to the ground in thousands of pieces. Sixty-eight people lost their lives in that DC–9.

One of the major factors in that accident was the alleged inability of the air traffic controllers to speak fluently in the English language with the Coronado's pilot. Recalling the earlier mentioned I.C.A.O. ruling, the Spanish mother tongued pilot of the CV–990 would have been obliged to converse with the controllers in either French or English. As he could not speak French, the language to be used on that occasion was English. But the military air traffic controller, on duty because of the strike of the civil operators, was allegedly not proficient in the English language. And besides this, even Air France's pilots, who would have been conversing with the controllers in their home tongue French, were also bitterly complaining that *even they* were having enormous difficulties with the military, to the extent of even accusing them of perpetrating dangerous situations, more than likely because of the different methods used in military and civil operations.

French, one of the fastest dying modern languages, is being guarded fanatically by those who use it as their only source of communication. And so in their eagerness, they have omitted to realize their agreement with the I.C.A.O. standard, or the international complications of such an industry as world air transportation. This therefore must be taken into account when trying to weigh up the situation in an overall unbiased fashion. And it should also be realized, that even though one does not and should not expect all baggage handlers to be university language graduates, if they are employed in such an "international" situation as has been described, they really should be made well aware, by their employers, of their weaknesses, so that when they are confronted with a situation that they do not fully understand, such as an unfamiliar foreign built and registered aircraft, they are instructed to get assistance from either someone who does know how the equipment is operated, or who is able to read and comprehend at least one of the languages used on the notices.

On the other hand, Turkish Airlines, which had elected to operate the machine through Paris airports, possibly should have been conversant with the French attitudes to foreign languages even though they evidently realized the international situation by including English on the notices. However, if they had taken the time to drive around the French roads, they would have quickly realized that the French tend to act first and then start thinking about the situation only when something goes wrong. This can readily be seen in the thousands of unnecessary road accidents, 99% of which were simply caused by someone doing something stupid, without beforehand, thinking about the possible consequences, or taking the most basic of precautions. Another example, which they do not like being reminded about, was the ridiculous *Maginot Line* fortifications in the last World War, built by the French along their border with Germany as the supposed "be all and end all" of stopping a German invasion. It of course, had one major fault that the French did not think about before outlaying millions upon millions of Francs on its construction, and it later proved to be the downfall of the whole country.

The baggage handler working in the aft cargo compartment of DC–10 registered number TC–JAV, did not report any difficulties in closing the compartment's door. This, of course, is a very frightening thing. It means that fatal design faults, still in these doors can exist without anyone knowing. He evidently omitted checking for the door being locked via the viewing window. This was not, of course, his responsibility.

There are a number of reasons for this rather peculiar attitude being taken to locking the door. Firstly, DC–10s are few and far between at most European airports, and therefore it would be highly unlikely that on each visit, a person who was proficient and understood the intricacies of the door closing operation would have been assigned to the job. Secondly, neither McDonnell Douglas nor Turkish Airlines showed any concern about the competence of personnel who would be closing the door according to the correct methods of operation. Thirdly, there were no notices in any language saying, "Do not swing on the handle". And fourthly, and possibly more important than all of the other reasons combined, the aft cargo compartment door on TC–JAV *had not been modified* by the McDonnell Douglas Corporation to the two most important Service Bulletins, namely 52–37 (issued on 3 July 1972 – 609 days prior to the disaster – and in the 167 days during which the DC–10 was standing around the Douglas factory before being delivered to Turkish Airlines), and Service Bulletin 52–40 (issued on 27 October 1972 – 493 days prior to the disaster – and with 51 days in which Douglas had the machine at its factory awaiting delivery). As will be recalled, Service Bulletin 52–37 should have made it impossible for the door's lock handle to be stowed if the door's locking mechanism was not properly positioned, and Service Bulletin 52–40 would have meant that a definite indication of the door not being locked, should have been *very evident* to the person operating it.

Anyhow, because of these inefficiencies, the Algerian baggage handler did manage to stow the door's locking handle once they had completely emptied the compartment of the luggage belonging to those getting off in Paris.

Early on that Sunday morning, a Mr. Erhan Ozer, an employee of Turkish Airlines who was stationed at their London offices, was sent over to Paris in order to join the flight when it arrived, because the engineer who would have normally been on board was on holidays, and his replacement had, during the early flight stages, reported that he was not feeling well. Turkish Airlines had specifically laid down in their flight instructions that the engineer on board had the responsibility of seeing and checking that the cargo doors were closed and secured properly. This could have been for two reasons (remembering that prior to the accident there were no mandatory regulations that the flight crew must check these doors); firstly, Turkish Airlines had experienced *on a large number of occasions* difficulties with these doors, and especially the aft cargo compartment door on TC–JAV. (One report that has reached me was from a Greater London Council employee who had returned from a business trip to Turkey, on board TC–JAV less than two weeks before the disaster. He said that on arriving at Heathrow, a public address announcement was made to the effect that baggage retrieval would be delayed for at least 30 minutes because of some difficulty – quite possibly, in hindsight, this difficulty could more than likely have been associated with problems of opening this door.) And secondly, the contracts between Turkish Airlines and both the *Aeroport de Paris* and the service company that was to handle Turkish aircraft during their stop-overs, specifically stated that it was Turkish Airline's responsibility to see and check that doors, panels and other openable sections of the aircraft were closed and properly secured prior to flight.

With all the passenger confusion being created by British Airways trying to load on board as many of their customers as the aircraft could take, and with the physical effort needed to get these people and their luggage into place on the wide-bodied jet, the time was ticking away. Turkish Airlines had evidently realized that a delay would occur. They must have been advised early on by British Airways of an impending influx of passengers, and they arranged for those arriving at Paris on the flight, whose journeys were not terminating there, to remain on the aircraft during the stop-over.

At approximately 1220 hours, twenty minutes after their scheduled departure time, Captain Nejat Berkoz and Co-pilot Orac Ulusman taxied the DC–10 out towards the end of runway 08.

For a departure from Paris *Orly* Airport, the pilots have a height/direction restriction. A take-off from runway 08, along the main cross-runway going from west to east, would mean heading initially towards the *Orly* Non-Direction Beacon situated some six miles east of the airport. The restriction then applied was that if the aircraft was *under* a height of 3,000 feet when passing over this beacon, the pilots must continue straight out to the next radio beacon located some 30 miles further on at *Coulommiers*. If, however, the machine was able to be above 3,000 feet at the *Orly* beacon, the pilots may then affect a turn to the left, heading northwards towards the beacon at *Montdidier*.

This restriction was enforced so as to keep low flying aircraft away from the city and suburbs of Paris.

Evidently, Captain Berkoz did not report to the *Orly* tower controllers that he would *not* be making the turn to the north after the *Orly* Beacon, but proceeding out towards *Coulommiers* instead. He lifted the huge DC–10 off the Paris runway just after 1231 hours that Sunday lunchtime.

Because of the size of the machine, the captain made a gentle sweeping turn around the township of *Coulommiers* and then headed north-northwest in the general direction of the *Montdidier* Beacon. All through this part of the flight everything appeared normal not only to the cockpit crew, but also the ground controllers, even though they were originally perplexed that no turn had been made soon after the *Orly* Beacon had been passed, recalling that the pilot had not reported that he intended to continue to the "further out" turning point.

Throughout this part of the journey, the DC–10 was gaining both speed and altitude. They flew around *Coulommiers* at just above 7,000 feet and at approximately 250 miles per hour. It would have been in the vicinity of this turning point that Captain Berkoz turned off the "Fasten Seat Belts" signs and presumably, because of the large number of passengers on board,

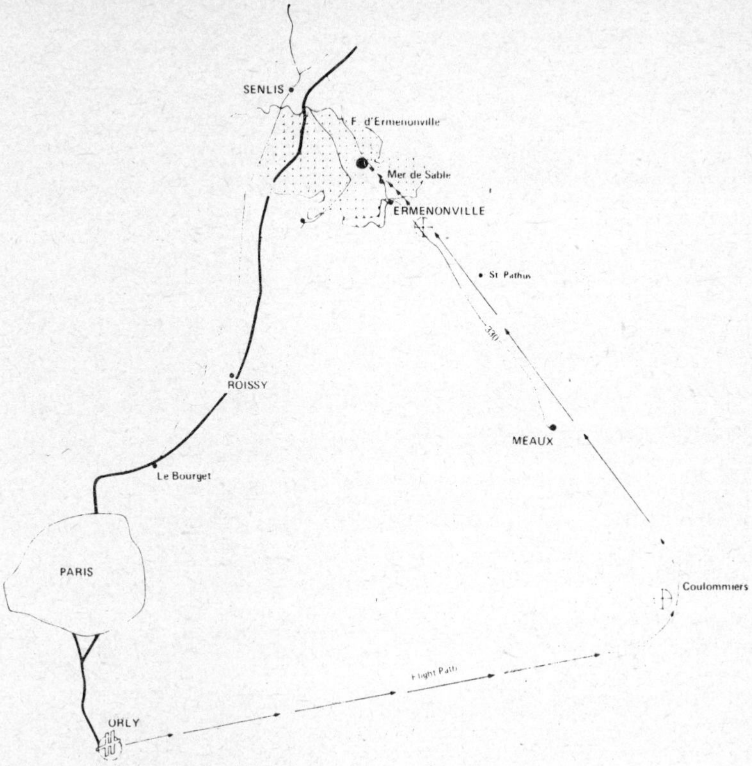

the stewardesses whose normal take-off stations were on the crew's jump-seats down at the rear of the cabin, had evidently moved forward in preparation for the serving of light refreshments during the 50 minutes hop. Remembering that there were just eight stewardesses to cope with the every whim of 330 adult passengers and five children, their time during the short trip would have been "cut out".

However at the same time, even though *Orly* Airport is one of the worst designed in the world for passenger comforts – there are no lavatories for instance, anywhere in the vicinity of the transit and boarding lounges – there did not seem to have been a sudden rush down to the 5 toilets in the rear of the DC–10. In fact, most of the passengers seemed to have remained in their seats or in their vicinity, and quite a large number of them had wisely not undone their seat-belts.

Once around the *Coulommiers* "dog-leg", the pilots levelled the aircraft off and resumed the auto-pilot programme of gaining both speed and height. At just under 9 minutes after lifting off the *Orly* runway, and in exactly similar fashion to the American Airlines DC–10 that was climbing over Windsor, Ontario, Canada on 12 June 1972, the cockpit voice recorder reveals that there was a sudden and explosive decompression throughout the cockpit and passenger cabin. At this exact point in the flight the aircraft was between the villages of *Oissery* and just to the south of *Saint Pathus*, heading roughly north at a height of 12,500 feet and at a speed of 345.5 miles per hour.

As the initial dust and cloudiness of the decompression abated, the pilots were of the impression that they also had been involved in a mid-air collision – yet the skies were fairly clear and no "second aircraft" could be seen. Practically simultaneously with the decompression, the three engine throttle handles returned to just forward of the "flight idle" position and the engines wound down; the auto-pilot disconnected because of the overload; a series of warning horns and sirens started up; and the aircraft commenced to bank to the left and went into a nose dive. The conversations also indicated that the flight control stick went "loose" and the rudder pedals gave an almighty jerk followed by an indication that they had become disconnected from the tail empennage.

By this time the DC–10 had assumed a steep nose-down angle – originally 20° below horizontal – which increased to just on 35° as the plane continued its journey downwards. It's speed was also increasing at an enormous rate as the machine dived towards the ground. The pilots realized that something drastic had happened and were checking the controls and instruments, most of which were evidently inoperable, when suddenly they noticed that the speed had increased at an alarming proportion so they tried to reduce this build-up by attempting to lower the aircraft's undercarriage and thus cause a "drag" which would hopefully slow it down. They operated the handle, but did not know whether the landing legs were in fact in the process of going down. After ten to fifteen seconds, once they realized that this in no way assisted them, they returned the handle to the "up" position and (unbeknown to them) the undercarriage legs started to raise.

One of the flight "peculiarities" – if it could be termed as such – of the DC–10, unrealized to that time, was that as the aircraft's speed was reduced, its nose would slowly go "down"; and as the aircraft's speed increased, the nose would slowly go "up". During the last few seconds of the uncontrollable flight, with the machine's speed nearing 475 miles per hour, the nose, even though no movements could be made to the aircraft's flight control panels on the wings and tail, gently rotated it from the 35° nose-down angle towards the horizontal, and this caused a number of the people on the ground in the vicinity, to think that the pilots were making their way towards one of the local airports at *Le Bourget* or *Roissy*. This was not so. The eyewitnesses had correctly seen the undercarriage lowered and then be raised again, and they had also seen the aircraft "flatten out" from its nose dive as it came nearer to the ground, but their reasoning behind these occurrences was completely wrong. The pilots had no control over the machine at all.

The part of the door that had broken away, was found by some local inhabitants and handed to the investigators late on the Monday evening after the disaster, 4 March 1974, in a wood between *Oissery* and *Saint Pathus*, but it was not identified as such until around midday the following day.

Back in the passenger cabin, one can only make presumptions on what actually happened.

With the aircraft climbing through 12,500 feet on its way to

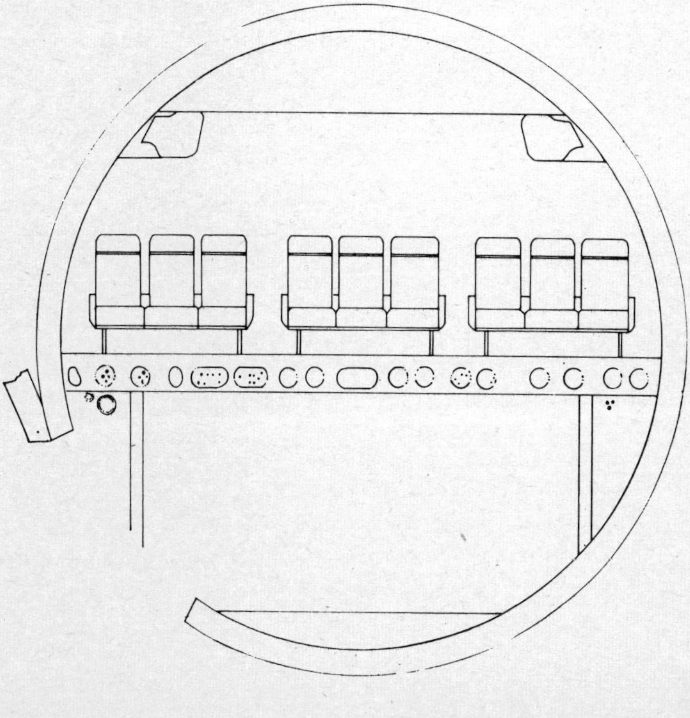

London, the aft cargo compartment door suddenly burst open and initially swung upwards in the direction of its hinge. The air currents then ripped it off after which it more than likely collided with the horizontal stabilizer on the tail before falling back to earth.

This would have caused an explosive decompression in the cargo compartment and then, because of the lack of air pressure underneath, the half-strength floor (compared to the DC–8) would have suddenly placed upon it not only the normal weight of the materials above, but also the pressurization in the passenger cabin.

Whereas, in the earlier American Airlines accident over Windsor, Ontario, the only weight on that section of the floor above the cargo hold was from a bar table and a small compartment dividing panel, on the Turkish Airlines DC–10 things were much different. There were six banks of passenger seats, each with three positions to a bank, and then immediately behind a narrow passageway that led to the rear emergency doors, there were two (one on either side adjacent to the doorways) cabin crew's jump-seats each with two positions.

As luck would have it, Turkish Airlines realized that because of a change in their seat manufacturer, the four rearmost banks of passenger seats on the left hand side of the cabin (marked "A" on the following plan) were of a different make to all the others on board. This permitted the investigators a positive identification.

The other parts of the aircraft's fuselage and cabin that now enter our story in detail, are the two cabin crew's jump-seats which are marked "B", and the position of the aft cargo compartment door in relation to the seats above, marked "C".

The aft cargo hold is located under the passenger cabin floor between fuselage stations 1748 and 1967. The actual collapse of the floor could have been anywhere between these points. However it is more than likely that roughly the same area of floor that collapsed over Windsor, Ontario, was also involved in this accident, which was between fuselage stations 1801 and 1942. (The actual numbering of fuselage stations are the inches in length of the aircraft commencing at a specific point, usually determined by the angles of certain design constants. In the DC–10, the very front of the radome on its nose is called fuselage station 239, being 239 inches behind this constant. The aft cargo compartment therefore was 18 feet 3 inches in length, starting at the point 125 feet 9 inches back from the machine's nose.)

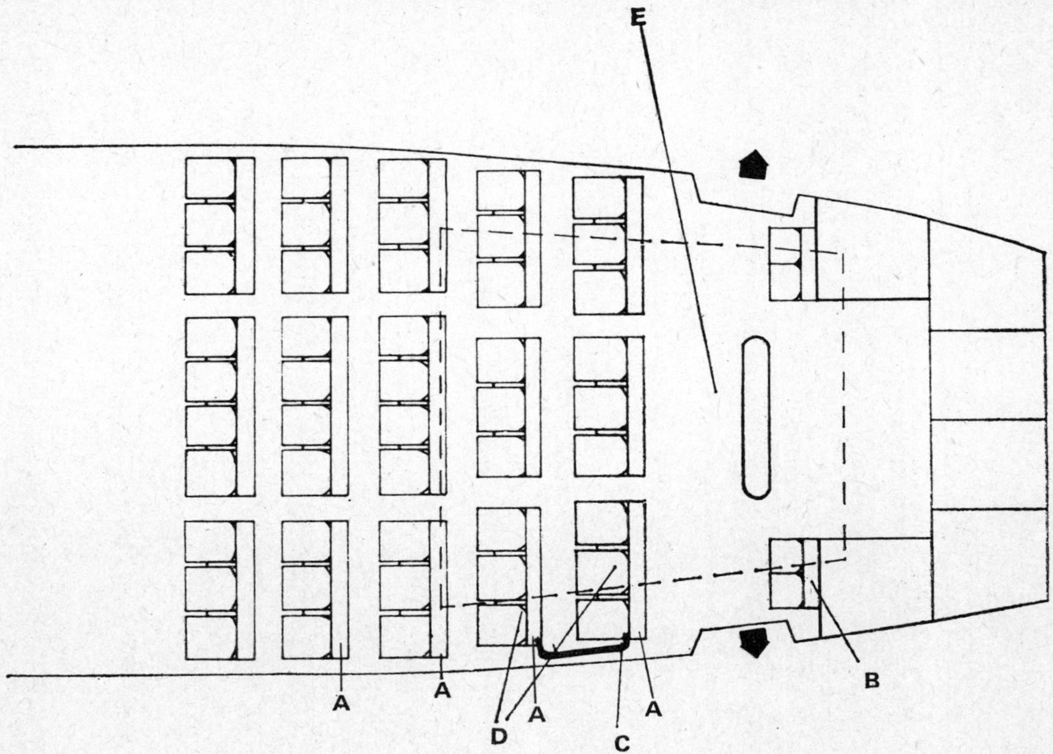
E
A
A
D
A
C
A
B

This area of possible collapse is indicated on the plan by the dotted line marked "E".

Once the decompression had occurred, the floor, with all the added weight on it, then suddenly collapsed. In contrast to the Windsor accident though, it is more than likely and practically certain that the whole area of the floor did completely cave in. When this happened, all of the control cables to the engine and the surfaces on the tail more than likely were snapped and jammed. Moreover, there seems to be a strong indication that when they were snapped, they jammed the aircraft's elevators in a nose-down attitude that caused the machine to go straight away into the dive.

As the floor collapse was occurring, the last two banks of seats on either side of the fuselage would have become detached from their mounts. Of the two on the left-hand side, their left legs which would have been in the seating runner incorporated into the section of the floor, next to the fuselage, that did not collapse, would have broken away. The right-hand legs of the same seats would have detatched from the other runner as that part of the floor collapsed. A similar occurrence, only to the opposite legs,

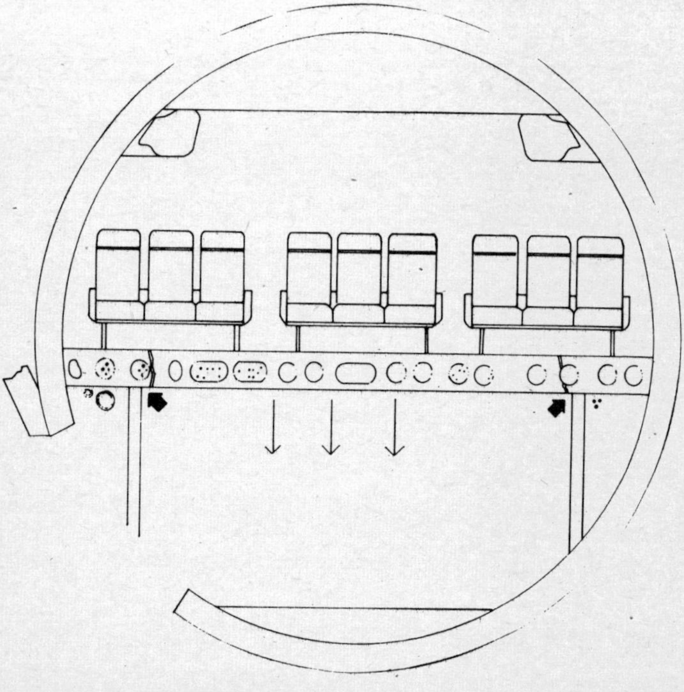

would have happened to the seats on the right-hand side of the fuselage.

The row of seats in the centre of the cabin would more than likely have remained attached to the floor structure because of the fairly evenness of the collapse, and this would have caused the two banks of seats on the right-hand side furthest away from the open cargo door to become jammed, more than likely at a crazy angle.

However, with the air still rushing out of the cargo compartment door, the two loose banks of seats that were at the back of the row on the left-hand side of the cabin, were sucked out of the hole where the door should have been, and the left-rear hostesses jump-seat, which also must have broken away, was itself sucked out as well. (The position of the seats that were found in the field near *Saint Pathus*, are marked "D" on the earlier seating plan.)

A careful examination of the seats found at *Saint Pathus* indicated that at the point when they were sucked out, the six passengers in them still had their seat-belts tightened around their waists. The positions in which the seats and bodies were

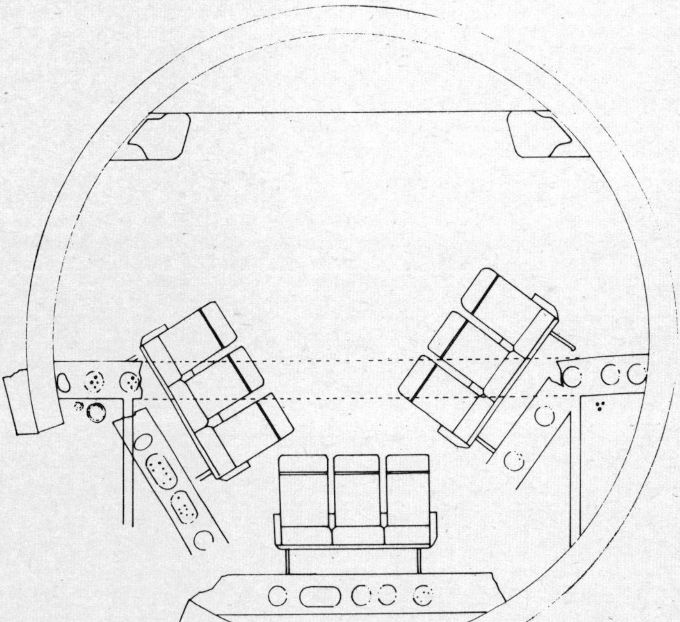

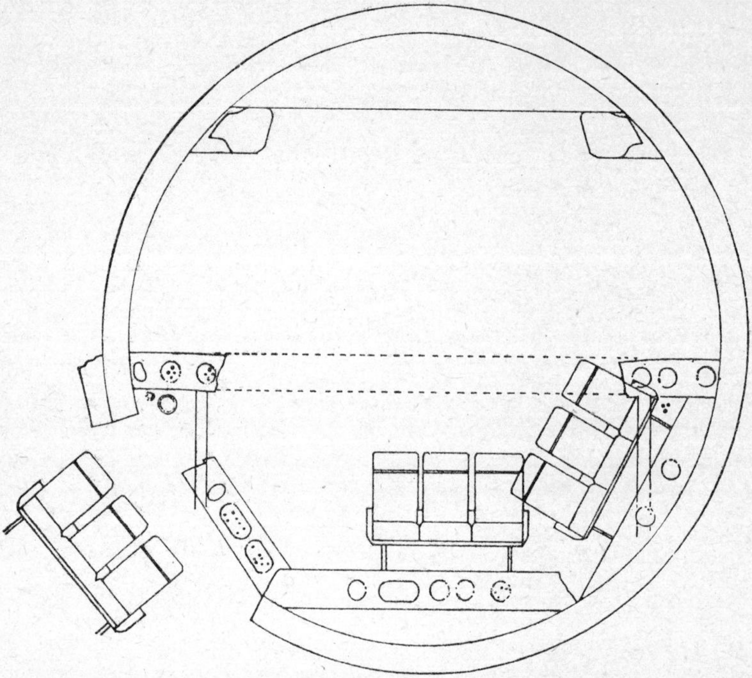

found, seems to indicate that the two banks of passenger seats left the aircraft first, closely followed by the crew's seat, which was not occupied. Once out, the seats and their contents would have fallen free in a trajectory course of about a half a mile in length, before impacting with the ground in Monsieur Dubois's field. It is more than likely that those six passengers lost consciousness on being ejected into the 345 miles an hour air currents through which the machine was flying. Their deaths would have been instantaneous on impact.

Everything that has been described up to this point in the accident would have happened in 2, possibly 3, seconds at the most. The whole sequence of events from the opening of the cargo compartment door to the ejection of the passengers and seats, and the aircraft suddenly leaning to the left and nosing over in the initial part of the steep dive would have happened so quickly that no one would have known quite what had occurred.

On conducting a number of surveys into past human factor experiences of accident situations, not as serious as, but quite possibly equally as scaring to the people on board, each of the

337 persons in the passenger cabin would have more than likely fallen into one of three possible categories.

For between the first 5 to 12 seconds after the decompression occurred, most of the people's minds would have been blank with the initial shock combined with the confusion of the situation.

Once this shock had temporarily worn off, say after the first 12 of the 77 seconds between the door coming open and the impact with the trees in the Forest of *Ermenonville*, everyone on board would have realized that a desperate situation existed. Those out of their seats or with the seat-belts undone, would possibly have been thrown heavily against the ceiling as the aircraft nosed over, dropping back as the dive progressed.

Nearly everyone on board that remained in their seats would have been unable to move around because of the gravity forces being built up by a combination of the increase in speed and the steep nose-down angle at which they were flying. Panic would certainly have set in as is usual in these cases, but there are two completely different types of panic that would have been experienced, with each conscious person on board falling into one or the other category.

Firstly, there is what is known as positive panic. The person realizes that something is wrong and their mind acts in terror at the situation confronting them, usually resulting in loud screaming and rapid, active movements in trying to shield themselves or move away from a point of danger. Secondly, there is negative panic where the person is, relatively speaking, still in a state of shock though he is somewhat unaware of the pending situation and remains static, motionless in his position, with any movement, and if one be made, being very slow and deliberate. In such a state, the person is unlikely to be fully aware of what is happening around him, though his mind does subconsciously tell him that a critical situation exists.

Because of the angle and speed of flight, there is not much doubt that it would have been nearly impossible for people to get up and start moving around, though it is more than likely that both persons and carry-on baggage and indeed anything else that was not tied down, would have been thrown to the ceiling before crashing back on top of passengers and seats once the machine's movements had relatively stabilized.

From then on, it was simply a matter of waiting for something to happen. And it did – 77 seconds after the door came off, and just on 600 seconds after the big machine had gently lifted off the *Orly* runway.

As the DC–10 "bellied" into the trees, its nose was pointing down between 4 and 7 degrees from horizontal, with it leaning between 10 and 15 degrees to the left-hand side. Its speed at the moment of its first impact with the more solid trees was recorded at 489.4 miles per hour and the machine evidently started breaking up from its underneath fuselage surfaces. For a distance of nearly three-quarters of a mile the forest was mowed down, progressing from the stage of just the tree-tops being missing to, within a distance of 150 metres further on, ground level contact near the crest of a small mound that ran at right angles to the crash path. The plane impacted very heavily with the earth at this point, semi bounced and sailed across a small gulley in which just the lower sections of the tree trunks remained standing. There was a further impact point on the top of the parallel mound on the other side of the gulley and this would have precipitated the complete disintegration of the parts of the aircraft that remained. It was after this point that clothes and other passenger belongings were found dangling from the branches of the trees on either side of the crash path.

Practically everything of substance that was on board the DC–10 was smashed into millions of small pieces. The disintegration was so complete, that even hardened crash investigators were appalled at the mess. In fact this could be the first time in civil aviation accident history that an aircraft and its contents have finished up in so many millions of small pieces, and the reason for not making it a definite "first" is because there have been a couple of past crashes into deep water, where all that was found were a few small bits and pieces floating on the surface. Even so, it is highly doubtful that these aircraft would have ended up in a complete disintegration otherwise possibly more buoyant pieces would have been located – but one simply does not know.

Ironically, back in 1968, on speaking with a senior designer at the Boeing Company's factory near Seattle, Washington, on this very same subject of the crashworthiness of these huge Jumbo type jets, the argument presented then in support of their safety was that, with all the cargo situated in the holds underneath the passenger cabin, this would act as a "buffer" between impact and persons.

Prior to 3 March 1974, it was impossible to comment on this theory: it could have been correct – and yet again, it need not necessarily have been so. No one was prepared to crash a $25 million aircraft just to find out.

Today, of course, the world knows otherwise.

APPENDICES

THE DOUGLAS AIRCRAFT MANUFACTURING HISTORY (IN SUMMARY)

APPENDIX 1

MODEL	FIRST FLIGHT	POWER UNITS	MAXIMUM SEATING	MAXIMUM SPEED	MAXIMUM RANGE
DT-1	1921	1 × 400 h.p. Liberty	2 crew	99 m.p.h.	274 miles
DT-2	1921	1 × 400 h.p. Liberty	2 crew	99 m.p.h.	274 miles
O-38B	1923	1 × 525 h.p. P & W R-1690	2 crew	150 m.p.h.	328 miles
O-43A	1923	1 × 675 h.p. Curtiss	2 crew	190 m.p.h.	unknown
TBD Devistator	1925	1 × 900 h.p. P & W R-1830	3 crew	237 m.p.h.	714 miles
B-18	1927	2 × 850 h.p. Wright R-1820	6 crew	215 m.p.h.	748 miles
SBD-5 Dauntless	1929	1 × 1200 h.p. Wright R-1820	2 crew	245 m.p.h.	1105 miles
DC-1	1933	2 × 875 h.p. Wright R-1820	12 pass.	190 m.p.h.	1163 miles
DC-2	1934	2 × 875 h.p. Wright R-1820	14 pass.	200 m.p.h.	783 miles
DC-3	1935	2 × 1200 h.p. P & W R-1830	32 pass.	190 m.p.h.	760 miles
DC-4	1938	4 × 1450 h.p. P & W R-2000	86 pass.	234 m.p.h.	1865 miles
DC-5	1939	2 × 1100 h.p. Wright R-1820	22 pass.	203 m.p.h.	864 miles
A-20G	1939	2 × 1600 h.p. Wright R-2600	4 crew	340 m.p.h.	1255 miles
A-26B Invader	1940	2 × 2000 h.p. P & W R-2800	3 crew	355 m.p.h.	1612 miles
AD2 Skyraider	1941	1 × 2700 h.p. Wright R-3350	1 crew	320 m.p.h.	915 miles
F3D Skynight	1941	2 × 3400 lb Westinghouse	2 crew	600 m.p.h.	1198 miles
B-66B Destroyer	1942	2 × 10000 lb Allison J71	3 crew	594 m.p.h.	1497 miles
F4D Skyray	1943	1 × 9700 lb P & W J57	1 crew	697 m.p.h.	1198 miles
A3D-2 Skywarrior	1944	2 × 12400 lb P & W J57	3 crew	610 m.p.h.	1048 miles
DC-6	1946	4 × 2500 h.p. P & W R-2800	86 pass.	313 m.p.h.	3178 miles
DC-6B	1951	4 × 2500 h.p. P & W R-2800	102 pass.	307 m.p.h.	3098 miles

PISTON/ JET	MILITARY CIVIL	OVERALL LENGTH	OVERALL WINGSPAN	MAXIMUM PAYLOAD*	NUMBERS BUILT	REMARKS
Piston	Military	37′ 8″	50′ 0″	*7293 lb	3	Torpedo carrier
Piston	Military	37′ 8″	50′ 0″	*7293 lb	63	Torpedo carrier
Radial	Military	32′ 0″	40′ 0″	*4458 lb	153	Observer/fighter/ bomber
In-line V	Military	33′ 11″	45′ 8″	*5300 lb	25	Observer/fighter
Radial	Military	35′ 0″	50′ 0″	*10194 lb	127	Fighter/bomber
Radial	Military	57′ 10″	89′ 6″	*27673 lb	350	Fighter/bomber
Radial	Military	33′ 0″	41′ 6″	*10855 lb	5,321	Fighter
Piston	Civil	62′ 0″	85′ 0″	3000 lb	1	Prototype only
Piston	Civil	62′ 0″	85′ 0″	3700 lb	218	Including 6 built in Japan
Piston	Civil	64′ 5″	95′ 0″	6500 lb	13,100	Including about 2,000 built in USSR and 487 in Japan
Piston	Civil	93′ 11″	117′ 6″	14200 lb	1,314	Including 72 DC-4Ms built in Canada
Radial	Military	48′ 0″	61′ 4″	*27200 lb	6,279	Fighter/bomber
Radial	Military	50′ 0″	70′ 0″	*35000 lb	unknown	Fighter/bomber
Radial	Military	38′ 2″	50′ 0″	*18263 lb	3,180	Transport
Jet	Military	45′ 6″	50′ 0″	*26850 lb	268	Fighter
Jet	Military	75′ 2″	72′ 6″	*83000 lb	294	Bomber
Jet	Military	45′ 8″	33′ 6″	*25000 lb	420	Fighter/transport
Jet	Military	76′ 4″	72′ 6″	*82000 lb	316	Fighter/transport
Piston	Civil	62′ 2″	78′ 0″	4326 lb	12	—
Piston	Civil	100′ 7″	117′ 6″	14800 lb	976	—
Piston	Civil	105′ 7″	117′ 6″	19200 lb	529	—

	FIRST FLIGHT	POWER UNITS	MAXIMUM SEATING	MAXIMUM SPEED	MAXIMUM RANGE
DC-7	1953	4 × 3250 h.p. Wright R-3350	99 pass.	360 m.p.h.	3846 miles
DC-7C	1955	4 × 3400 h.p. Wright R-3350	105 pass.	357 m.p.h.	4767 miles
DC-8-10	1958	4 × 13500 lb P & W JT5C	177 pass.	593 m.p.h.	4330 miles
DC-8-50	1962	4 × 18000 lb P & W JT3D	177 pass.	593 m.p.h.	7139 miles
DC-8-63	1967	4 × 19000 lb P & W JT3D	251 pass.	596 m.p.h.	4888 miles
DC-9-10	1965	2 × 14500 lb JT8D P & W	90 pass.	582 m.p.h.	2188 miles
DC-9-30	1969	2 × 14500 lb P & W JT8D	155 pass.	582 m.p.h.	1382 miles
DC-10-10	1971	3 × 40000 lb GE CF6-6	345 pass.	590 m.p.h.	2825 miles

PISTON/ JET	MILITARY CIVIL	OVERALL LENGTH	OVERALL WINGSPAN	MAXIMUM PAYLOAD	NUMBERS BUILT	REMARKS
Piston	Civil	108′ 11″	117′ 6″	20600 lb	215	—
Piston	Civil	112′ 3″	127′ 6″	19500 lb	121	—
Jet	Civil	150′ 6″	142′ 5″	37315 lb	62	—
Jet	Civil	150′ 6″	142′ 5″	34360 lb	230**	Still being manufactured
Jet	Civil	187′ 5″	148′ 5″	67735 lb	254†	Includes 61/62 numbers made Still being manufactured
Jet	Civil	104′ 5″	93′ 4″	2800 lb	135	—
Jet	Civil	119′ 4″	93′ 4″	23138 lb	475***	Includes DC-9-40s made. Still being manufactured
—	Civil	181′ 5″	155′ 4″	101700 lb	103‡	—

NOTES: * denotes gross maximum take-off weights for military aircraft.
† denotes figures built and ordered to the end of 1970.
‡ denotes numbers ordered by October 1973.

APPENDIX 2

The following are the names of those who died in the Turkish Airlines holocaust during the early afternoon of Sunday, 3 March 1974. (An asterisk preceding the name, indicates that more than one member of a family was involved.)

BRITISH SUBJECTS (179 dead):
*Abrahamson – Mr. Harry
*Abrahamson – Mrs. Florrie
Addis – Mr. Thomas William
*Arditi – Master Julian Edgar
Arthur – Mr. Brian Lucius
Backhouse – Mr. John Steven
Bacon – Mr. Roy
Barrow – Mrs. Jean Edith Teresa
*Beasley – Mr. Ronald Charles Fredrick
*Beasley – Mrs. May Florence
*Beckwith – Dr. Michael George
*Beckwith – Mrs. Patricia Ann
Black – Mrs. Claire Papier
Bland – Mrs. Christina Carmen
Bowley – Mr. Charles Christopher
Bradley – Mr. Ivor John
Bressloff – Mr. Sydney Maurice
Breton – Mr. Robert Anthony
Brigstocke – Mr. Geoffrey Reginald William
Brown – Mrs. Bessie Victoria
Burn – Miss Susan Mary
Burris – Mrs. Jean Mary
Burtonwood – Mr. Jack
Byatt – Mrs. Georgina Evelyn
Cain – Mr. David John Steven
Colleen – Miss Esther Winifred
Connelly – Mrs. Diana
Conway – Mr. James
Cooper – Mr. John Hugh
Cornish – Mr. Arthur Laurance
Coult – Mr. Richard Tomlin
Cowell – Mr. David Lewis
Dack – Mr. William Claude
Dangerfield – Mr. Trevor Vincent
Davies – Mr. Hubert James
Davies – Mrs. Margaret Eileen
*Deane – Mr. William Harry
*Deane – Mrs. Brenda Winifred
Delaney – Mr. Ronald Williams
Dickman – Mrs. Marjorie Magdalen
*Doran – Mr. Kenneth Christopher
*Doran – Mrs. Phyllis Gertrude
Dottridge – Mr. Mark Dunsford
Dullforce – Mr. Eric George
Ellis – Mr. Bryan Graham
*Fairburn – Mr. Brian Edgerton
*Fairburn – Mrs. Joy Deidre
Fillery – Mrs. Jean Constance
Fontana – Mr. Carlo Alberto John Erco
Freer – Mr. Joseph Jack
Fripp – Mr. Nicholas Jeremy Steuart
Fuller – Mr. Ian Roger
Gillman – Mr. Robert James
Green – Mr. Peter Jeremy
Griffin – Mr. Ernest John
Griffin – Mrs. Isobel Kennedy
Hannah – Mr. Michael
Hayes – Mr. William James Patrick
Hill – Mr. Francis Eustace
Hillman – Mr. Paul Anthony
Hope – Mr. Francis Michael
Hope – Mrs. Joyce Marjorie
Hunt – Mr. George Desmond
Hurdle – Mrs. Sheelagh
*Hutton – Mr. Patrick John
*Hutton – Mrs. Anne Elizabeth
Jackson – Mr. Malcolm Henry
Jones – Master Nicholas Paul
*Kelsey – Mr. Eric
*Kelsey – Mrs. Catherine Esther
*Kelsey – Miss Ann Mary Olga
*Kelsey – Miss Helen Esther Ann

Kendall – Mr. Christopher Mather
King – Mr. Stuart Frank
*Kohnstamm – Dr. Otto Sigmund
*Kohnstamm – Mrs. Winifred Walton
*Kween – Mr. David
*Law-Wright – Mr. John Anthony
*Law-Wright – Mrs. Susan Vanessa
Levet – Mr. Graham Cecil
Lewis – Mr. Morton Lloyd
Lloyd – Mr. Elfryn
Marriage – Mr. Thomas Roger
McCaig – Mrs. Anna Philomena
McClinton – Mr. John Patrick
*McDonald – Mr. Peter Graham
*McDonald – Mrs. Wendy Elizabeth
McMeekin – Mr. James Wilson
*Middleton – Mr. Jack
*Middleton – Mrs. Joan Margaret
*Milne – Mr. Robert Arthur
*Milne – Mrs. Pauline Florence
Morley – Mr. Rex
Morley – Lady Hedwig Catherina
Morris – Mr. Peter Graham
Morris – Mr. Maurice Clive
Neale – Miss Helen Victoria Rumsey
Newton – Mrs. Mabel Ada Doris
Nicholls – Mr. David John Russell
Okuniewski – Mr. Tadeusz
Osborne – Mrs. Gaynor Mary
*Overton – Mr. James Allan
*Overton – Mrs. Susan Brenda
Paine – Mr. Leslie George
Parish – Captain Julian Glynne Woodbine
Paul – Mr. Terence Fredrick
Payman – Mr. Herbert Saul
Perry – Mr. Martin Robert
*Portch – Mr. Malcolm Ronald
*Portch – Mrs. Edin Delysia
*Portch – Miss Trudi Melinda
*Powell – Mr. Ronald Edward
*Powell – Mrs. Ann
Pratt – Miss Prudence Hedley
Price – Mr. Anthony
*Qunta – Mr. Anders Sipo
*Qunta – Mrs. Margaret Mary
Rawlinson – Mr. Geoffrey
Raymond – Mr. Paul
Reay – Mr. Desmond Armstrong McCabe
Riley – Mr. Richard Anthony
Roberts – Mr. Nigel John Northcliffe
Roberts – Miss Lynne Barbara
Roch – Mrs. Sylvia Mary Beaumont Ormond
Rogers – Miss Patricia Margaret
*Rogers – Mr. George Thomas
*Rogers – Mrs. Ivy Maud Letitia
Rosher – Mrs. Sarah Elizabeth
Rowlandson – Mrs. Gladys Mary
Ryan – Mr. Anthony Colin
Rynsard – Mr. Gregory Robert Anthony
Sanders – Mr. Fredrick Ernest George
Sanders – Mr. Ronald Mark
*Sanders – Mrs. Dorothy May
Sargent – Mr. Charles Henry
Savidge – Mr. Robert Briggs
Scott – Mr. David George
Seager – Mrs. Rosalind Vivienne
Sheel – Miss Pamela
*Slater – Mr. John Frederick
*Slater – Mrs. Gloria Josephine
*Slater – Miss Nadine Nicola
Smith – Mr. Lawrence Congleton
*Sorkin – Mr. Malcolm
*Sorkin – Mrs. Ruth Yvone
*Sturgess – Mr. Edward William
*Sturgess – Mrs. Edith Amelia
Sworder – Miss Erica Ann
Sykes – Mrs. Sarah Elizabeth
Thomas – Mrs. Hannah Marianne
Tilbrook – Mr. Michael
Tomlin – Mr. David Alexander
Townsend – Mr. Michael Fredrick
Wallis – Mr. Robert William
Walsh – Mr. Peter John
Wardley – Mr. Edwin William
Ware – Mr. Lionel James
Warnett – Mr. Peter Frederick
Waterhouse – Mr. Sydney
*Watkins – Mr. John Victor
*Watkins – Mrs. Mary
Weld-Forester – Mrs. Georgina
Wheal – Mrs. Wendy Lynn
White – Mrs. Glenda Jean
Whitehead – Mr. Michael Rodney
Wilkinson – Mrs. Isabella
Williams – Mr. Owen James
*Wilson – Mr. John North Lewis
*Wilson – Mrs. Joyce
*Winterton – Mr. John Winston
*Winterton – Mrs. Winifred Clarice
Withers – Mr. Peter John
*Wood – Mr. Graham Milton
*Wood – Mrs. Margaret Rosemary
Woods – Miss Linda
Woolvett – Mr. Peter William
Zuelchaur – Mr. Peter Walter Leslie

JAPANESE SUBJECTS (48 dead):
Awakura – Mr. Takanori
Fumihiro – Mr. Fujita
Goto – Mr. Junichi
Hayashi – Mr. Masaki
Higuchi – Mr. Masaki
Higuchi – Mr. Takehiro
Hiramatsu – Mr. Masamachi
Hokiya – Mr. Atsuko
Hosoi – Mr. Shiego
Imazeki – Mr. Masanori
Ishibashi – Mr. Hiroshi

Ishikawa – Mr. Tomoo
Ito – Mr. Ikuo
Iwanaga – Mr. Nobuo
Kameyama – Mr. Makoto
Kamiya – Mr. Koji
Kanai – Mr. Bieko
Kato – Mr. Junzo
Kawabata – Mr. Kazumaza
Kawaguchi – Mr. Kenichi
Kenbo – Mr. Kazuyuki
Kichitake – Mr. Tsukuru
Kinugasa – Mr. Motohisa
Kobayashi – Mr. Hiromasa
Kojima – Mr. Yoshitami
Kumesawa – Mr. Yoshio
Matsume – Mr. Tabanori
Mori – Mr. Yasuyuki
Morikawa – Mr. Norikazu
Nagao – Ms. Fumie
Ohama – Mr. Yasushi
Ohama – Mr. Yomei
Okamura – Mr. Yuki
Okura – Mr. Tetsuo
Okura – Mr. Takashi
Sakata – Mr. Tadaharu
Sakurai – Mr. Kenichiro
Sato – Mr. Kakuichi
Sato – Mr. Tetsuo
Shimizu – Mr. Masami
Shinmi – Mr. Takeo
Shiraki – Mr. Hiroshi
Takashima – Mr. Yasunori
Tani – Mr. Shoji
Tominaga – Mr. Yasuhiro
Tsuchiya – Mr. Hiroshi
Tsukada – Mr. Kimio
Yano – Mr. Mitsuo

TURKISH SUBJECTS (52 dead):
Afir – Mr. Faruk Omer
*Akbulut – Mr. Mehmet
*Akbulut – Mrs. Serife Tulga
Altinay – Miss Fatma Rona
Arditi – Mrs. Klarita
Barka – Hostess Fatma
Beller – Mr. Zeki Dogan
*Benghi – Mr. Halil
*Benghi – Mrs. Yildiz
Berkoz – Captain Nejat
Birgill – Hostess Ayse
Bircan – Mr. Mehmet
Cebeci – Mr. Oktay
Celebi – Mr. Mustafa Ihsan
Cetin – Mrs. Selvinas
Ciril – Master Meral
Cizravi – Mrs. Fatma Gonul
Emeksiz – Miss Sevim
Erdin – Mr. Ilgun
*Gurel – Mr. Fikri
*Gurel – Mrs. Hasibe
Hidir – Hostess Semra Cazibe
Ime – Miss Fani
*Inan – Miss Hatice Burein
*Inan – Master Mehmet Engin
*Kapur – Mr. Aliye Filiz
*Kapur – Miss Basak Kerime
*Kohen – Mr. Hayim
*Kohen – Mrs. Coya
Koray – Mr. Engin
*Koz – Mr. Nihat
*Koz – Miss Miuhibe
Mihlar – Mr. Sukran
*Oner – Mr. Metin
*Oner – Mrs. Seniha Gonul
Oral – Mr. Ahmet
Ozer – Flight Engineer Erhan
Ozturk – Mr. Mustafa Serafettin
Pekerol – Mr. Mehmet Sedettin
Sahin – Hostess Selime Sibel
Simsek – Mr. Ahmed Cemalttin
Sonmez – Hostess Culay Naciye
Temizkan – Mr. Islam
Tezean – Chief Steward Hayri
Ucok – Mr. Hasan Engin
Ulusman – Co-pilot Oral
*Uziyel – Mr. Muse
*Uziyel – Mrs. Rasel
*Uziyel – Master Liat
Yardimic – Mr. Uildan
Yilmazer – Hostess Niloun
Yuruker – Miss Gunseli

AMERICAN SUBJECTS (26 dead):
Blair – Mr. Penn
Chard – Mr. Gary Ray
Filipkiewicz – Mr. Roman
Hanessain – Mr. John Jnr.
*Hart – Mr. Laurence Mell
*Hart – Mrs. Alice Joan
*Hart – Miss Diane
*Hart – Miss Nancy Jean
*Hart – Master Robert Raymond
Kalinsky – Miss Nancy Beth
*Kween – Miss Phyllis
Lemmer – Miss Carol J.
Lux – Miss Clara
Peterson – Mr. Rehn
Safran – Mr. Milton
Smith – Master Ronald Lee Jnr.
*Wilcox – Mr. Wayne Ayres
*Wilcox – Mrs. Ouida Neill
*Wilcox – Master Clark Neill
*Wilcox – Miss Kailan
*Wright – Mr. Thomas Prescott
*Wright – Mrs. Fay
*Wright – Miss Carlaine
*Wright – Master Jackson Prescott
*Wright – Miss Sheryl

FRENCH SUBJECTS (16 dead):
Bernard – Mlle. Françoise Elisabeth
Donnat – M. Abel Roger

Druesne – M. François Marie Jean
*Lerendu – M. Robert Andre Jean
*Lerendu – Mme. Josette
*Lerendu – Mlle. Corinne Elise Simone
*Lerendu – Mm. Jaques
Lezay – Mme. Moniquet
Mayer – M. Ladislas
Loubenzow – Mlle. Lucie Ludmilla
Nguyen Thi – Mlle. Cam Ha
Pauly – Mlle. Françoise
Pellegars – M. François Marie Gabriel
Poisson – Mlle. Martine Janive Françine
Reynaud – Mme. Robert
Wormser – Mlle. Betty Françoise

BRAZILIAN SUBJECTS (5 dead):
Compenella dos Santos – Sr. Roberto Antonio
*Lins – Sr. Nelson
*Lins – Sra. Nailka
Pumar – Sr. Anadey
Souza – Sr. Mario Jorge

INDIAN SUBJECTS (3 dead):
Kothary – Mr. Rasiklal
Lakhana – Mr. Abdul Razzak
Shah – Mr. Ramahlal

ARGENTINIAN SUBJECTS (2 dead):
*Pernias – Mrs. Olda Lina Elisa
*Pernias – Miss Alexandra Maria

MOROCCAN SUBJECTS (2 dead):
Bouhadana – Mlle. Tamar
Chaoavi Benaodulla – M. Mohamed Saad

SPANISH SUBJECTS (2 dead):
Justo – Sr. Ernesto
*Pernias – Sr. Pedro

YUGOSLAV SUBJECTS (2 dead):
Sen – Mr. Nazif
Senyurt – Mr. Neomettin

AUSTRALIAN SUBJECT (1 dead):
Cowie – Miss Rachel Margaret

AUSTRIAN SUBJECT (1 dead):
Deister – Herr Rainer

BELGIAN SUBJECT (1 dead):
Coussens – M. Jean Jaques

CYPRIOT SUBJECT (1 dead):
Mustafa – Mr. Nebil

GERMAN SUBJECT (1 dead):
Muller – Fln. Christiane

IRISH SUBJECT (1 dead):
Macdonald – Fr. James